Springer Theses

Recognizing Outstanding Ph.D. Research

Aims and Scope

The series "Springer Theses" brings together a selection of the very best Ph.D. theses from around the world and across the physical sciences. Nominated and endorsed by two recognized specialists, each published volume has been selected for its scientific excellence and the high impact of its contents for the pertinent field of research. For greater accessibility to non-specialists, the published versions include an extended introduction, as well as a foreword by the student's supervisor explaining the special relevance of the work for the field. As a whole, the series will provide a valuable resource both for newcomers to the research fields described, and for other scientists seeking detailed background information on special questions. Finally, it provides an accredited documentation of the valuable contributions made by today's younger generation of scientists.

Theses are accepted into the series by invited nomination only and must fulfill all of the following criteria

- They must be written in good English.
- The topic should fall within the confines of Chemistry, Physics, Earth Sciences, Engineering and related interdisciplinary fields such as Materials, Nanoscience, Chemical Engineering, Complex Systems and Biophysics.
- The work reported in the thesis must represent a significant scientific advance.
- If the thesis includes previously published material, permission to reproduce this must be gained from the respective copyright holder.
- They must have been examined and passed during the 12 months prior to nomination.
- Each thesis should include a foreword by the supervisor outlining the significance of its content.
- The theses should have a clearly defined structure including an introduction accessible to scientists not expert in that particular field.

More information about this series at http://www.springer.com/series/8790

Jason Sanders

Dynamics of the Milky Way

Tidal Streams and Extended Distribution Functions for the Galactic Disc

Doctoral Thesis accepted by
the University of Oxford, UK

Springer

Author
Dr. Jason Sanders
Institute for Astronomy
University of Cambridge
Cambridge
UK

Supervisor
Prof. James Binney
Department of Physics
Oxford University
Oxford
UK

ISSN 2190-5053 ISSN 2190-5061 (electronic)
Springer Theses
ISBN 978-3-319-18771-6 ISBN 978-3-319-18772-3 (eBook)
DOI 10.1007/978-3-319-18772-3

Library of Congress Control Number: 2015940733

Springer Cham Heidelberg New York Dordrecht London

Printed on acid-free paper

Springer International Publishing AG Switzerland is part of Springer Science+Business Media (www.springer.com)

Publications Related to This Thesis

This thesis has used material from the following first-authored publications.
Chapter 2 uses material from Sanders J., 2012a, MNRAS, 426, 128.
Chapter 3 uses material from Sanders J. L., Binney J., 2014, MNRAS, 441, 3284.
Chapter 4 uses material from Sanders J. L., Binney J., 2015, MNRAS, 447, 2479.
Chapter 5 uses material from Sanders J. L., Binney J., 2013a, MNRAS, 433, 1813.
Chapter 6 uses material from Sanders J. L., Binney J., 2013b, MNRAS, 433, 1826 and Sanders J. L., 2014, MNRAS, 443, 423.
Chapter 7 uses material from Sanders J., 2012b, MNRAS, 425, 2228.

This thesis is dedicated to Hannah and my parents for their constant love and support.

Supervisor's Foreword

In the 1990s our current picture of galaxy formation was established. In this picture small companions are constantly falling into the gravitational potential wells of larger galaxies such as ours. As they orbit in the force field of the larger galaxies, they are disrupted by tidal forces. The upshot is that twin streams of stars (and presumably undetected dark matter) are pulled from opposite points of the small companion. Ibata et al. 1994 discovered such a companion, the Sagittarius dwarf spheroidal, quite close to the centre of our Galaxy but lying on the opposite side of the Galactic Centre from us. Stars striped from the Sagittarius dwarf were soon after identified at angular distances >10° from the dwarf, and by 2003 the tidal streams formed by stars stripped from the Sagittarius dwarf had been traced right around the sky. In the new millenium data from the Sloan Digital Sky Survey, which contained accurate multi-colour photometry of stars down to 20th magnitude, dramatically increased the rate at which new tidal streams were discovered. First individual globular clusters were shown to have extensive tidal tails comprising stars that could no longer be bound to the cluster (Odenkirchen et al. 2003; Belokurov et al. 2006). Then streams were discovered that had no evident progenitor (Belokurov et al. 2007; Grillmair and Dianotos 2006).

A fundamental task of Galactic dynamics is to map the Galaxy's gravitational field, and hence its distribution of dark matter. From very early on it was recognised that tidal streams have enormous potential in this connection (Lynden-Bell and Lynden-Bell 1995) but it was not clear how this potential should be realised. The crudest approach rested on the assumption that at large radii the gravitational field is essentially spherical, so each stream would be confined to a plane. A more sophisticated approach assumed that the stars of a stream were essentially on the same orbit as the stream's progenitor. In 2008 I showed that proper motions and distances could be predicted from measured sky positions, line-of sight velocities and the assumption that all stream stars were on the same orbit. A student, Andy Eyre, took up this idea, extended it to the case in which proper motions rather than line-of-sight velocities had been measured, and tested the reconstructions with N-body simulations of cluster disruption. These simulations revealed that the approximation that all stars are on the same orbit becomes increasingly problematic as the errors on the measured quantities increase to realistic values. Following Tremaine (1999), Eyre examined the formation of N-body streams in angle-action coordinates and showed that in any gravitational potential streams have a characteristic structure in angle-action space. In particular, there is a systematic shift in the actions of the orbits of stream stars as one proceeds down each of the tidal tails. If this structure is ignored when searching for the correct potential seriously incorrect results can arise.

Eyre left the field after being awarded his doctorate because he felt called to study medicine. So Jason Sanders took up where Eyre had left off. Given the forensic power of angle-action coordinates, Jason gave significant time throughout his thesis work to devising more powerful methods for the determination of angle-action coordinates from ordinary phase-space coordinates, and a critical comparison of the cost-effectiveness of these techniques is a useful feature of his thesis.

For the then known tidal streams Jason quantified the errors in the parameters of the recovered potential that would arise from assuming all stream stars were on the same orbit. Since these errors could be of order one, he proceeded to introduce a new principle for the selection of the true potential. The new principle precisely exploits the systematic variation of orbit along a stream, taking the form of linear relationships between each angle variable and the associated orbital frequency.

The mapping between observed quantities such as distance, line-of-sight velocity, and proper motion and angle-action variables is very nonlinear. Consequently, when observational errors are significant one should avoid computing the angles and actions of individual stars and instead project model streams into the space of observable and adjust the parameters of the models to find the range which is consistent with the observational data. Jason implemented this approach for some mock data. His implementation is, however, not the definitive one because it involves integrating the probability of the stream over the error ellipsoid of each star. For realistic errors in the data, the probability of a given stream is near zero through most of the error ellipsoid of the star, so it would be much more efficient to integrate the star's error ellipsoid through the volume in which the stream has significant probability. This is yet to be done.

In a series of papers from 2010 we showed that kinematic data for stars near the Sun can be modelled with considerable precision by assuming that the distribution function of the stars is an analytic function of the actions. In this work we acknowledged that stars have different ages, but not that they have different chemical compositions. In the last section of his thesis Jason seeks an analytic function of age, metallicity and actions that accounts for recent survey data. He is not entirely successful, but a paper that will soon appear in Monthly Notices of the Royal Astronomical Society shows that two small but important modifications to the assumed history of star formation and the way radial migration is handled yield an extended distribution function that provides good fits to all the data except the density of higher-metallicity stars far from the plane. It is not improbable that this conflict between model and data will be resolved by changes to the data rather than to the model.

Oxford, UK
May 2015

Prof. James Binney

References

Belokurov V, Evans NW, Irwin MJ, Hewett PC, Wilkinson MI (2006) ApJL 637:L29
Belokurov V (2007) APJ 658:337
Grillmair CJ, Dionatos O (2006) APJL 643:L17
Ibata RA, Gilmore G, Irwin MJ (1994) Nature 370:194
Lynden-Bell D, Lynden-Bell RM (1995) MNRAS 275:429
Odenkirchen M (2003) AJ 126:2385
Tremaine S (1999) MNRAS 307:877

Abstract

One of the key goals of Milky Way science is measuring the distribution of dark matter in the Galaxy. Through the study of Galactic dynamics, inferences can be made about the structure of the Galaxy, and hence the dark matter distribution. To this end, we present a study of methods useful for modelling and understanding dynamical systems in the Galaxy.

A natural choice of coordinate system when studying dynamical systems is the canonical system of angle-action coordinates. We present methods for estimating the angle-actions in both axisymmetric and triaxial potentials. These fall into two categories: non-convergent and convergent. The non-convergent methods are fast approaches, mostly based on approximations to Stäckel potentials. We investigate the accuracy of these methods for realistic Galactic potentials. The slower convergent methods operate by constructing generating functions to take us from simple analytically-tractable potentials to our target potential.

Tidal streams should prove useful for constraining the large-scale dark matter distribution in the Galaxy. Armed with our new angle-action tools, we investigate the properties of known streams in a realistic Galactic potential. We present a simple algorithm for constraining the Galactic potential using a tidal stream, which exploits the expected structure of a stream in the angle-frequency space of the true potential. We expand this approach into a fully probabilistic scheme that allows for handling of large errors, missing data and outliers.

We close by discussing another tool useful for modelling the dynamics of the Galaxy: extended distribution functions for the Galactic disc. We present a simple extension of an action-based distribution function from Binney (2010) that includes metallicity information, and compare the model predictions with current data. These models are essential for incorporating the selection effects of any survey, and reveal the important chemo-dynamic correlations that expose the history and evolution of the Galaxy.

Reference

Binney J (2010) MNRAS 401:2318

Acknowledgments

First, I would like to thank Prof. James Binney for his supervision during my D.Phil. He is very generous with his time and ideas, and I have been very lucky to have such an enthusiastic supervisor. James gives thought to any problem sent his way, and always seems to know an appropriate course of action to work through the problem. He has also spent considerable time reading drafts of this thesis, for which I am very grateful. James has kindly provided code used as a basis of part of the work in this thesis.

Thanks also go to the current and past members of the Oxford Galaxy Dynamics group. The regular coffee time discussions were a crucial addition to my astrophysics education, and our Monday meetings were a welcome start to the week. I would like to thank, in particular, Paul McMillan, whose provision of the torus construction code and Walter Dehnen's GalPot code were essential for much of the work in this thesis. Much of the tidal streams work in this thesis is based on the work of Andy Eyre. I thank Andy for supplying initial advice and code crucial in kick-starting my D.Phil.

I acknowledge the Science & Technology Facilities Council for funding my D.Phil for the past three years.

Room 2.9 Rudolf Peierls Centre has been a second home for the past three years. It is an awful office (too cold in winter and too warm in summer), but it is made by the people, both past and present, who have inhabited it. In particular, I would like to thank Francesco Fermani and Mattia Sormani for their friendship and advice.

My time is Oxford has been wonderful, and this is largely due to some great friends. Regular dinners with Roly and Emily, and Grace and Henry have been a brilliant distraction from work and have provided many a moment of hilarity.

I never would have reached where I am today without the continued support of my family. My parents, Rose and Richard, have always encouraged and enabled me to do well. Their regular visits with offerings of red wine have been greatly appreciated! My sister, Thea, and brother, Adam, always know the right thing to say, whether it is a joke or some kind words of support.

Finally, and most importantly, I thank Hannah. I love her, and her support over the past three years has been invaluable. She has been so patient and kind, even when things have been tough. Our life together in Oxford has been fantastic, and I can't wait to start the next chapter together in Cambridge.

Contents

Chapter 1
Introduction

1.1 The Galaxy

The Universe has structure on a vast range of scales, from the elements and molecules found on the Earth, to huge cosmic webs of matter that extend over millions of light-years. Our Galaxy, the Milky Way, is a complex structure that sits at the interface of these extreme scales. Gas is fed along the cosmic webs into the Galaxy, cools and forms stars, which go on to produce the important elements that fuel life on Earth.

The nature of our Galaxy has intrigued humans for many generations. Clearly visible with the naked eye, the Milky Way sits as a huge murky swathe across the sky. The Ancient Greeks were the first to suggest that the Galaxy is actually composed of many stars, but it wasn't until 1610, when Galileo Galilei used his recently-invented telescope to resolve these stars, that this hypothesis was confirmed. We now know that there are approximately 10^{11} stars in our Galaxy. These stars are gravitationally bound and located mainly in a thin disc. The Sun resides in the disc of the Galaxy such that our view of the Galaxy is a narrow band in the sky. Inspired by the work of Thomas Wright and the structure of the Solar system, Kant concluded that such a structure could occur if a systematic rotation of the Milky Way balanced the inward gravitational forces. The Galaxy also has a significant proportion of gas and dust, which causes our view of the Galaxy to be obscured and altered (e.g. Berry et al. 2012; Schlafly et al. 2014).

Kant also had the insight that many of the fuzzy *nebulae* first observed by Abd al-Rahman al-Sufi in 964 may be very distant stellar systems similar to our own Galaxy, as opposed to being part of our Galaxy. This hypothesis remained controversial until the early 20th century and was the subject of the *Great Debate* between Shapley and Curtis. Kant's idea gained evidence with improved observations by Messier in 1781 and Herschel in 1786, but the first clear evidence came in 1845 when Lord Rosse resolved substructure in the faintest *nebulae*. He discovered that they fell into two distinct categories: elliptical and spiral. The structure of the spiral galaxies reinforced Kant's hypothesis that these systems had some systematic rotation similar to the Milky Way.

J. Sanders, *Dynamics of the Milky Way*, Springer Theses,
DOI 10.1007/978-3-319-18772-3_1

Determining distances to astrophysical objects has long remained an issue in astrophysics, and it proved to be the solution to the debate over the nature of the *nebulae*. To find distances to nearby stars, Herschel assumed the absolute magnitude of all stars is identical and so constructed the first 3D map of the Milky Way. To pin down the nature of the *nebulae*, it was necessary to estimate their distances. In the 1920s Edwin Hubble made two important contributions in this respect. His first contribution was to use the 100-inch telescope on Mount Wilson to resolve the outer parts of spiral galaxies. He found that they consisted of many point-like faint objects very similar to stars, and concluded that the spiral galaxies must lie at very large distances. Secondly, when observing the Andromeda galaxy, Hubble found that some of these objects had a characteristic periodic luminosity similar to local Cepheid variable stars. Based on these local stars, it was already known that there is a tight relationship between absolute magnitude and period, such that Hubble was able to estimate the distance to the Andromeda galaxy. He found a distance of $\sim$300 kpc—much larger than estimates of the size of the Milky Way. Therefore, the *nebulae* must be distinct from our own Galaxy.

The work of these early astronomers has led us to the view that the Universe is populated with galaxies, and that our Galaxy is just one of many. The evolution and dynamics of these impressive structures is still an ongoing and highly active research area. By all accounts our Galaxy is a typical star-forming galaxy, and as such is somewhat of a representative sample of all galaxies in the Universe. However, our observation point is unique and we are able to observe our own Galaxy in far greater detail than we could ever hope to observe other galaxies. Therefore, the Milky Way is an ideal test-bed for understanding the fundamental properties and behaviour of all galaxies. In this thesis we seek to further understand the nature of the Milky Way, and hence *all* galaxies.

1.2 Dark Matter

Despite our view of the Galaxy, and all galaxies, being dominated by stars and gas, by far the most dominant component of any galaxy is the dark matter. This poorly-understood component is hypothesised to exist based on the effects of its gravitational influence. In the Milky Way it is believed that the dark matter contributes about ten times the mass of the baryonic matter (Wilkinson and Evans 1999; Xue et al. 2008; Watkins et al. 2010; McMillan 2011a).

The nature of dark matter is one of the biggest questions in modern physics. Its existence was first posited in the early thirties by Oort (1932) and Zwicky (1933). Both Oort and Zwicky found that the motions of astrophysical objects could not be explained by the visible matter, and so it was necessary to invoke some additional dark matter to explain the observations. Oort found that the stars in the Milky Way had velocities inconsistent with the visible matter in the Galactic disc, whilst Zwicky found that the velocities of galaxies in clusters were inconsistent with the visible cluster mass. Later, Begeman (1989) used the Westerbork 21 cm line observations to find the rotation curve of the gas in NGC 3198 is approximately flat. This observation could not be accounted for by the visible mass, and hence implied the existence of a dark halo. Invoking extra invisible matter is clearly a large leap, but it could explain the observations. Subsequently dark matter has become an accepted, yet still not understood, component of the Universe.

Dark matter is a crucial component in the currently accepted Λ-CDM cosmology (Liddle 2003) and the theory for the formation of galaxies. Dark matter makes up approximately 85 % of the total matter in the Universe, and plays a central role in explaining the power spectrum of the Cosmic Microwave Background radiation (CMB), as well as enabling galaxies to form from the small fluctuations we observe in the CMB. Early in the Universe, due to its lack of interaction, the dark matter clumped into overdensities that were relics of some early quantum fluctuations in the density field. The CMB shows the relics of these early density perturbations in the baryons (Planck Collaboration et al. 2013). With baryons alone, the fluctuations are not large enough to account for the structure seen in the Universe today, so dark matter is required to produce sufficiently large overdensities. These overdensities accreted more dark matter until they became small haloes of dark matter. Baryonic matter was slowly accreted into the centre of these dark matter haloes, and cooled into smaller self-gravitating structures that formed into galaxies. It is therefore believed that all galaxies, including the Milky Way, are surrounded by a halo of dark matter.

This hierarchical formation picture of the Universe is reinforced by large-scale cosmological simulations (e.g. Springel et al. 2005; Boylan-Kolchin et al. 2009) in which initial perturbations in a uniform dark matter density field are allowed to grow for the age of the Universe. These simulations produce large overdensities connected by web-like structures with each overdensity also surrounded by many smaller *subhaloes*. Navarro et al. (1996) found that the structure of the haloes from these simulations follows a universal profile.

Currently, many resources are devoted to understanding the nature of dark matter. A direct detection of dark matter is the ultimate goal. These particle physics experiments are the necessary route for understanding the true nature of dark matter. However, astrophysical insight is also required to form a full picture of the nature of dark matter. The density and the velocity distribution of the dark matter at the Sun are essential ingredients to understanding the results from a direct detection experiment. Additionally, the shape of the dark matter halo in our Galaxy provides clues to the physics that formed the halo (Sackett 1999; Bailin and Steinmetz 2005), and will provide more universal insights into dark matter distributions in the Universe.

Although widely accepted, dark matter is not the sole explanation for many of the observations mentioned here. Alternative theories of gravity such as MOND (Milgrom 1983) seek to explain many of the observations by altering the gravitational field equations. Newton's gravity, or indeed Einstein's Theory of General Relativity, is well tested on Solar system scales, but its application to galactic and cluster scales is still debated (Uzan 2009; Reyes et al. 2010).

Measuring the distribution of dark matter is a key goal of Milky Way science. This thesis discusses methods and approaches that take steps towards achieving this goal.

1.3 Galactic Surveys

The most promising route to understanding dark matter is through measuring its gravitational influence, either on the visible matter or the photons emitted by this visible matter. Gravitational lensing of the light from distant galaxies around galaxy clusters reveals the presence of large dark matter haloes surrounding the clusters (see Massey et al. 2010, for a review). In the Galaxy, we are able to see the influence of the dark matter on the visible matter, in particular the stars. We are now in a golden age of large-scale Galactic surveys of stars, with many hundreds of thousands of stars being accurately mapped in position, velocity and chemistry space. For understanding the Galaxy as a whole, we require large number statistics, such that we can measure densities, velocity dispersions etc., assuming that the population is in dynamical equilibrium. However, within these large numbers of stars, there will be a subset of stars that are members of non-equilibrium structures. These should provide powerful diagnostics on the Galactic potential (see Sect. 1.6). For modelling the Galaxy as a whole, we require large representative samples from large-scale Galactic surveys.

In the last thirty years, increasingly sophisticated technology has allowed large Galactic surveys to become a reality. Large-scale photometric surveys, such as 2MASS (Skrutskie et al. 2006, observing in the infra-red) and the Sloan Digital Sky Survey (Eisenstein et al. 2011), have measured the magnitudes of several hundred million stars in the Galaxy. These are complemented with large proper motion catalogues such as PPMXL (Roeser et al. 2010) and UCAC-2 (Zacharias et al. 2010), such that we can map much of the Galaxy in 4D. To fully understand the Galaxy, it is necessary to view the Galaxy in full 6D, which involves combining this 4D information with line-of-sight velocities and distances.

In 1989 the Hipparcos satellite was launched (Perryman et al. 1997) with the aim of recovering accurate distances to many stars. Hipparcos measured the parallaxes of around 10^5 nearby stars. Its successor, *Gaia* (Perryman et al. 2001), was launched in late 2013 and will measure parallaxes for around 10^9 stars, or one % of all stars in the Galaxy. Parallaxes provide the most accurate distances to nearby stars. For more distant stars we must rely on (spectro-)photometric distances (e.g. Burnett et al. 2011).

Spectroscopic surveys are required to measure line-of-sight velocities. These are expensive surveys, so we can only hope to survey a small fraction of the hundreds of millions of stars in the photometric catalogues. However, in addition to measuring the line-of-sight velocity, we are also in a position to measure many other properties of the stars from a spectrum, such as effective temperature, surface gravity and chemical abundances. The first big effort to make a large spectroscopic survey was the Geneva-Copenhagen survey (Nordström et al. 2004, GCS) which took spectra for a fraction of the stars with Hipparcos parallaxes. For the first time, a large sample of stars were viewed in 6D phase-space plus chemistry space etc. We will return to the GCS later in this thesis so will not discuss it further here. The Sloan Digital Sky Survey launched a spectroscopic survey: the Sloan Extension for Galactic Understanding and Exploration (SEGUE) (Yanny et al. 2009). SEGUE is a low-resolution spectroscopic survey of stars fainter than 14th magnitude, complemented by SDSS photometry. As such, it provides a view of the outer parts of the disc, dominated by the thick disc, and the stellar halo of the Galaxy, and so complements the more local GCS sample. The RAVE survey has collected medium resolution spectra (R $\sim$ 7500) for around 5×10^5 stars brighter than $I = 12$, enabling the accurate determination of line-of-sight velocities of stars in a large volume of the Galaxy. The selection volume of this survey means that it is dominated by the thin and thick discs of the Galaxy, so acts as a complement to the more extended thick-disc-dominated SEGUE surveys.

Gaia will also measure line-of-sight velocities for many stars accurate to $0.5\,\mathrm{km\,s^{-1}}$ for $V = 15$ (Munari et al. 2001). However, *Gaia* is not able to make accurate measurements of the stellar properties such as chemical abundances. Gaia-ESO is an ongoing spectroscopic survey measuring line-of-sight velocities chemical abundances of around 10^5 stars (Gilmore et al. 2012). It promises to complement the *Gaia* data set.

1.4 Dynamics

The clear way to exploit all the available data from large surveys is through modelling the dynamics. A natural time-scale for the evolution of the Galaxy is $\sim$200 Myr. This is the time taken for a star near the Sun to complete a single revolution around the Galactic centre. This implies two things. The first is that stars near the Sun have orbited the Galaxy approximately 50 times since the formation of the Galaxy. Therefore, it is reasonable to assume that the Galaxy has reached an approximately steady state. Secondly, the time taken for a star to orbit is significantly longer than any human time-scale, so we have no hope of measuring the acceleration of many stars. The time-scales in the Solar System are short enough for detailed observations of the entire planetary orbits to be made. From this the acceleration, and hence the masses of the Sun and the other planets, can be inferred. We are not nearly as lucky when working with the Galaxy.[1]

The Solar System has been studied in great detail, and a great many tools have been developed for this study. The motion of each of the planets around the Sun can be simply described to zeroth order as Keplerian motion about a point mass. However, the other planets, most importantly Jupiter, provide small corrections to this behaviour. The most natural way to approach this problem is with a perturbative scheme, and the variables most suited to perturbative solutions of dynamical systems are the angle-action variables. These variables form an important part of the work in this thesis so we will introduce them here.

[1] The exceptions to this are the S stars at the Galactic centre (Genzel et al. 2010), which orbit the central Galactic black hole on time-scales of order 10 yr. These stars have been used to infer the presence of a central black hole.

1.4.1 Angle-Action Coordinates

An n-dimensional dynamical system can be described by the Hamiltonian, H, expressed in terms of canonical coordinates, $(\boldsymbol{q}, \boldsymbol{p})$. These coordinates satisfy the $2n$ Hamilton's equations given by

$$\frac{\mathrm{d}\boldsymbol{p}}{\mathrm{d}t} = -\frac{\partial H}{\partial \boldsymbol{q}}, \tag{1.1}$$

$$\frac{\mathrm{d}\boldsymbol{q}}{\mathrm{d}t} = +\frac{\partial H}{\partial \boldsymbol{p}}. \tag{1.2}$$

If the Hamiltonian is integrable, it permits n integrals of the motion (Arnold 1978). Any function of the integrals of the motion is also an integral of motion. We choose the action variables denoted by $\boldsymbol{J}$. These coordinates are *in involution* i.e. they have zero Poisson bracket $[J_i, H] = 0$. Therefore, their corresponding conjugate coordinates, the angles $\boldsymbol{\theta}$, obey the equations $\partial H/\partial\theta_i = 0$, and the Hamiltonian can be expressed purely as a function of $\boldsymbol{J}$. Hamilton's equations for these canonical coordinates are

$$\frac{\mathrm{d}\boldsymbol{J}}{dt} = -\frac{\partial H}{\partial \boldsymbol{\theta}} = 0, \tag{1.3}$$

$$\frac{\mathrm{d}\boldsymbol{\theta}}{dt} = +\frac{\partial H}{\partial \boldsymbol{J}} \equiv \boldsymbol{\Omega}(\boldsymbol{J}), \tag{1.4}$$

where we have introduced the frequencies, $\boldsymbol{\Omega}$. Therefore, we can write the equations of motion simply as

$$\boldsymbol{J} = \text{const.}, \tag{1.5}$$

$$\boldsymbol{\theta} = \boldsymbol{\Omega}(\boldsymbol{J})t + \boldsymbol{\theta}(0). \tag{1.6}$$

An additional detail is that the angles are 2π-periodic, such that the actions $\boldsymbol{J}$ describe an n-torus over which a particle moves. These equations already reveal the power of the angle-action formalism—motion that is complicated in $(\boldsymbol{q}, \boldsymbol{p})$ becomes simple when viewed in angle-actions. The angle-action variables also provide a basis for the development of a perturbative solution to the equations of motion. In fact, the angle-action coordinates of the Kepler problem were invented to explore the role played by planet-planet interactions in the dynamics of the Solar System. These coordinates will form the base of much of the work described in this thesis. For non-integrable Hamiltonians and resonant orbits, the angle-action coordinates do not exist.

1.4.2 Angle-Action Coordinates in Galactic Dynamics

The simplicity of the equations of motion for angle-action coordinates make them attractive variables to work with. They also possess many other useful properties, which are outlined in Binney and Tremaine (2008) and will be discussed later in this thesis. Their use, however, remains awkward due to the difficulty of their calculation in a general potential. To find the angles and actions, the potential must produce separable Hamilton-Jacobi equations. In 1D this is a triviality and we are able to express the radial action as an integral over the radial potential. However, only in a few cases can this integral be performed analytically. The isochrone potential (Henon 1959) is the most general spherical potential for which we can find analytic relations to take us between $(\boldsymbol{x}, \boldsymbol{v})$ and $(\boldsymbol{J}, \boldsymbol{\theta})$. The Kepler and spherical harmonic oscillator potential are two limits of the isochrone potential. In all other spherical potentials, the actions and angles must be found numerically.

The most general class of axisymmetric and triaxial potentials that produce separable Hamilton-Jacobi equations are the Stäckel potentials (Stäckel 1893; de Zeeuw 1985). There are no cases where the actions can be found analytically. Instead, the actions must be found numerically in all Stäckel potentials by evaluating 1D integrals. This attractive property of the Stäckel potentials has led several authors (Dejonghe and de Zeeuw 1988; Famaey and Dejonghe 2003) to consider whether realistic Galactic potentials could be fitted with a Stäckel potential. In Chap. 2, we pursue this idea further to estimate the actions via locally fitting a Stäckel potential to a realistic Galactic potential.

The most general method for relating $(\boldsymbol{x}, \boldsymbol{v})$ and $(\boldsymbol{J}, \boldsymbol{\theta})$ is through the torus machinery developed in a series of papers by McGill and Binney (1990), Binney and Kumar (1993), Kaasalainen and Binney (1994) and McMillan and Binney (2008). This machinery proceeds by constructing a generating function that maps a simple toy torus to the target torus. At a given $\boldsymbol{\theta}$ on this torus, $(\boldsymbol{x}, \boldsymbol{v})$ may be simply found using the constructed generating function. This approach suffers from the disadvantage of supplying $(\boldsymbol{x}, \boldsymbol{v})$ given $(\boldsymbol{J}, \boldsymbol{\theta})$ which for many purposes is not ideal. For instance, McMillan and Binney (2013) found such a routine was inhibitive to constraining the Milky Way potential using *Gaia*-like data.

To take advantage of the attractive properties of angle-action coordinates, it is necessary to produce algorithms for their approximation in general, realistic Galactic potentials. In recent years various approximation schemes have been developed for quickly estimating actions and angles in axisymmetric and triaxial potentials (Sanders 2012a; Binney 2012a; Sanders and Binney 2014). The first three chapters of this thesis detail this work.

1.5 Dynamical Modelling of the Galaxy

Instead of working with individual stars, we must use collections of stars from large-scale Galactic surveys to get some handle on the Galactic potential, and hence the mass distribution. The collective behaviour of the stars will only be consistent with known physics if we have correctly identified the Galactic potential. In this thesis, we describe two different structures in the Galaxy that may be used to measure the Galactic potential.

The first of these is the Galactic disc. The long-range force of gravity dominates the behaviour of the Galaxy, and the force on an individual star is dominated by forces from distant stars as opposed to the few nearby objects. As such, we can model the stars in the Galaxy as being a collisionless fluid with some distribution function (DF), $f(\boldsymbol{x}, \boldsymbol{v})$, that obeys the collisionless Boltzmann equation (Binney and Tremaine 2008)

$$\frac{\mathrm{d}f}{\mathrm{d}t} = 0. \tag{1.7}$$

By Jeans' theorem, if we make f a function of the integrals of the motion, $\boldsymbol{I}$, we naturally satisfy this equation. One choice of integrals are the previously mentioned actions, $\boldsymbol{J}$. The actions depend on the potential of the Galaxy, $\Phi(\boldsymbol{x})$, such that we can describe the Galaxy with a combination of the two functions f and Φ. The collisionless Boltzmann equation ties together the spatial and velocity structure of the DF. Therefore, only when we have chosen the correct potential will the DF produce a consistent picture of the 6D phase-space structure of the Galaxy. This physically-motivated restriction on the model allows us to constrain the potential of the Galaxy.

Action-based distribution functions for the Galactic disc have been constructed in a series of papers by Binney (2010, 2012b), and fitted to both local data from the Geneva-Copenhagen survey and predicted more extended data from RAVE (Binney et al. 2014). Similarly, Bovy and Rix (2013) have used these models to constrain the mass scale-length of the Galactic disc using SEGUE data.

The real power from DF modelling comes from the complementary constraints of many populations. Each distinct population of stars has its own DF but crucially all populations must reside in the same potential. For instance, Walker and Peñarrubia (2011) divide stars in a globular cluster into two colour bins and simultaneously construct distribution functions for these populations to constrain the potential. A similar approach is certainly possible for the Galactic disc. Bovy et al. (2012) find that the Galactic disc can be considered as a series of spatially simple populations when separated into bins in the chemical abundance plane, ([α/Fe], [Fe/H]). When chemistry is added to dynamics it provides a powerful combination for constraining the Galactic potential.

Additionally, chemistry of the stars reveals important information regarding the history of the Galaxy. Essentially each star retains memory of its birth place through its internal composition. However, the actions of a star can change throughout its lifetime through dynamical processes, such as radial migration (Sellwood and Binney 2002). Schönrich and Binney (2009) constructed a full chemo-dynamical model of the Galaxy that included the effects of radial migration and found that the Geneva-Copenhagen survey was well fitted with such a model. By fitting these models to the data, the evolution of the Galaxy can be uncovered. At the end of this thesis we present an application of DF modelling of the Galaxy, and we explore the possibility of constructing an extended distribution function, which allows for chemo-dynamical modelling of the Galactic disc. The aim is to produce a consistent picture both of the current dynamical structure of the disc and also the past history and evolution of the Galaxy.

1.6 Tidal Streams

In the previous section, we have stressed the importance of smooth equilibrium models for studying the Galaxy. We know from our own Galaxy (e.g. Hou et al. 2009; Belokurov et al. 2006), as well as images of external galaxies, that substructures, such as spiral waves and infalling satellites, are important structures on top of these smooth models. Indeed one of the first observations of an external galaxy by Lord Rosse was of the Whirlpool galaxy, which exhibits spiral patterns and an infalling satellite. Just as with the Solar System, the incorporation of perturbative effects is made simpler through the use of angle-action variables. For instance, the velocity distribution in the Solar neighbourhood shows significant non-equilibrium structure (Dehnen 1998; Famaey et al. 2005). One such structure, the Hyades, can be accounted for by a perturbation due to the inner/outer Lindblad resonance of a spiral wave, which can be simply modelled in angle-action space (Sellwood 2010; McMillan 2011b, 2013). Additionally, the angle-action variables provide the ideal tool to find non-equilibrium structure in our Galaxy. Clumps in action-space, or non-uniformity in the angles, is clear evidence of substructure (McMillan and Binney 2008; Sanderson et al. 2014).

When we look into the halo of the Galaxy, it is rich in non-equilibrium substructure (Belokurov et al. 2006). This is entirely to be expected in the Λ-CDM picture of the Universe, where a large galaxy, such as our own, is accompanied by many smaller structures. The Large and Small Magellanic Clouds, first observed by Abd al-Rahman al-Sufi in 964, and the Sagittarius Dwarf Galaxy (Ibata et al. 1995) are the largest such structures, but there are many smaller globular clusters and satellites orbiting out in the halo (Koposov et al. 2008). Any satellite orbiting in the Galaxy will naturally experience strong tidal forces from the gravitational field of the Milky Way, which will cause the stars to be tidally stripped from the satellite. The stripped stars form long filamentary structures called tidal streams. Substructure associated with the Large Magellanic cloud was first observed by Mathewson et al. (1974), but this is not believed to be associated with tidal stripping. Not long after its discovery, Johnston et al. (1995) predicted that the Sagittarius dwarf would be heavily disrupted, and that debris might be observed in the solar neighbourhood. The tidal tails of the Sagittarius dwarf were found in the 2MASS data by Majewski et al. (2003). Since then, many streams have been discovered, mostly through matched filtering techniques on the Sloan Digital Sky Survey data, and the famous 'field of streams' image (Belokurov et al. 2006) showed that, as anticipated, the halo of our Galaxy is rich in such objects. We review the known streams of the Milky Way more fully in Chap. 5.

These structures are also present in external galaxies, such as M31 (Ibata et al. 2001), though we only observe high-mass satellites due to their greater luminosity. The number and distribution of these structures are interesting for comparison with cosmological predictions of galaxy formation. However, here we are interested in tidal streams as a probe of the Galaxy's potential.

1.6.1 Constraining the Galactic Potential with Tidal Streams

By its nature, a tidal stream is a correlated system of stars, and, as such, we can construct dynamical models for its formation and structure. Around each pericentric passage, or disc passage, some stars are stripped from the progenitor, and can then be considered as moving freely in the external gravitational potential of the Galaxy. Unlike the Galactic disc, we cannot consider the stars as being in a dynamical equilibrium—quite the opposite, the distribution function of the stream is a strong function of the angles. Importantly, the angles of the stars increase along the stream. The structure of the stream reflects the underlying potential of the Galaxy, such that inferences about the Galactic potential can be made from stream observations (McGlynn 1990; Johnston et al. 1996, 1999). Additionally, they prove highly complementary to the local measurements of the dark matter possible through disc modelling, and instead probe the global dark matter distribution. As such, tidal streams may be the best way to constrain the large-scale distribution of dark matter in our Galaxy.

Naively, we could consider the stars in the stream as a series of snapshots in time of a single star orbiting in the Galaxy. For a low-mass progenitor, the actions for each of the stars in the stream will be approximately the same, whilst the angles will increase linearly along the stream at a rate governed by the frequency of the progenitor. With this assumption, we can essentially measure the acceleration due to the Galaxy. This naive picture indicates how constraining the Galactic potential using streams operates, but in detail it is wrong.

There are many proposed methods for constraining the potential using tidal streams. The most primitive of these use the assumption that a stream delineates an orbit (Jin and Lynden-Bell 2007; Binney 2008). Due to the simplistic nature of this modelling approach, it has been employed successfully on real data, notably the GD-1 stream (Grillmair and Dionatos 2006) by Koposov et al. (2010). The validity of the simple stream-orbit approximation has been brought into doubt by several authors (Dehnen et al. 2004; Choi et al. 2007; Eyre and Binney 2011). Importantly, it is clear that a stream can never follow an orbit as, for a stream to form, the particles must be on different orbits. If all particles were on the same orbit, the 'stream' would not stretch in time but instead oscillate in length as it passed from apocentre to pericentre and back. We require the particles in a stream to follow slightly different orbits for the stream to stretch in time. Eyre and Binney (2011) demonstrated that assuming a stream delineates an orbit can lead to biases when estimating the parameters of the Galactic potential. Additionally, these authors provided a clear and thorough discussion of the structure of streams in angle-action space. Angle-action coordinates are very attractive for viewing the formation of streams, and, in the correct potential, there are very clear correlations between the angles and the actions of a stream. However, the discussion in Eyre and Binney (2011) was limited to potentials that have separable Hamilton-Jacobi equations (Stäckel potentials, of which spherical potentials are limiting cases) for which the angle-action coordinates are numerically tractable.

The angle-action approach allows for a clearer understanding of the structure of streams, as well as presenting a route to improving the modelling of the streams in order to constrain the potential. This approach more correctly models the formation and dynamics of the stream. Similarly, many authors have proposed approaches to constraining the potential without assuming that the stream delineates an orbit. Varghese et al. (2011) correct the orbit track of the progenitor by releasing stars at pericentric passage assuming some model for the progenitor. This approach can rapidly generate stream models but relies on strong modelling assumptions. Similarly, Küpper et al. (2012) use a *streak-line* method for constructing streams by continually releasing particles from the progenitor at the tidal radius. This approach has been used to constrain properties of the Via Lactea II simulation from its constituent streams by Bonaca et al. (2014). Johnston et al. (1999) and Price-Whelan and Johnston (2013) use an 'orbit-rewinding' approach that integrates the orbit of each observed star backwards in time until the particles comes within some phase-space distance of the progenitor. This method requires knowledge of the progenitor, which is often not available (e.g. GD-1), but can be treated as an unknown parameter to marginalize over (Price-Whelan et al. 2014). The method also requires one to introduce some arbitrariness in the choice of phase-space distance measure.

Despite not delineating an orbit, a stream still forms a very tight clump in action-space. Peñarrubia et al. (2012) show that in the correct potential the spread in the integrals, or actions, is minimized, and this spread increases when using the wrong potential. This entropy minimization technique is also used by Sanderson et al. (2014). It certainly avoids the stream/orbit question but only by neglecting the phase information of the stream. The high degree of correlation in the angle structure of the stream is indicative of its constraining power, so to throw out this information is perhaps not wise. Finally, for high-mass progenitors such as Sagittarius, perhaps the only approach to using the tidal tails to constrain the potential is through full N-body modelling (e.g. Law and Majewski 2010). Such an approach models the full dynamical effects of a cluster in a tidal field, but (perhaps crucially) neglects the effects of dynamical friction that must affect such a high-mass progenitor (Jiang and Binney 2000).

Following Eyre and Binney (2011) the key questions were: Is the stream = orbit approximation valid for realistic Galactic potentials? How can we exploit the insights of Eyre and Binney (2011) to constrain a potential using tidal streams which improves on orbit fitting? How can we find the angle-action coordinates in a general potential? Chapters 5 and 6 of this thesis will aim to answer these questions.

1.7 Overview of Thesis

This thesis discusses approaches to modelling the dynamics of the Milky Way. The common aim is to understand the structure of the Galaxy, particularly the distribution of dark matter, by various dynamical probes. We focus on two dynamical probes to achieve this goal: extended distribution function modelling of the Galactic disc provides constraints on the local structure of the Galaxy, which is complemented by the more global constraints provided by tidal streams. Both of these approaches rely on the construction of appropriate angle-action-based models. This thesis consists of three key parts divided over seven main chapters. Firstly, Chaps. 2, 3 and 4 develop algorithms for estimating angle-action coordinates in general axisymmetric and triaxial potentials. Chapters 5 and 6 go on to discuss tidal streams and how we may use them to constrain the Galactic potential. We discuss distribution functions for the Galactic disc in Chaps. 7 and 8. Finally in Chap. 9, we present the conclusions of the thesis, along with a look to the future and the importance of the presented work. Here we briefly summarize the content of each chapter.

1.7.1 Angle-Action Estimation in a General Axisymmetric Potential

The usefulness of angle-action variables in galaxy dynamics is well known, but their use is limited due to the difficulty of their calculation in realistic galaxy potentials. In Chap. 2, we present methods for estimating angle-action variables in realistic axisymmetric Milky Way potentials. The presented methods are split into two categories: convergent and non-convergent methods. The first non-convergent method proceeds by locally fitting a Stäckel potential over the region an orbit probes, and estimating the true actions as those in the fitted potential. The quality of the method is assessed by estimating the angle-action variables of a range of disc and halo-type orbits, and we present a brief application by projecting the Geneva-Copenhagen survey into angle-action space. We further present a new method for estimating the actions based on the adiabatic approximation of Schönrich and Binney (2012). This new method improves on the assumption of separability in cylindrical polar coordinates and uses a more general assumption of separability in spheroidal coordinates. We compare the accuracy and computational time of these new methods with other known non-convergent methods (Binney 2012a). Finally, we present a slower convergent method for estimating the actions. The method operates by iteratively constructing tori using action estimates from the non-convergent approaches. The work in this chapter is based on that published by Sanders (2012a).

1.7.2 Actions, Angles and Frequencies from Numerically Integrated Orbits

In Chap. 3, we present a method for extracting actions, angles and frequencies from an orbit's time series. The method recovers the generating function that maps an analytic phase-space torus to the torus to which the orbit is confined, by simultaneously solving the constraints provided by each time step. We test the method by recovering the actions and frequencies of tori in a triaxial Stäckel potential, and use it to investigate the structure of orbits in a triaxial potential that has been fitted to our Galaxy's Sagittarius stream. We show how the method operates when working with near-resonant and chaotic orbits. The method promises to be useful for analysing N-body simulations. The work in this chapter is based on that published by Sanders and Binney (2014).

1.7.3 Action Estimation Using a Triaxial Stäckel Approximation

In Chap. 4, we present a fast method for estimating the actions in a general triaxial potential. The method is an extension of the axisymmetric approach presented by Binney (2012a), and operates by assuming that the true potential is sufficiently close to a Stäckel potential. We require rapid action evaluation in order to find the moments of triaxial distribution functions. The method uses only the initial phase-space point to estimate the classical integrals and hence does not involve any time-consuming orbit integration. We briefly inspect the accuracy of the method by estimating the actions of orbits in a triaxial Navarro-Frenk-White potential (Navarro et al. 1996). We go on to show how the method can be used to construct triaxial distribution functions that satisfy the Jeans equations, and so can be used effectively to build models of external galaxies, as well as triaxial components of our own Galaxy.

1.7.4 Stream-Orbit Misalignment

Tidal streams don't, in general, delineate orbits. A stream-orbit misalignment is expected to lead to biases when using orbit-fitting to constrain models for the Galactic potential. In Chap. 5, we discuss the expected magnitude of the misalignment and the resulting dangers of using orbit-fitting algorithms to constrain the potential. We summarize data for known streams that should prove useful for constraining the Galactic potential, and compute their actions in a realistic Galactic potential. We go on to discuss the formation of tidal streams in angle-action space, and explain why, in general, streams do not delineate orbits. The magnitude of the stream-orbit misalignment is quantified for a logarithmic potential and multi-component Galactic potentials. Specifically, we focus on the expected misalignment for the known streams. By introducing a two-parameter family of realistic Galactic potentials, we demonstrate that assuming these streams delineate orbits can lead to order one errors in the halo flattening and halo-to-disc force ratio at the Sun. We present a discussion of the dependence of these results on the progenitor mass, and demonstrate that the misalignment is mass-independent for the range of masses of observed streams. Hence, orbit-fitting does not yield better constraints on the potential if one uses narrower, lower-mass streams. The work in this chapter is based on that published by Sanders and Binney (2013a).

1.7.5 Stream Modelling in Angle-Frequency Space

In Chap. 6, we present new techniques for constraining the Galactic potential using tidal streams, which crucially do not rely on the assumption that a stream delineates an orbit. The key idea is that a stream takes on a very simple structure when viewed in angle and frequency space in the correct potential. We present a simple algorithm for constraining the Galactic potential using these observations and show that we are able to recover the parameters of a logarithmic potential from a simulation. We investigate the impact of observational errors on the algorithm and conclude that the simple approach is not robust enough to large observational errors. We proceed by writing down a fully probabilistic model for a stream in the space of observables using the expected structure in angle-frequency space. We show that we are able to recover the parameters of the potential by sampling from this model, and that the probabilistic model can handle contamination of the data with non-stream members. The work in this chapter is based on that published by Sanders and Binney (2013b) and Sanders (2014).

1.7.6 Determining the Velocity Dispersion of the Thick Disc

A recent study by Moni Bidin et al. (2012a) used a sample of 412 red giants to determine the vertical velocity dispersion of the thick disc as a function of Galactic height. The results of this study were then used in Moni Bidin et al. (2012b) to measure the mass density in the Galactic disc, and it was concluded that there is much less dark matter in the solar neighbourhood than anticipated. In Chap. 7, we present a study of the method used in Moni Bidin et al. (2012a) by attempting to recover the velocity dispersion of the thick disc from the realistic Galactic distribution function of Binney (2012b). We show that the errors reported by Moni Bidin et al. (2012a) are far too small, as two significant sources of error are neglected. We also demonstrate that the method leads to biases in the vertical velocity dispersion gradient with Galactic height. This work is based on that published by Sanders (2012b).

1.7.7 Extended Distribution Functions for the Galactic Disc

In Chap. 8, we present an extension to the action-based distribution functions of Binney (2012b) which includes metallicity information. Purely dynamical distribution functions have limited use when attempting to model data sets, due to the presence of selection effects. Surveys naturally select in the colour and magnitude of the stars, which in turn affects the observed metallicity distribution. Metallicity and age are believed to be correlated with velocity as, for instance, older metal-poor stars are members of hotter populations. Therefore, inclusion of metallicity in a distribution function is necessary for modelling the dynamics of a survey. Additionally, it provides a probe of the formation and history of the Galaxy, and useful leverage on constraining the Galactic potential. We demonstrate that our extended distribution functions provide good fits to data from the Geneva-Copenhagen survey, and show the effects of simple selection functions on metallicity and velocity distributions. We go on to construct a mock SEGUE G dwarf sample from our model and compare to the data.

References

Arnold VI (1978) Mathematical methods of classical mechanics. Springer, New York
Bailin J, Steinmetz M (2005) ApJ 627:647
Begeman KG (1989) A & A 223:47
Belokurov V et al (2006) ApJL 642:L137
Berry M et al (2012) ApJ 757:166
Binney J (2008) MNRAS 386:L47
Binney J (2010) MNRAS 401:2318
Binney J (2012a) MNRAS 426:1324
Binney J (2012b) MNRAS 426:1328
Binney J et al (2014) MNRAS 439:1231
Binney J, Kumar S (1993) MNRAS 261:584
Binney J, Tremaine S (2008) Galactic dynamics, 2nd edn. Princeton University Press, New Jersey
Bonaca A, Geha M, Kuepper AHW, Diemand J, Johnston KV, Hogg DW (2014) ArXiv e-prints
Bovy J et al (2012) ApJ 759:131
Bovy J, Rix H-W (2013) ApJ 779:115
Boylan-Kolchin M, Springel V, White SDM, Jenkins A, Lemson G (2009) MNRAS 398:1150
Burnett B et al (2011) A & A 532:A113
Choi J-H, Weinberg MD, Katz N (2007) MNRAS 381:987
de Zeeuw T (1985) MNRAS 216:273
Dehnen W (1998) AJ 115:2384
Dehnen W, Odenkirchen M, Grebel EK, Rix H-W (2004) AJ 127:2753
Dejonghe H, de Zeeuw T (1988) ApJ 329:720
Eisenstein DJ et al (2011) AJ 142:72
Eyre A, Binney J (2011) MNRAS 413:1852
Famaey B, Dejonghe H (2003) MNRAS 340:752
Famaey B, Jorissen A, Luri X, Mayor M, Udry S, Dejonghe H, Turon C (2005) A & A 430:165
Genzel R, Eisenhauer F, Gillessen S (2010) Rev Mod Phys 82:3121
Gilmore G et al (2012) Messenger 147:25
Grillmair CJ, Dionatos O (2006) ApJL 643:L17
Henon M (1959) Annales d'Astrophysique 22:126
Hou LG, Han JL, Shi WB (2009) A & A 499:473
Ibata R, Irwin M, Lewis G, Ferguson AMN, Tanvir N (2001) Nature 412:49
Ibata RA, Gilmore G, Irwin MJ (1995) MNRAS 277:781
Jiang I-G, Binney J (2000) MNRAS 314:468

Jin S, Lynden-Bell D (2007) MNRAS 378:L64
Johnston KV, Hernquist L, Bolte M (1996) ApJ 465:278
Johnston KV, Spergel DN, Hernquist L (1995) ApJ 451:598
Johnston KV, Zhao H, Spergel DN, Hernquist L (1999) ApJL 512:L109
Kaasalainen M, Binney J (1994) MNRAS 268:1033
Koposov S et al (2008) ApJ 686:279
Koposov SE, Rix H-W, Hogg DW (2010) ApJ 712:260
Küpper AHW, Lane RR, Heggie DC (2012) MNRAS 420:2700
Law DR, Majewski SR (2010) ApJ 714:229
Liddle A (2003) An introduction to modern cosmology, 2nd edn. Wiley, New York
Majewski SR, Skrutskie MF, Weinberg MD, Ostheimer JC (2003) ApJ 599:1082
Massey R, Kitching T, Richard J (2010) Rep Prog Phys 73:086901
Mathewson DS, Cleary MN, Murray JD (1974) Ap J 190:291
McGill C, Binney J (1990) MNRAS 244:634
McGlynn TA (1990) ApJ 348:515
McMillan PJ (2011a) MNRAS 414:2446
McMillan PJ (2011b) MNRAS 418:1565
McMillan PJ (2013) MNRAS 430:3276
McMillan PJ, Binney JJ (2008) MNRAS 390:429
McMillan PJ, Binney JJ (2013) MNRAS 433:1411
Milgrom M (1983) ApJ 270:365
Moni Bidin C, Carraro G, Méndez RA (2012) ApJ 747:101
Moni Bidin C, Carraro G, Méndez RA, Smith R (2012) ApJ 751:30
Munari U, Agnolin P, Tomasella L (2001) Balt Astron 10:613
Navarro JF, Frenk CS, White SDM (1996) ApJ 462:563
Nordström B et al (2004) A & A 418:989
Oort JH (1932) Bull Astr Inst Neth 6:249
Peñarrubia J, Koposov SE, Walker MG (2012) ApJ 760:2
Perryman MAC et al (2001) A & A 369:339
Perryman MAC et al (1997) A & A 323:L49
Planck Collaboration et al (2013) ArXiv e-prints
Price-Whelan AM, Hogg DW, Johnston KV, Hendel D (2014) ArXiv e-prints
Price-Whelan AM, Johnston KV (2013) ApJL 778:L12
Reyes R, Mandelbaum R, Seljak U, Baldauf T, Gunn JE, Lombriser L, Smith RE (2010) Nature 464:256
Roeser S, Demleitner M, Schilbach E (2010) AJ 139:2440
Sackett PD (1999) In: Merritt DR, Valluri M, Sellwood JA (eds.) Astronomical society of the pacific conference series, galaxy dynamics—a Rutgers symposium, vol 182, p 393
Sanders J (2012a) MNRAS 426:128
Sanders J (2012b) MNRAS 425:2228
Sanders JL (2014) MNRAS 443:423
Sanders JL, Binney J (2013a) MNRAS 433:1813
Sanders JL, Binney J (2013b) MNRAS 433:1826
Sanders JL, Binney J (2014) MNRAS 441:3284
Sanderson R, Helmi A, Hogg DW (2014) In: American astronomical society meeting abstracts, vol 223, p. 346.10
Schlafly EF et al (2014) ApJ 789:15
Schönrich R, Binney J (2009) MNRAS 396:203
Schönrich R, Binney J (2012) MNRAS 419:1546
Sellwood JA (2010) MNRAS 409:145
Sellwood JA, Binney JJ (2002) MNRAS 336:785
Skrutskie MF et al (2006) AJ 131:1163
Springel V et al (2005) Nature 435:629

Stäckel P (1893) Math Ann 42:537
Uzan J-P (2009) SSR 148:249
Varghese A, Ibata R, Lewis GF (2011) MNRAS 417:198
Walker MG, Peñarrubia J (2011) ApJ 742:20
Watkins LL, Evans NW, An JH (2010) MNRAS 406:264
Wilkinson MI, Evans NW (1999) MNRAS 310:645
Xue XX et al (2008) ApJ 684:1143
Yanny B et al (2009) AJ 137:4377
Zacharias N et al (2010) AJ 139:2184
Zwicky F (1933) Helv Phys Acta 6:110

Chapter 2
Angle-Action Estimation in a General Axisymmetric Potential

2.1 Introduction

In the study of dynamical systems, it is becoming increasingly important to be able to process and understand large multi-dimensional data sets efficiently. The stars in our own Galaxy, the Milky Way, are being increasingly observed in full six-dimensional phase-space through the combination of astrometry and radial velocity measurements. Full 6D phase-space information is currently available for stars in the solar neighbourhood from the Geneva-Copenhagen and RAVE surveys (Nordström et al. 2004; Zwitter et al. 2008) and this is to be greatly expanded on by the space mission *Gaia* (Perryman et al. 2001). Beyond our Galaxy, the advent of integral-field spectroscopy has led to projects such as SAURON (Bacon et al. 2001 and subsequent papers), which mapped the kinematics of a representative sample of 72 nearby elliptical and spiral galaxies, and subsequently ATLAS3D (Cappellari et al. 2011 and subsequent papers), which combined SAURON observations with CO and HI observations to study the kinematics of a complete volume-limited sample of 260 local early-type galaxies.

Although no galaxy is ever in perfect dynamical equilibrium, equilibrium dynamical models are central to the interpretation of observations of both our Galaxy and external galaxies. N-body simulations of cosmological clustering yield a picture in which dark-matter haloes are far from dynamical equilibrium only during short-lived and quite rare major mergers. In general a dark-matter halo can be well approximated by a dynamical equilibrium that is mildly perturbed by accretion. A major reason for the importance of equilibrium models is that we can infer a galaxy's gravitational potential, and thus its dark-matter distribution, only to the extent that the galaxy is in equilibrium. Moreover, equilibrium models are the simplest models and more complex configurations, involving spiral structure or an on-going minor merger for example, are best modelled as perturbations of an equilibrium model. Schwarzschild modelling (Schwarzschild 1979) constructs equilibrium models by describing the configuration of a model by a weighted set of orbits. However, this approach is not the most natural as each orbit is characterised by its initial phase-space coordinates.

J. Sanders, *Dynamics of the Milky Way*, Springer Theses,
DOI 10.1007/978-3-319-18772-3_2

Additionally, observational data is often understood by performing large N-body simulations. Whilst such models are straightforward to produce, the configurations of the models are difficult to control and characterise. It is necessary that techniques are developed which can simplify both observational and simulation data without losing the richness of the phase-space information.

2.1.1 Angle-Action Variables

The natural way to model a dynamical equilibrium is via Jeans theorem, which assures us that the system's distribution function (DF) can be assumed to be a non-negative function of isolating integrals. Since one expects a smooth time-independent gravitational potential to admit up to three functionally independent isolating integrals, Jeans theorem states that we should be able to represent an equilibrium stellar system by the density of stars in a three-dimensional space of integrals rather than in full six-dimensional phase space. This reduction in dimensionality makes the system very much easier to comprehend and model.

Since any function of integrals is itself an integral, infinitely many different integrals may be used as arguments of the DF. However, the action integrals J_i stand out as uniquely suited to be used as arguments of the DF. Along with the angles, the actions form a set of canonical coordinates that can be used to express the equations of motion in a trivial form: the actions are integrals of the motion whilst the angles increase linearly with time. Such a formulation instantly reduces the complexity of any dynamical data set. Along an orbit, the six phase-space dimensions are reduced to three angle coordinates. Angle-action variables can be defined for any quasi-periodic orbit. Initially introduced to study celestial mechanics, angle-action variables now have great potential for galaxy dynamics due to their attractive properties. It is particularly convenient to use actions as arguments of the DF as (i) they are adiabatic invariants, (ii) the zero-point of an action is well defined and (iii) the range of values an action may take is independent of the other actions. The angle-action variables also provide a basis for the development of a perturbative solution to the equations of motion (see Binney and Tremaine 2008 for a much fuller discussion of the merits of angle-action variables). Finally, we mention that angle-action variables can be used to hunt for substructure in the Galaxy (e.g. McMillan and Binney 2008), and will be used in Chap. 6 to model tidal streams.

Despite the aforementioned advantages, angle-action variables remain awkward to work with in practical applications due to the difficulty of their calculation in a general potential. They are easily calculated when the potential is spherical and with more work can be numerically calculated when the potential is of Stäckel form. However, neither of these approaches is satisfactory when working with realistic galaxy potentials, because such potentials do not satisfy these conditions. The development of methods to estimate angle-action variables in a general potential is crucial if we are to benefit from the advantages of angle-action variables and the wealth of techniques that utilise them. For some applications, such as stream modelling, we require accurate actions for a few data points, such that we can afford to spend a long time evaluating the actions. However, for other applications, such as disc modelling, we require a fast algorithm for estimating the actions of many data points. Therefore, a variety of algorithms are required.

In this chapter, we present methods for estimating angle-action variables in a general axisymmetric potential. The methods presented in this chapter fall into two categories: slower convergent algorithms that produce very accurate actions, and faster non-convergent algorithms that produce less accurate actions. The majority of these methods are based on the analytic calculation of the actions in an axisymmetric Stäckel potential, so, in Sect. 2.2, we give a brief overview of the determination of angle-action variables in this class of potential. We continue by presenting a series of non-convergent methods based on the Stäckel potential. In Sect. 2.3, we present a method for estimating the angle-actions by locally fitting a Stäckel potential to the region a given orbit explores in an axisymmetric potential. The actions and angles in the general axisymmetric potential are estimated as those in the fitted Stäckel potential. The results of the method are examined by analysing artificial data in Sect. 2.3.3, and we demonstrate the practical application of the method by inspecting the Geneva-Copenhagen Survey in angle-action space in Sect. 2.3.4. In Sect. 2.4, we present the adiabatic approximation from Schönrich and Binney (2012) that is based on separability in cylindrical polar coordinates. In Sect. 2.5, we present a new method that improves on the adiabatic approximation by assuming separability in a general prolate spheroidal coordinate system. In Sect. 2.6, we present the method from Binney (2012) that assumes the potential is close to a Stäckel potential. Finally, we show how these non-convergent methods can be combined with torus construction (McMillan and Binney 2008) to produce a convergent method for finding the actions from a phase-space point $(\boldsymbol{x}, \boldsymbol{v})$. The results of all of these methods are compared in Sect. 2.8. The work on which this chapter is based was published by Sanders (2012).

2.2 Actions and Angles in a Stäckel Potential

The most general class of potentials in which we are able to calculate the angle-action variables analytically is that of Stäckel potentials. In a confocal ellipsoidal coordinate system, these potentials produce separable Hamilton-Jacobi equations. A full discussion of Stäckel potentials is given in de Zeeuw (1985). Here we limit the discussion to oblate axisymmetric Stäckel potentials which are associated with prolate spheroidal coordinates (λ, ϕ, ν). A specific prolate spheroidal coordinate system is defined by two constants (a, c). These coordinates are related to cylindrical polar coordinates (R, ϕ, z) by

$$\frac{R^2}{\tau - a^2} + \frac{z^2}{\tau - c^2} = 1, \tag{2.1}$$

where λ and ν are the roots of τ such that $c^2 \leq \nu \leq a^2 \leq \lambda$. Surfaces of constant λ are prolate spheroids and surfaces of constant ν are two-sheeted hyperboloids of revolution that intersect the spheroids orthogonally. Given λ and ν we can find expressions for R and z as a function of (λ, ν),

$$\begin{aligned} R^2 &= \frac{(\lambda - a^2)(\nu - a^2)}{c^2 - a^2}, \\ z^2 &= \frac{(\lambda - c^2)(\nu - c^2)}{a^2 - c^2}. \end{aligned} \tag{2.2}$$

To convert from the canonical coordinates, (R, z, p_R, p_z), to the coordinates (τ, p_τ), we introduce the generating function $S(\lambda, \nu, p_R, p_z)$ given by

$$S(\lambda, \nu, p_R, p_z) = p_R R(\lambda, \nu) + p_z z(\lambda, \nu). \tag{2.3}$$

We can then find $p_\tau = \partial S/\partial \tau$ as

$$\begin{aligned} p_\lambda &= \frac{1}{2} p_R \sqrt{\frac{\nu - a^2}{(c^2 - a^2)(\lambda - a^2)}} + \frac{1}{2} p_z \sqrt{\frac{\nu - c^2}{(a^2 - c^2)(\lambda - c^2)}}, \\ p_\nu &= \frac{1}{2} p_R \sqrt{\frac{\lambda - a^2}{(c^2 - a^2)(\nu - a^2)}} + \frac{1}{2} p_z \sqrt{\frac{\lambda - c^2}{(a^2 - c^2)(\nu - c^2)}}. \end{aligned} \tag{2.4}$$

Similarly, differentiating Eq. (2.2) we find $p_\lambda = P_\lambda^2 \dot{\lambda}$ and $p_\nu = P_\nu^2 \dot{\nu}$ where the dot denotes differentiation with respect to time and

$$\begin{aligned} P_\lambda^2 &= \frac{\lambda - \nu}{4(\lambda - a^2)(\lambda - c^2)}, \\ P_\nu^2 &= \frac{\nu - \lambda}{4(\nu - a^2)(\nu - c^2)}. \end{aligned} \tag{2.5}$$

Inversion of Eq. (2.4) allows us to write the Hamiltonian as

$$\begin{aligned} H &= \frac{1}{2}(p_R^2 + p_z^2) + \frac{L_z^2}{2R^2} + \Phi(R, z) \\ &= \frac{1}{2}\left(P_\lambda^2 \dot{\lambda}^2 + P_\nu^2 \dot{\nu}^2 + \frac{L_z^2}{R^2(\lambda, \nu)}\right) + \Phi(\lambda, \nu) \\ &= \frac{1}{2}\left(\frac{p_\lambda^2}{P_\lambda^2} + \frac{p_\nu^2}{P_\nu^2} + \frac{L_z^2}{R^2(\lambda, \nu)}\right) + \Phi(\lambda, \nu), \end{aligned} \tag{2.6}$$

where L_z is the z-component of the angular momentum. A potential, Φ_S, is of Stäckel form in a particular prolate spheroidal coordinate system if

$$\Phi_S = -\frac{f(\lambda) - f(\nu)}{\lambda - \nu}. \tag{2.7}$$

Φ_S is fully defined by a single function $f(\tau)$. A single function may be used as λ and ν take different ranges of values except at $\lambda = a^2, \nu = a^2$, where we require f to be continuous so the potential remains finite. As in all axisymmetric potentials, the energy, E, and z-component of the angular momentum, L_z are isolating integrals. In a Stäckel potential we are in the fortunate position of being able to find analytically a third isolating integral, I_3. To do this we solve the Hamilton-Jacobi equation. We introduce the generating function, $W(\lambda, \nu, \boldsymbol{J})$ for the transformation between the prolate spheroidal coordinates and the as yet unknown actions and corresponding angles. With this generating function we write $p_\tau = \partial W/\partial\tau$ and set the Hamiltonian at fixed $\boldsymbol{J}$ to the energy E. We make the Ansatz that $W = \sum_\tau W_\tau(\tau)$ and find that

$$\begin{aligned}\lambda E + f(\lambda) - 2(\lambda - a^2)(\lambda - c^2)\Big(\frac{\partial W_\lambda}{\partial\lambda}\Big)^2 + \Big(\frac{c^2 - a^2}{\lambda - a^2}\Big)\frac{L_z^2}{2} \\ = \nu E + f(\nu) - 2(\nu - a^2)(\nu - c^2)\Big(\frac{\partial W_\nu}{\partial\nu}\Big)^2 + \Big(\frac{c^2 - a^2}{\nu - a^2}\Big)\frac{L_z^2}{2}.\end{aligned} \tag{2.8}$$

We introduce the separation constant

$$K = c^2 E - \frac{L_z^2}{2} + I_3, \tag{2.9}$$

where I_3 is the third integral. We now have expressions for the momenta as a function of τ and the classical integrals i.e.

$$2(\tau - a^2)(\tau - c^2)p_\tau^2 = (\tau - c^2)E - \Big(\frac{\tau - c^2}{\tau - a^2}\Big)\frac{L_z^2}{2} - I_3 + f(\tau). \tag{2.10}$$

This expression allows us to find the third isolating integral as

$$I_3 = (\tau - c^2)E - \Big(\frac{\tau - c^2}{\tau - a^2}\Big)\frac{L_z^2}{2} + f(\tau) - 2(\tau - a^2)(\tau - c^2)p_\tau^2. \tag{2.11}$$

We define the action variables, J_λ and J_ν, as

$$J_\tau = \frac{1}{2\pi}\oint p_\tau \mathrm{d}\tau, \tag{2.12}$$

where the integration is over all values of τ for which $p_\tau^2 \geq 0$. As $p_\tau = p_\tau(\tau, E, I_2, I_3)$, the actions are solely functions of the isolating integrals and thus constants of the motion. The third action, J_ϕ, is simply L_z. Therefore, given a Cartesian phase-space point $(\boldsymbol{x}_0, \boldsymbol{v}_0)$, we can find the three isolating integrals,

$\boldsymbol{I} = (E, L_z, I_3)$, using the coordinate transformation and perform the 1D integrals in Eq. (2.12) to find the actions.

The actions give an absolute measure of the extent of the oscillations of the orbit in each of the coordinates. At large radii, the prolate spheroidal coordinate system becomes spherical such that $\lambda \approx R^2 + z^2$. Therefore, we can think of J_λ as a measure of the radial oscillations. The ν coordinate increases as we move away from the $z = 0$ plane, so we may think of J_ν as a measure of the vertical oscillations.

The corresponding angle coordinates, θ_τ and θ_ϕ, are calculated using the generating function, $W(\lambda, \phi, \nu, J_\lambda, L_z, J_\nu)$ for the canonical transformation from $(\lambda, \phi, \nu, p_\lambda, L_z, p_\nu)$ to $(\theta_\lambda, \theta_\phi, \theta_\nu, J_\lambda, L_z, J_\nu)$. The angles are found by differentiating the generating function with respect to the respective action such that

$$\theta_\tau = \frac{\partial W}{\partial J_\tau} \text{ for } \tau = \{\lambda, \nu\}; \; \theta_\phi = \frac{\partial W}{\partial L_z}. \tag{2.13}$$

A full list of formulae, as well as a discussion of how to perform the quadratures numerically, is given in Appendix A. In this appendix, we also give expressions for the frequencies, $\boldsymbol{\Omega} = \partial H / \partial \boldsymbol{J}$, which will be useful in later chapters.

2.3 Estimating Actions in a Fitted Stäckel Potential

We have seen how to find actions in a Stäckel potential. We now go on to show how we can use these insights to estimate actions in a general potential. Given the ease with which we can calculate actions and angles in a Stäckel potential, it seems sensible to investigate how well a Stäckel potential can fit a Galaxy model so that we may estimate the actions and angles as those calculated in this best-fitting potential. It has been known for some time that Stäckel potentials do not give a good fit to the potential of the Galaxy globally due to the rigid conditions they must fulfil. Dejonghe and de Zeeuw (1988) outline a method for fitting a general axisymmetric potential with a Stäckel potential, which can be applied both globally and locally. These authors produced global fits for the Bahcall-Schmidt-Soneira Galaxy model (Bahcall et al. 1982) with errors in the potential nowhere exceeding 3 %, and Jasevicius (1994) carried out a similar analysis on a broader range of Milky Way potential models with similar results. As expected, the fits are worst in the central 0.5 kpc of the Galaxy. De Bruyne et al. (2000) sought to fit axisymmetric potentials locally using a set of Stäckel potentials in order to calculate the third isolating integral, I_3. When applied to a Miyamoto-Nagai potential, I_3 was found to vary by approximately 10 % along an orbit. Here we follow the method presented by Dejonghe and de Zeeuw (1988).

Suppose we have an axisymmetric potential $\Phi(R, z)$ that we wish to fit by a Stäckel potential Φ_{fit}. We begin by choosing a prolate spheroidal coordinate system by specifying (a, c). The coordinate system is fully specified by the combination $(a^2 - c^2)$, so we are free to set $c^2 = 1$, which reduces numerical difficulties. We determine a by using a property of an axisymmetric Stäckel potential (de Zeeuw 1984). It follows from Eq. (2.7) that for a Stäckel potential, Φ_S,

$$\frac{\partial^2}{\partial\lambda\partial\nu}[(\lambda-\nu)\Phi_S]=0. \tag{2.14}$$

Therefore, for a general potential Φ we use this equation as a definition for the coordinate system. We find this gives an estimate for a at a point (R, z)

$$a^2-c^2=R^2-z^2-\left[3z\frac{\partial\Phi}{\partial R}-3R\frac{\partial\Phi}{\partial z}+Rz\left(\frac{\partial^2\Phi}{\partial R^2}-\frac{\partial^2\Phi}{\partial z^2}\right)\right]\Big/\frac{\partial^2\Phi}{\partial R\partial z}. \tag{2.15}$$

Later, we calculate a sufficiently accurate value of a by evaluating this expression at multiple positions along an orbit and averaging.[1] With this choice of a we transform $\Phi(R, z)$ to $\Phi(\lambda, \nu)$ and specify the fitting region: $\lambda_-\leq\lambda\leq\lambda_+$, $\nu_-\leq\nu\leq\nu_+$. A global fit corresponds to $\nu_-=c^2$, $\nu_+=\lambda_-=a^2$, $\lambda_+=\infty$. We define the *auxiliary function*

$$\chi(\lambda,\nu)\equiv-(\lambda-\nu)\Phi(\lambda,\nu). \tag{2.16}$$

If the potential Φ is of Stäckel form, this auxiliary function is simply $\chi(\lambda,\nu)=f(\lambda)-f(\nu)$. We seek the function f that makes $\Phi_{\rm fit}$ most like Φ by minimising the square difference of the potential auxiliary function and the fitting potential auxiliary function, $\chi_{\rm fit}$, over the fit region. Therefore, we minimise the functional

$$F[f]=\int_{\lambda_-}^{\lambda_+}\mathrm{d}\lambda\int_{\nu_-}^{\nu_+}\mathrm{d}\nu\,\Lambda(\lambda)N(\nu)(\chi(\lambda,\nu)-f(\lambda)+f(\nu))^2, \tag{2.17}$$

where $\Lambda(\lambda)$ and $N(\nu)$ are weighting functions allowing us to acquire a better fit in certain areas. These functions must be finite when integrated over the fitting region. We choose the normalised weighting functions

$$\Lambda(\lambda)=4\lambda^{-5}(\lambda_-^{-4}-\lambda_+^{-4})^{-1},\quad N(\nu)=(\nu_+-\nu_-)^{-1}. \tag{2.18}$$

This choice of weighting functions gives preferential weight to smaller values of λ where the potential is harder to fit. Analytic minimisation of the functional F results in a best fit function

$$f(\lambda)=\bar{\chi}(\lambda)-\frac{1}{2}\bar{\bar{\chi}},\quad f(\nu)=-\bar{\chi}(\nu)+\frac{1}{2}\bar{\bar{\chi}}, \tag{2.19}$$

[1] For an oblate Stäckel potential, this expression gives the exact a^2-c^2 at every point. However, for a general oblate potential, this expression can give negative values such that $a^2<0$. This implies that the best-fitting coordinate system is oblate (see Sect. 2.3.5). However, this only seems to occur when the equipotentials are near spherical (e.g. at high angular momentum in McMillan potential). For the logarithmic potential explored later this formula selects prolate coordinate systems for $q<1$ and oblate for $q>1$.

where

$$\begin{aligned}\bar{\chi}(\lambda) &= \int_{\nu_-}^{\nu_+} \mathrm{d}\nu\, \chi(\lambda, \nu) N(\nu), \\ \bar{\chi}(\nu) &= \int_{\lambda_-}^{\lambda_+} \mathrm{d}\lambda\, \chi(\lambda, \nu) \Lambda(\lambda), \\ \bar{\bar{\chi}} &= \int_{\nu_-}^{\nu_+} \int_{\lambda_-}^{\lambda_+} \mathrm{d}\lambda \mathrm{d}\nu\, \chi(\lambda, \nu) \Lambda(\lambda) N(\nu).\end{aligned} \tag{2.20}$$

The derivation of these equations is given in Appendix B. The quality of the fit we have achieved is then measured by $F[f]$.

2.3.1 Procedure

Combining the above two sections, we can estimate the actions and angles of a phase point $(\boldsymbol{x}, \boldsymbol{v})$ by first fitting a Stäckel potential to the given potential over the region the orbit probes, and then calculating the angle-action variables in this fitted potential. Therefore, given a point $(\boldsymbol{x}, \boldsymbol{v})$, we follow this procedure:

1. We begin by calculating the z-component of the angular momentum, L_z, and the energy, E, in the 'true' potential.
2. We then integrate the orbit in the 'true' potential. We use the initial time-steps of the orbit integration to find the best-fitting coordinate system: at several points along the orbit we evaluate Eq. (2.15) and average to find a sufficiently accurate value for a. With the coordinate system found, we continue integrating to find the edges of the orbit, λ_+, λ_- and ν_+, that define the fitting region. The edges of the orbit are given approximately by the points where $\dot{\tau} = 0$ in the best-fitting prolate spheroidal coordinate system. The minimum and maximum τ edges are distinguished by inspecting the sign of $\ddot{\tau}$. For all realistic potentials every orbit crosses the $z = 0$ plane so we set $\nu_- = c^2$.
3. We can now find a best-fitting Stäckel potential over this region. Using Eq. (2.19) we tabulate $f(\lambda)$ and $f(\nu)$ for 40 points in (λ_-, λ_+) and (ν_-, ν_+) respectively so that we may interpolate these smooth functions. Any call outside the ranges is calculated fully using Eq. (2.19) with a full re-computation of $\bar{\chi}(\tau)$.
4. With the best fit potential now calculated, we find I_3 using Eq. (2.11) for three points on the boundary of the orbit (on the minimum λ edge, the maximum λ edge and the maximum ν edge) and take an average. We have already found these three points when determining the edges of the orbit, so this choice involves minimum additional computational effort and provides a fair estimate for I_3 over a large region of the orbit. However, this choice of I_3 can lead to the initial phase-space point $(\boldsymbol{x}, \boldsymbol{v})$ being forbidden. Therefore, with this choice of I_3, we check whether $p^2(\nu) > 0$ and $p^2(\lambda) > 0$ for the initial phase-space point using Eq. (2.11), and,

if not, then we calculate I_3 from Eq. (2.10) using only the initial phase-space point. This procedure[2] reduces the numerical noise around the turning points, particularly in R.

5. With the three isolating integrals calculated, we are in a position to estimate the actions and angles using the method outlined in Sect. 2.2. The limits of the orbit are redetermined by finding from Eq. (2.10) the points where $p_\tau^2 = 0$ using Brent's method, and are not given by $\tau_\pm$. We do this because using $\tau_\pm$ may result in including points in the integration where $p_\tau^2 < 0$.

Table 2.3 quantifies the efficiency of this procedure.

2.3.2 Discussion

Kent and de Zeeuw (1991) propose several methods for estimating I_3 by assuming the potential is close to separable. For general disc orbits in realistic Galactic potentials, they found that the most accurate of the techniques was the 'least-squares method'. This method sought to minimise

$$\int \mathrm{d}t \left[(\tau(t)-c^2)E-\left(\frac{\tau(t)-c^2}{\tau(t)-a^2}\right)\frac{L_z^2}{2}+f(\tau(t))-2(\tau(t)-a^2)(\tau(t)-c^2)p_\tau^2(t)-I_3\right]^2$$

with respect to the coordinate system and I_3, where the integral is in reality a sum over phase-space points from an orbit integration. In a general potential, we require some approximation for $f(\tau)$. The method of Kent and de Zeeuw (1991) does not require an explicit fit of a Stäckel potential to the true potential. The authors chose to approximate $f(\lambda) \approx -(\lambda - c^2)\Phi(R(\lambda, c^2), 0)$. In Sect. 2.6 and Chap. 4 we employ a similar approach to estimating $f(\tau)$ from a general potential. As we have already performed an orbit integration as part of our procedure, we could choose to estimate the best-fitting coordinate system and I_3 in the manner presented by Kent and de Zeeuw. With a best-fitting coordinate system chosen we could then find the best-fitting Stäckel potential hopefully to further refine our estimate of I_3. Then with estimates of both $f(\lambda)$, $f(\nu)$, I_3 and the coordinate system we could use Eq. (2.12) to find the actions.

The calculation of the actions is sensitive to the value of the momenta throughout the orbit. Therefore, it is important that we find a good approximation to both the coordinate system and $f(\tau)$. Later we will show that, from the proposed methods, the estimate of $f(\tau)$ that produces the smallest spread in the action estimates is obtained by fitting a Stäckel potential as done here. The method of Kent and de Zeeuw (1991), whilst appropriate for finding a good choice of coordinate system and estimate for I_3, is probably not optimal for action estimation.

---

[2] Later we will adopt a slightly different procedure (see Chap. 6 and Appendix G for details).

2.3.3 Application

We now investigate how successful the above routine is for estimating the angle-action variables in a general axisymmetric potential. To demonstrate the applicability of the method to data, we choose a realistic Milky Way potential from McMillan (2011a). This potential consists of two exponential discs for the thick and thin discs of the Galaxy and two spheroids for the bulge and dark matter halo. We select the 'best' model from this paper, which had parameters for the components of the potential set by current constraints. The equipotential contours for this model are plotted in Fig. 2.1. It is clear that, as we move out from the centre, the contours become more circular so we anticipate that they are better fit by surfaces of constant λ and ν. Therefore, we expect more accurate estimates of the angle-action variables for orbits at larger radii. Also, orbits that probe a large range of R and/or z should have less accurate angle-action variable estimates as these orbits probe a large range of curvature of the equipotential contours. Therefore, we expect the method to work best for small J_λ and J_ν but large L_z.

de Zeeuw et al. (1986) show that for the corresponding density of a Stäckel potential to be everywhere non-negative, the density cannot fall off faster that r^{-4}, where r is the spherical radius. The best potential from McMillan (2011a) has a flat rotation curve, and an Navarro-Frenk-White halo (Navarro et al. 1996). Therefore, the density distribution towards the centre goes approximately as r^{-2} and falls to r^{-3} in the outskirts. We, therefore, expect the best-fitting Stäckel potentials to be suitable. However, as we are only locally fitting Stäckel potentials, we can produce

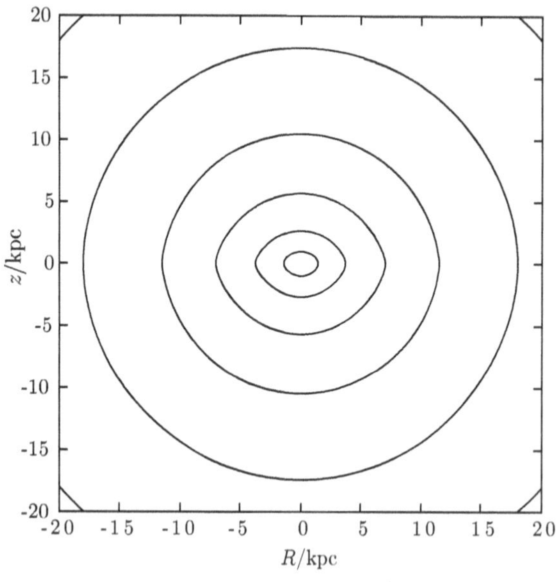

Fig. 2.1 Contours of $\ln(\Phi(R,z)/\Phi(0,0))$ for McMillan's best-fitting Milky Way potential. The contours increase from the centre in equally space units of 0.15 with the central contour at -0.15

unphysical global Stäckel potentials i.e. $\rho(\boldsymbol{x}) < 0$ for some $\boldsymbol{x}$, that are still suitable locally.

We assess the validity of the method by comparing the results with the 'exact' angle-action variables calculated using the 'torus machine' (McMillan and Binney 2008). Orbital tori are three-dimensional surfaces characterised by the three actions $\boldsymbol{J} = (J_R, L_z, J_z)$ obtained as the images of analytic tori under a canonical transformation. The strength of the torus machine lies in constructing a torus given a set of actions, $\boldsymbol{J}$, such that the phase-space coordinates, $(\boldsymbol{x}, \boldsymbol{v})$, may be obtained as functions of the angles, $\boldsymbol{\theta}$, over the surface of the torus. Therefore, a simple test for the Stäckel potential fitting procedure is first to produce a list of phase space coordinates with fixed actions but randomly chosen angles using the torus machine. The success of the method is then measured by how accurately the angle-action variables can be reproduced. We note that the canonical transformation produced by the torus machine maps J_R into J_λ and J_z into J_ν. From now on, we will use the more intuitive notation for the actions, J_R and J_z, and similarly for the angles, θ_R and θ_z.

The errors in the actions of a given torus from the torus machine may be estimated from the residuals of the Hamiltonian over the surface of the torus. The error in the Hamiltonian, ΔH, is related directly to the error in the actions by

$$\Delta H = \frac{\partial H}{\partial \boldsymbol{J}} \Delta \boldsymbol{J} = \boldsymbol{\Omega} \cdot \Delta \boldsymbol{J} \tag{2.21}$$

where we find the frequency $\boldsymbol{\Omega}$ directly from the torus machine. Assuming the errors in J_R and J_z are approximately equal and uncorrelated, the error in the actions may be estimated as

$$\Delta J \approx \frac{\Delta H}{\sqrt{\Omega_R^2 + \Omega_z^2}}. \tag{2.22}$$

The true angles of an orbit in a potential increase linearly with time. The errors in the torus angles are estimated by the residuals of the angles away from this expected straight line. Clearly, we require the angle and action errors from the torus machine to be smaller than the errors from the Stäckel-fitting procedure in order to state anything meaningful about the systematic errors from our method.

2.3.3.1 Single Torus

Here we discuss the results of applying the procedure to 10,000 randomly generated points from the torus[3] $\boldsymbol{J} = (J_R, L_z, J_z) = (0.078, 1.9, 0.097)\,\mathrm{kpc^2 Myr^{-1}}$. This torus was chosen to be representative of the actions of a disc star in the solar neighbourhood. For this torus, the errors in the actions and angles are $\Delta J/J = 0.01\,\%$ and $(\Delta\theta_R, \Delta\theta_\phi, \Delta\theta_z) = (1.0, 0.2, 1.0) \times 10^{-5}$ rad. The orbit in the (R, z) plane is

[3]Throughout this chapter, the actions are stated in units of $\mathrm{kpc^2 Myr^{-1}} = 977.8\,\mathrm{kpc\,km\,s^{-1}}$.

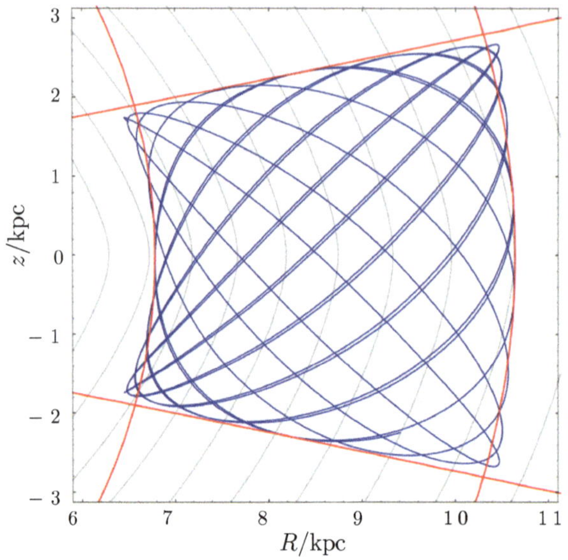

Fig. 2.2 Fit region for a single orbit—the *blue line* shows the orbit with actions $\boldsymbol{J} = (J_R, L_z, J_z) = (0.078, 1.9, 0.097)\,\text{kpc}^2\text{Myr}^{-1}$. The *black lines* are the equipotential contours of McMillan's best-fitting Milky Way potential. The *red lines* show the lines of constant λ and ν that define the region over which the potential is fitted

shown in Fig. 2.2. This orbit has apses at $R \approx (6.5, 10.5)$ kpc and $z_{\text{max}} \approx 2.8$ kpc. Also shown in the figure are the curves defining the fit region and equipotential contours for McMillan's best-fitting potential.

The residuals in the fitted potential over the fitting region defined in Fig. 2.2 are plotted in Fig. 2.3. Everywhere within the fitting region, the error in the potential is less than 0.2 % of the maximum difference in the potential across the fitting region. We note here that a good fit for the potential does not necessarily correlate with an accurate calculation of the actions. Small changes in the potential can cause large changes in the motion of a particle, so, whilst a good fit for the potential is necessary, we don't expect the errors in the actions to be of similar order.

The 10,000 phase-space points are shown in scatter plots of (R, θ_R) and (z, θ_z) in Fig. 2.4. We can see that R and z are periodic in the angles. We define the zero-point of θ_R such that the radial periapsis and apoapsis correspond to $\theta_R = 0$ and $\theta_R = \pi$ respectively. θ_z is defined such that $z = 0$ corresponds to $\theta_z = 0, \pi$ and $z = \pm z_{\text{max}}$ corresponds to $\theta_z = \pi/2, 3\pi/2$. The zero-point of θ_ϕ is defined such that $\theta_\phi = \phi$ at periapsis. The spread of the z coordinates of the points at a given angle is much larger than the spread in the R coordinates.

When the Stäckel fitting method is applied to this set of phase-space points, we find that the root-mean-square (RMS) deviations of the actions, ΔJ, are given by $\Delta J_R/J_R \approx 4.9\,\%$ and $\Delta J_z/J_z \approx 4.2\,\%$. We also find that there is a very tight anticorrelation between ΔJ_R and ΔJ_z. All phase-space points have the same energy

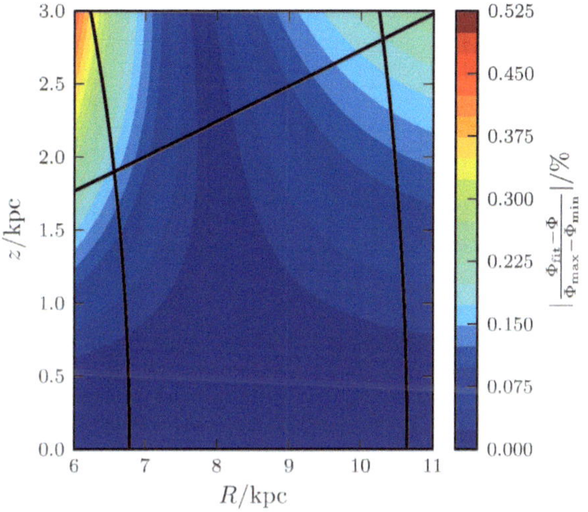

Fig. 2.3 Filled contour plot of the percentage difference between the best-fitting Stäckel potential and McMillan's best-fitting Milky Way potential. $\Phi_{\rm min}$ and $\Phi_{\rm max}$ give the values of the potential on the minimum and maximum λ edges respectively. Also plotted in black are the curves of constant λ and ν which define the region over which the potential is fitted. This is the fit region corresponding to the orbit shown in Fig. 2.2, with actions $\boldsymbol{J} = (J_R, L_z, J_z) = (0.078, 1.9, 0.097)\,\mathrm{kpc^2Myr^{-1}}$. We see that, within the fitting region, the difference between the fitted potential and the potential we are attempting to fit is less than 0.2 % of the maximum potential difference across the fitting region

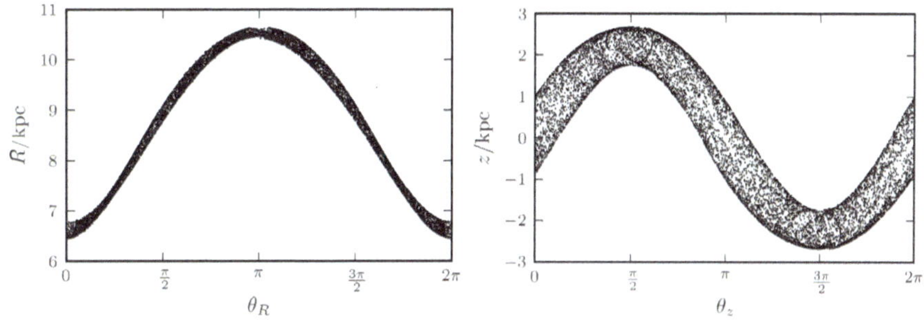

Fig. 2.4 Scatter plot of R against θ_R and z against θ_z for 10,000 randomly selected phase-space points from the torus detailed in Sect. 2.3.3.1

as we are using the potential that was used to integrate the orbit to find the energy. Therefore, all the points lie along the intersection of the surface of constant energy with the (J_R, J_z) plane. If we overestimate J_R, we must underestimate J_z in order to have the correct energy. The RMS deviations in the angles for the 10,000 phase-space points are $(\Delta\theta_R, \Delta\theta_\phi, \Delta\theta_z) = (4.1, 1.1, 5.1) \times 10^{-2}$ rad.

Errors as a function of angle: It is informative to investigate how the errors in the derived actions and angles vary with true angle around the torus. The derived actions are approximately independent of the true angles as they depend only on

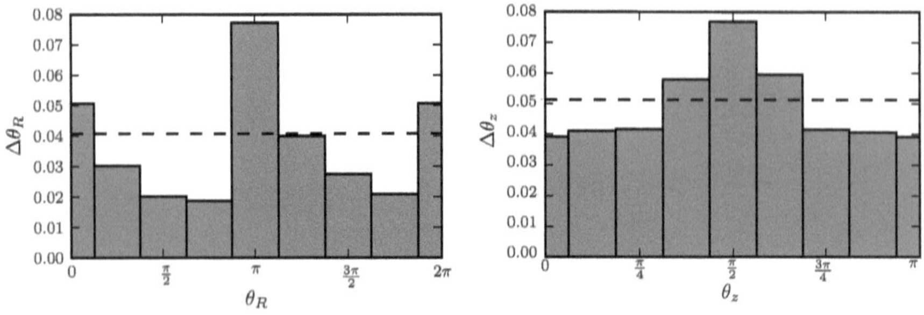

Fig. 2.5 RMS error in the angles binned as a function of angle for the torus in Sect. 2.3.3.1. The *dashed line* shows the total RMS error from all the points on the torus. For the bottom panel, we have taken advantage of the symmetry in the $z = 0$ plane and mapped the $\theta_z = (0, 2\pi)$ interval onto $\theta_z = (0, \pi)$ such that $\theta_z = \pi/2$ corresponds to $\pm z_{\max}$ etc. The largest error occurs at the apses for both cases

the path of the orbit, which is determined by the fitted potential and not the initial point on the orbit. Any small variation is due to the choice of prolate spheroidal coordinate system and variations in the fitted potential. However, we find that the error in the derived angle varies with true angle. In Fig. 2.5 we plot the RMS errors in the angles binned as a function of true angle for both θ_R and θ_z. Maximum errors occur at the turning points in the (R, θ_R) and (z, θ_z) plots shown in Fig. 2.4. For the radial and vertical angle, the largest error occurs at apoapsis. In a Stäckel potential, the momenta, p_τ, depend on τ and the isolating integrals, which, once determined, are taken to be constant. Therefore, at a given location in the orbit, the angle is solely a function of the position coordinates and the velocity information is essentially ignored. Around turning points in the orbit, the velocity coordinates contain the majority of the information, whilst the position coordinates are changing very slowly. Therefore at turning points the errors in the angles are large as the angle coordinates are estimated using this reduced phase-space information. In general, the errors in θ_z are larger than the errors in θ_R.

2.3.3.2 Multiple Tori

We have seen that the method gives reasonable estimates for the actions for a particular torus, but, in order to use the method with confidence, we need to see how the errors depend on the torus. Here we repeat the above procedure for a range of different tori that probe the different regions of the potential. We work with two groups of tori: those with low actions and torus machine errors less than $\Delta J/J = 0.01\,\%$ and $(\Delta\theta_R, \Delta\theta_\phi, \Delta\theta_z) = (27.0, 5.1, 990) \times 10^{-6}$ rad and those with high actions and torus machine errors less than $\Delta J/J = 1\,\%$ and $(\Delta\theta_R, \Delta\theta_\phi, \Delta\theta_z) = (2.0, 1.7, 1.7) \times 10^{-2}$ rad. The low-action group consists of 100 tori with actions $J_R = (0.001, 0.005, 0.01, 0.05, 0.1)\,\mathrm{kpc^2Myr^{-1}}$, $J_z = (0.001, 0.005, 0.01, 0.05, 0.1)\,\mathrm{kpc^2Myr^{-1}}$ and $L_z = (1.0, 2.0, 3.0, 4.0)\,\mathrm{kpc^2Myr^{-1}}$.

These tori probe the region $3\,\text{kpc} < R < 22\,\text{kpc}$, $|z| < 5\,\text{kpc}$ and are chosen to be representative of disc-type tori. The high-action group consists of 36 tori with actions $J_R = (0.5, 1.0, 5.0)\,\text{kpc}^2\text{Myr}^{-1}$, $J_z = (0.5, 1.0, 5.0)\,\text{kpc}^2\text{Myr}^{-1}$ and $L_z = (1.0, 2.0, 3.0, 4.0)\,\text{kpc}^2\text{Myr}^{-1}$. These tori probe the region $2\,\text{kpc} < R < 120\,\text{kpc}$, $|z| < 100\,\text{kpc}$. We include the second group to demonstrate that the method can deal with orbits that deviate very far from the plane and probe a very large region of the potential. We would like to be able to apply the method to halo stars and tidal streams, so it is important to understand the errors for these high-action tori.

Actions: As mentioned previously, we expect the errors in J_R and J_z will be large when $J_R/|L_z|$ and/or $J_z/|L_z|$ are large. In this regime, the orbit probes a large central region of the potential so we anticipate the potential fit will be poorer. In Fig. 2.6, the RMS deviations in the actions for the complete orbit sample are plotted against the combination of the actions $(J_R + J_z)/|L_z|$. We can see that, as anticipated, the absolute errors correlate with this action combination. In fact, the correlation is much tighter than the individual correlations with $J_R/|L_z|$ and $J_z/|L_z|$, so the errors in the method are dependent on the sum of the actions $(J_R + J_z)$. It is this measure that tells us how much an orbit strays from a circular orbit and thus how much of the potential it explores.

We also note from Fig. 2.6, that, at a given value of $(J_R + J_z)/|L_z|$, the errors in J_R and J_z are of similar magnitudes. As explained above, the errors in J_R and J_z compensate for each other to recover the correct energy. In Fig. 2.7, we plot this correlation between the RMS errors in J_R and J_z. A consequence of this tight correlation is that when one action is much greater than the other, the relative error in the smaller action will be much greater than the relative error in the larger action. However, it is worth noting that the absolute error is far more important than the relative error. Given a distribution function for a steady-state galaxy, $f(\boldsymbol{J})$, the absolute error in f is given by

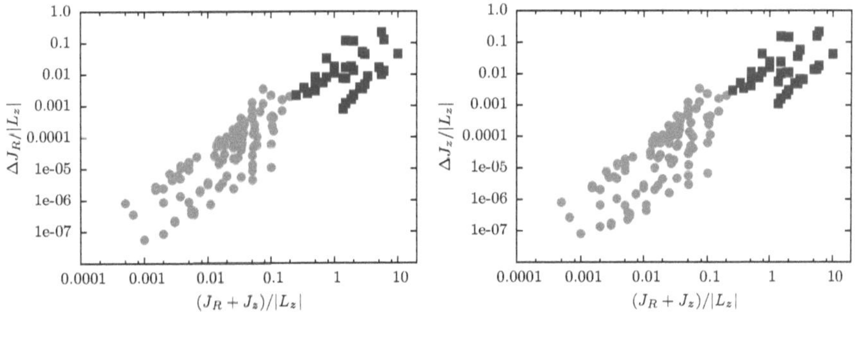

Fig. 2.6 Absolute RMS deviations of the actions for the sample of tori detailed in Sect. 2.3.3.2. The *blue circles* are data for which the relative errors in the actions from the torus machine are less than 0.01 % and the red squares are those with errors less than 1 %. The errors correlate loosely with both J_R and J_z separately, but there is a much tighter correlation between the errors and $(J_R + J_z)$

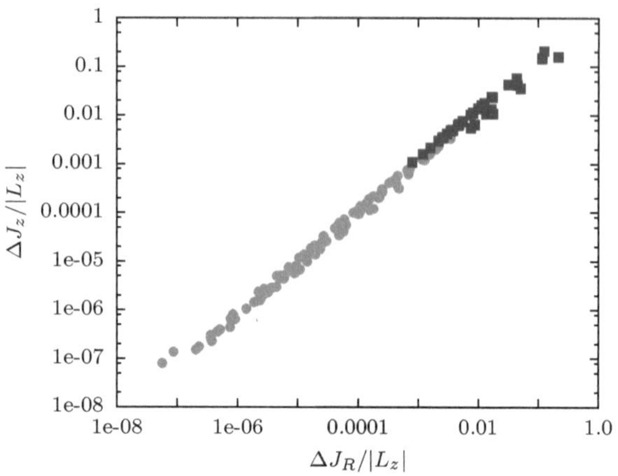

Fig. 2.7 RMS deviations in J_R and J_z for the sample of tori detailed in Sect. 2.3.3.2. The *blue circles* are data for which the relative errors in the actions from the torus machine are less than 0.01 % and the red squares are those with errors less than 1 %

$$(\Delta f)^2 = \sum_{i,j} \frac{\partial f}{\partial J_i}\frac{\partial f}{\partial J_j}\text{cov}(J_i, J_j), \tag{2.23}$$

where $\text{cov}(X, Y)$ is the covariance between variables X and Y. In the case of uncorrelated errors between the actions, this simply becomes

$$(\Delta f)^2 = \sum_i \left(\frac{\partial f}{\partial J_i}\Delta J_i\right)^2. \tag{2.24}$$

The distribution function for the Milky Way is approximately exponential in the actions (Binney 2010 and Chap. 8):

$$f(\boldsymbol{J}) \sim \prod_i \mathrm{e}^{a_i J_i}, \tag{2.25}$$

where a_i is independent of the action J_i. Therefore, the absolute error in f is given by

$$(\Delta f)^2 = \sum_i (f a_i \Delta J_i)^2. \tag{2.26}$$

Similarly, the relative error in the distribution function is given by

$$\left(\frac{\Delta f}{f}\right)^2 = \sum_i \left(\frac{\partial \ln f}{\partial \ln J_i}\frac{\Delta J_i}{J_i}\right)^2 = \sum_i (a_i \Delta J_i)^2. \tag{2.27}$$

Both the absolute and relative error in the distribution function are determined by the absolute errors in the actions, so we need not be overly concerned that the relative error in one action is much larger than the relative error in another.

From the relationship illustrated in Fig. 2.6, we can estimate the error in a given estimate of J_R and J_z. Performing a linear fit to both sets of data points independently, we find that, for $(J_R + J_z)/|L_z| \lesssim 10$, a good fit for the RMS errors in both J_R and J_z is given by

$$\Delta J \approx 0.01 \frac{(J_R + J_z)^{\frac{3}{2}}}{|L_z|^{\frac{1}{2}}}. \tag{2.28}$$

The errors in the actions have a weak dependence on L_z as orbits with higher L_z explore regions of the potential that are more spherical and hence easier to fit with a Stäckel potential (see Fig. 2.1).

Angles: We present the RMS deviations in the angles for the 100 tori in the low-action group in Fig. 2.8. In general, the relative errors in the angles are larger than the errors in the actions. The calculation of an action involves only a single integral, whereas the corresponding angle calculation involves nine integrals. Each integral folds in more error from the fitting method, so we expect the relative errors in the angle variables to be significantly larger than the errors in the actions. From Fig. 2.8, we see that the errors in the angles correlate with the relative error in the actions. The error in θ_ϕ has been plotted against $\Delta J_R/L_z$, whilst the other two angles have been plotted against the relative error in their respective action. As the errors in J_R and J_z are approximately equal (Fig. 2.7), the three sets of points are essentially θ_i against $\Delta J/J_i$. As with the errors in the actions, we may estimate the error in a given calculation of the angles by fitting the data in Fig. 2.8. We find the errors are approximately given by

$$\Delta\theta_R \approx \left(\frac{\Delta J}{J_R}\right)^{0.75}, \quad \Delta\theta_z \approx \left(\frac{\Delta J}{J_z}\right)^{0.75}, \quad \Delta\theta_\phi \approx \left(\frac{\Delta J}{L_z}\right)^{0.5}. \tag{2.29}$$

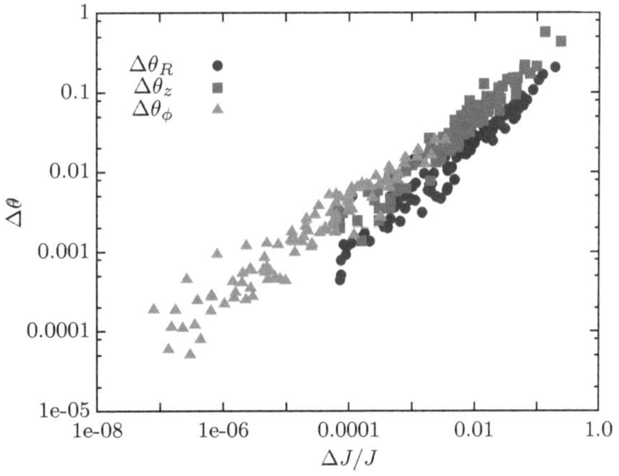

Fig. 2.8 RMS deviations in the angles for the 100 low-action tori as a function of the relative error in their respective action. For θ_ϕ, we have plotted the error against $\Delta J_R/L_z$

2.3.4 Geneva-Copenhagen Survey

Now that we have understood the systematic errors of the method, we apply it to real data. The Geneva-Copenhagen Survey (GCS) (Nordström et al. 2004) is a sample of 16682 nearby F and G stars, and is perhaps the best data set with full 6D phase-space information. It provides us with a platform to motivate a discussion of errors involved in a practical calculation. The GCS has been analysed by many authors and specifically looked at in angle-action space by Sellwood (2010) and McMillan (2011b). From the table produced by Holmberg et al. (2009), we select the 13,518 objects that have full 6D phase-space information. We correct the data for the solar velocity with respect to the local standard of rest as calculated by Schönrich et al. (2010) i.e. $(U, V, W)_{\odot} = (11.1, 12.24, 7.25)\,\mathrm{km\,s^{-1}}$.[4] Using the 'best' model Galactic potential from McMillan (2011a) sets the solar radius as $R_0 = 8.29\,\mathrm{kpc}$ and the velocity of the local standard of rest as $v_{\mathrm{LSR}} = 239.1\,\mathrm{km\,s^{-1}}$. The results of estimating the actions and angles using the Stäckel-fitting procedure are shown in Fig. 2.9. These are very similar to the equivalent plots from Sellwood (2010) and McMillan (2011b). We see that the plot of J_R against L_z has a markedly parabolic shape. The minimum of this parabola corresponds to the circular orbit at the solar radius. The shaded red region is inaccessible by stars that are at the solar position and have $J_z = 0$. We can understand this using the epicyclic approximation (Binney and Tremaine 2008): a star with a given L_z has some guiding-centre radius R_c. For $J_R = 0$, the star follows a circular orbit of radius R_c at an angular rate L_z/R_c^2, such that $\phi = \theta_\phi = tL_z/R_c^2$. Non-zero J_R produces radial oscillations about this circular orbit, such that a star circulates the guiding centre of the $J_R = 0$ orbit. A star requires enough radial action to produce a radial oscillation large enough to pass through the solar neighbourhood, hence the parabolic shape. The angles θ_R and $\theta_\phi - \phi$ describe the location of the star with respect to the guiding centre. They are correlated as, whilst a star leads the guiding centre, $\theta_\phi < \phi$ and $\theta_R < \pi$, whilst, when trailing, $\theta_\phi > \phi$ and $\theta_R > \pi$. At the solar position $\phi = \pi$, hence the observed selection effect in Fig. 2.9. From the plot of θ_ϕ against θ_R, we see that the majority of stars are at $\theta_R = 0, \pi$ corresponding to the apses of their radial motion. However, there is still a lot of structure in between these extrema: the peak at $\theta_R/\pi \approx 0.54$ corresponds to the Hyades moving group.

2.3.4.1 Structure in the GCS

Since Dehnen (1998) investigated the kinematics of the solar neighbourhood using data from the *Hipparcos* satellite, it has been known that, when viewed in the (U, V) velocity plane, the local distribution of stars consists of a series of groups and clusters. These structures were classified by Famaey et al. (2005) and are all thought to have a dynamical origin (De Simone et al. 2004; Antoja et al. 2010). From the GCS sample, we can identify several of the larger structures from the peaks in the (U, V)

[4] We work in a right-handed Galactocentric Cartesian coordinate system with the positive x direction pointing towards the Galactic centre.

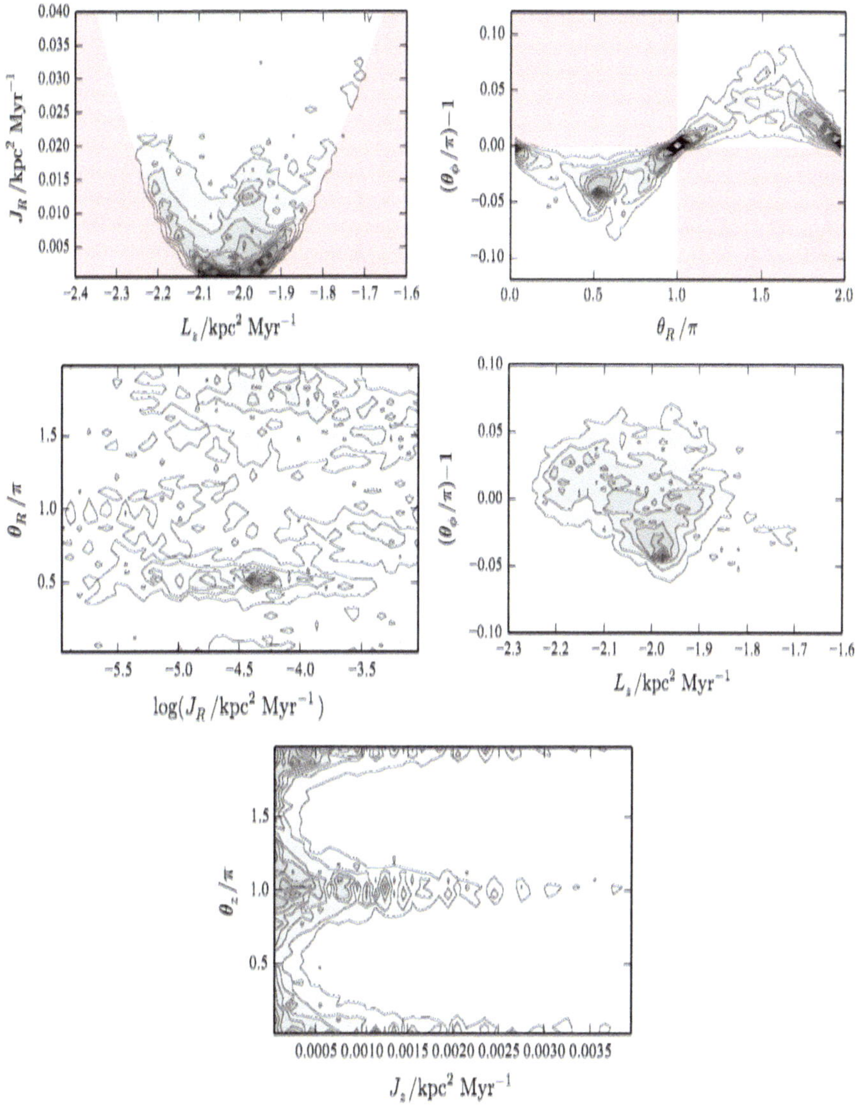

Fig. 2.9 Density contours for the actions and angles of 13,518 objects from the Geneva-Copenhagen Survey. The contours are linearly spaced. The *red shaded region* is inaccessible by stars that are at the solar position and have $J_z = 0$. The strong peak at $\theta_R/\pi \approx 0.54$ is due to the Hyades. We also see the Hercules stream at $\log J_R/\,\mathrm{kpc}^2\mathrm{Myr}^{-1} \approx -3.3$ and $\theta_R/\pi \approx 0.85$

distribution. This is done by binning the stars in U and V and then performing a wavelet transform. In Fig. 2.10, we show the results of the wavelet transform for structure on scales $\sim$12 km s^{-1}. The peaks in this plot are identified with the known

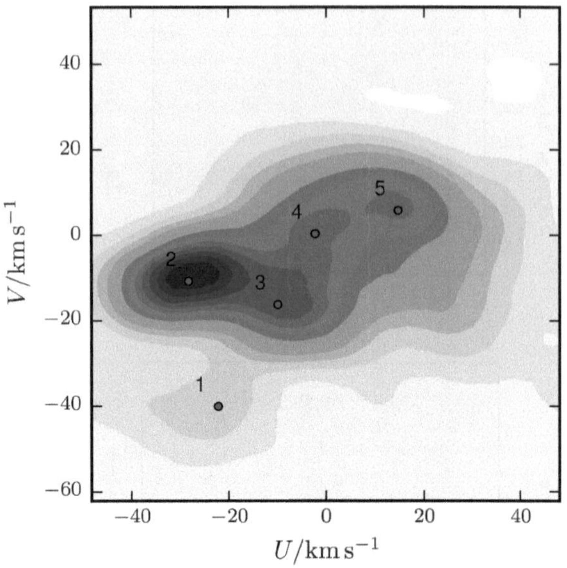

Fig. 2.10 Density contours for the (U, V) plane of 13,518 objects from the Geneva-Copenhagen Survey smoothed to a scale of $\sim$12 km s^{-1} via a stationary wavelet transform. The five major structures are identified by red points: *1* Hercules, *2* Hyades, *3* Pleiades, *4* Coma Berenice, *5* Sirius

groups, specifically the Hercules, Hyades, Pleiades, Coma Berenice and Sirius. As discussed by McMillan (2011b), for a star located at the solar position, each point in the (U, V) plane represents a unique point in the $(J_R, L_z, \theta_R, \theta_\phi)$ space. Lines of constant L_z and θ_ϕ form an approximately Cartesian grid in the (U, V) plane, whilst lines of constant J_R and θ_R form an approximately polar grid. Therefore, we can find a reduced set of angle-actions (excluding any vertical action and angle) for each of the peaks identified in Fig. 2.10. From inspecting the (U, V) plane, we can see that the distribution deviates from axisymmetric equilibrium, so estimating the actions and angles assuming an axisymmetric potential is clearly an oversimplification. However, this is a necessary first step to create a basis on which we can include non-axisymmetric perturbations.

We represent each group by a 6D phase-space point placed at the solar position with zero vertical velocity but U and V determined by the identification from Fig. 2.10. We then estimate the actions and angles corresponding to this point, again using McMillan's 'best' potential. The results are shown in Table 2.1. This gives us an opportunity to discuss the various sources of errors in a realistic use of the method. There are two sources of error in this calculation—the systematic errors introduced by the Stäckel fitting method and the errors in the input coordinates. Holmberg et al. (2009) estimated the space-velocity errors for each star as 1.5 km s^{-1}, which when combined with the errors estimated in Schönrich et al. (2010) for the solar motion gives velocity errors of $(\Delta U, \Delta V) = (1.7, 1.6)$ km s^{-1}. The majority of this error arises from the uncertainty in the distances. We convolve each peak position by these errors to calculate 10,000 Gaussianly distributed points in (U, V) space about these peaks, and then estimate the errors in the output actions and angles by the RMS scatter in the resulting $(\boldsymbol{J}, \boldsymbol{\theta})$ coordinates. These errors can then be compared and combined with the known systematic errors from Sect. 2.3.3.2 and are shown in Table 2.1.

Table 2.1 Velocities with respect to the local standard of rest and angle-action coordinates for density peaks of known structures in the Geneva-Copenhagen Survey

	$U/\,\mathrm{km\,s^{-1}}$	$V/\,\mathrm{km\,s^{-1}}$		
Hercules	−22.1	−40.1		
Hyades	−28.2	−10.8		
Pleiades	−9.8	−16.3		
Coma Berenice	−2.2	0.3		
Sirius	14.7	5.7		
	$J_R/\,\mathrm{kpc^2 Myr^{-1}}$	$L_z/\,\mathrm{kpc^2 Myr^{-1}}$	θ_R	$\theta_\phi - \pi$
Hercules	$(4.0 \pm 0.3^{0.006}_{0.3}) \times 10^{-2}$	-1.69 ± 0.01	$2.65 \pm 0.04^{0.008}_{0.036}$	$-(9.6 \pm 0.7^{0.05}_{0.68}) \times 10^{-2}$
Hyades	$(1.2 \pm 0.1^{0.001}_{0.1}) \times 10^{-2}$	-1.94 ± 0.01	$1.91 \pm 0.07^{0.005}_{0.07}$	$-0.118 \pm 0.007^{0.0001}_{0.0066}$
Pleiades	$(7.3 \pm 0.1^{0.005}_{0.1}) \times 10^{-3}$	-1.89 ± 0.01	$2.68 \pm 0.08^{0.004}_{0.077}$	$-(4.2 \pm 0.7^{0.006}_{0.7}) \times 10^{-2}$
Coma Berenice	$(1.7 \pm 1.4^{0.0002}_{1.4}) \times 10^{-4}$	-2.03 ± 0.01	$1.5 \pm 1.1^{0.0009}_{1.1}$	$-(9.3 \pm 6.4^{0.0008}_{6.4}) \times 10^{-3}$
Sirius	$(3.5 \pm 0.8^{0.001}_{0.8}) \times 10^{-3}$	-2.07 ± 0.01	$5.28 \pm 0.13^{0.003}_{0.13}$	$0.06 \pm 0.01^{0.00002}_{0.006}$

The errors in U and V are $1.7\,\mathrm{km\,s^{-1}}$ and $1.6\,\mathrm{km\,s^{-1}}$ respectively. The errors in the angles and actions are presented as a^b_c where a is the total error and b and c are the contributions from the systematic errors of the Stäckel fitting method and the errors in the space-velocities respectively, such that $a^2 = b^2 + c^2$. In all cases, the observational error dominates the systematic error

We find that, for all the coordinates, the error in the data dominates the systematic error introduced by the method. As noted by McMillan (2011b) the relationship between errors in $(\boldsymbol{x}, \boldsymbol{v})$ and $(\boldsymbol{J}, \boldsymbol{\theta})$ is non-trivial. Specifically, at very low radial actions, any error in the velocity can introduce a 2π error in the θ_R coordinate. We see this occurring for the Coma Berenice peak—the error in the θ_R coordinate is large as the peak is positioned very close to the origin of the (U, V) plane. We also see that the errors in J_R and θ_ϕ are of order one for Coma Berenice.

We note that we have not included any error for the size of the structures in phase-space, nor any error due to the choice of smoothing from the wavelet transform, nor any error for the assumption that all the stars are situated at the solar position, nor any error in the choice of gravitational potential. Investigating the error in the actions due to the range of viable potentials for the Milky Way is beyond the scope of this investigation. Even with this underestimated error, the errors from the data dominate the systematic errors. We conclude that, given the accuracy of the current data, the determination of the angle-action coordinates using the Stäckel-fitting method is not limited by the well-understood systematic errors.

Table 2.2 Deviations between actions and angles for stars in the Geneva-Copenhagen Survey sample estimated with the Stäckel fitting method and the data from McMillan (2011b)

	RMS difference	Expected RMS error
$\Delta J_R/10^{-4}\,\mathrm{kpc^2 Myr^{-1}}$	5.4	4.9
$\Delta J_z/10^{-4}\,\mathrm{kpc^2 Myr^{-1}}$	7.6	4.9
$\Delta\theta_R/0.01$ rad	2.6	0.9
$\Delta\theta_\phi/0.01$ rad	1.2	0.6
$\Delta\theta_z/0.01$ rad	7.7	40.1

The second column gives the RMS of the expected systematic errors from the Stäckel fitting method

2.3.4.2 Comparison with McMillan (2011b)

McMillan (2011b) calculated the angles and actions of the GCS sample by using the torus machine iteratively. Here we compare the results to our own estimates of the angle-actions. The potential used by McMillan (2011b) was the 'convenient' potential detailed in McMillan (2011a), which places the Sun at a Galactocentric radius $R_0 = 8.5\,\mathrm{kpc}$ with a velocity of the local standard of rest $v_{\mathrm{LSR}} = 244.5\,\mathrm{km\,s^{-1}}$. We calculate the RMS deviations between McMillan's data and ours, and present the results in Table 2.2. We also show the expected RMS errors in the angles and actions estimated using the Stäckel fitting method. The largest discrepancy between our data and McMillan's occurs for the θ_z coordinate. The expected error for this coordinate is also large, and very much larger than the actual error. The largest error in θ_z is produced by stars near their turning points. At low z the majority of stars are well away from their turning points, and the resulting errors in the angles are smaller than if we observed a sample of stars that were uniformly sampled in θ_z. Only those stars with very low vertical action have corresponding large errors in θ_z. For all the other variables, the discrepancy between our data and McMillan's data seems to be in agreement with the expected systematic errors of our method.

2.3.5 Prolate Axisymmetric Potentials

Later we will need to use the above algorithm on prolate axisymmetric potentials, so here we extend the algorithm for this case. Oblate spheroidal coordinates are associated with prolate axisymmetric potentials. These are defined by the roots

$$\frac{R^2}{\tau - a^2} + \frac{z^2}{\tau - c^2} = 1, \tag{2.30}$$

where λ and ν are the roots of τ such that $a^2 \leq \lambda \leq c^2 \leq \nu$. Note that now surfaces of constant ν are oblate spheroids and surfaces of constant λ are two-sheeted hyperboloids of revolution. A general prolate axisymmetric potential accommodates two classes of orbit: inner and outer long-axis loops. An inner long-axis loop crosses $z = 0$ inside the focus $\sqrt{c^2 - a^2}$ and has $\nu_0 = c^2$. These orbits are approximately radially bound by surfaces of constant λ and vertically by surfaces of constant ν. An outer long-axis loop crosses outside this focus and has $\lambda_1 = c^2$. These orbits are approximately radially bound by surfaces of constant ν and vertically by surfaces of constant λ. We choose to set $c^2 = 1$ and set a^2 using Eq. (2.15).[5] The equations for the actions, angles and frequencies are identical to the oblate case. We, therefore, adopt the same procedure as outlined in Sect. 2.3: we begin by integrating the orbit to find a best choice of coordinate system, find the boundaries of the orbit, fit a Stäckel potential to this region using Eqs. (2.19) and (2.21), and then estimate the actions, angles and frequencies as those in this best-fitting potential. We must ensure that, when a coordinate bounces instead of oscillating, we include an appropriate factor of 2 and we must add an additional π to the appropriate angle when $z < 0$ to remove the degeneracy in the choice of coordinates. Note that, for an outer long-axis loop orbit, $J_R = J_\nu$ and $J_z = J_\lambda$, whereas for an inner long-axis loop orbit $J_R = J_\lambda$ and $J_z = J_\nu$.

As an example, we look at loop orbits in the axisymmetric logarithmic potential described by

$$\Phi(R, z) = \frac{V_c^2}{2} \ln\left(R^2 + \frac{z^2}{q^2}\right), \tag{2.31}$$

where $V_c = 220\,\mathrm{km\,s^{-1}}$ and q controls the flattening. We generate tori with the actions $(J_R, L_z, J_z) = (0.29, 3.8, 0.45)\,\mathrm{kpc^2 Myr^{-1}}$ using the torus machine for $q = (0.6, 0.9, 1.1, 1.9)$. On each torus we generate an initial condition $(\boldsymbol{x}, \boldsymbol{v})$ with $(\theta_R, \theta_\phi, \theta_z) = (0.1, 0.0, 0.0)$ rad, and integrate this initial condition for 500 adaptive time-steps. In Fig. 2.11 we show the recovered actions, angles and frequencies for this orbit, along with the input actions and frequencies from the torus machine. For $q = 0.6, 0.9$ the orbit is a short-axis loop, for $q = 1.1$ the orbit is an outer long-axis loop, and for $q = 1.9$ the orbit is an inner long-axis loop. The actions for the orbits in the near-spherical potentials ($q = 0.9, 1.1$) are accurate to ~2 %, whilst for the flatter potentials ($q = 0.6, 1.9$) the errors increase to ~5 %. We see the anti-correlation between the radial and vertical actions. As the flattening is increased ,we see the expected decrease in the vertical frequency, Ω_z. The frequency recovery has small biases.

---

[5]Note that a^2 can be less than zero, making a imaginary. However, all physical quantities depend on $\sqrt{c^2 - a^2}$, so this is irrelevant.

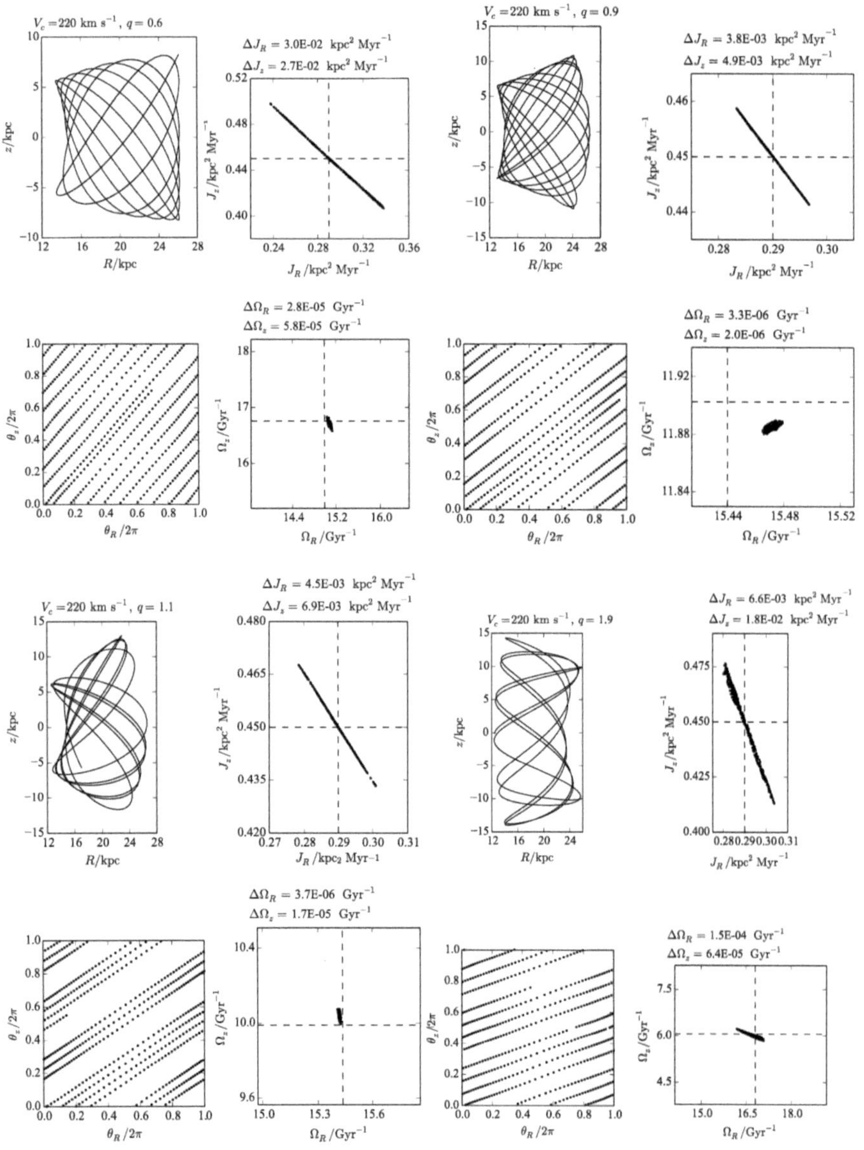

Fig. 2.11 Actions, angles and frequencies for orbits integrated in the logarithmic potential with differing flattening, q. Each set of four panels corresponds to a different flattening. The *top left* is $q = 0.6$, *top right* $q = 0.9$, *bottom left* $q = 1.1$, and *bottom right* $q = 1.9$. Within each set of four panels, we show the orbit in the meridional plane in the *top left panel*, the J_R and J_z estimates in the *top right* (with the respective standard deviations given above each plot), the angle recovery in the *bottom left*, and Ω_R and Ω_z estimates in the *bottom right*. The *dashed lines* are the true values from the torus machine

2.4 Polar Adiabatic Approximation

We now move on to discuss other methods for estimating the actions in an axisymmetric potential. The adiabatic approximation provides an alternative method for estimating the actions. In its simplest form (Binney 2010), the adiabatic approximation is based on the observation that the oscillation of a star perpendicular to the Galactic plane is much more rapid than the oscillation in the plane, so we may consider the vertical motion to be determined by a slowly varying potential. This approximation assumes that the motion in the plane is unaffected by the motion perpendicular to the plane. The absence of energy transfer between the radial and vertical motion leads to an underestimate of the centrifugal potential for the radial motion and hence an underestimate of the maximum radius of an orbit. Binney and McMillan (2011) attempted to resolve this issue by replacing L_z with $(|L_z| + J_z)$ in the effective radial potential. Schönrich and Binney (2012) improved on this by including a correction to the radial energy due to the changes in the vertical energy along an orbit. It is this final approach that we present here.

Following Schönrich and Binney (2012) we assume that the vertical motion at a given radius, R_0, is governed by the potential $\Psi_z(z) = \Phi(R_0, z) - \Phi(R_0, 0)$ such that the vertical energy, E_z, is

$$E_z = \frac{1}{2}v_z^2 + \Psi_z(z). \tag{2.32}$$

Then the vertical action is estimated to be

$$J_z = \frac{2}{\pi}\int_0^{z_{\max}} \mathrm{d}z\, v_z, \tag{2.33}$$

where $z_{\max}$ is the vertical height where the vertical velocity, v_z, is zero. By linear interpolation we may reverse this calculation such that, for a given pair of J_z and R_0, we may calculate $E_z(J_z, R_0)$. Over the course of an orbit we take J_z to be constant but the vertical energy will be changing as the orbit explores different radii. For overall energy conservation, this energy must be transferred from the vertical motion into the radial motion. Therefore, the radial motion is governed by the one-dimensional potential

$$\Psi_R(R) = \Phi(R, 0) + \frac{L_z^2}{2R^2} + E_z(J_z, R) - E_z(J_z, R_c), \tag{2.34}$$

where R_c is the guiding-centre radius (the radius of a circular orbit with z-component of angular momentum L_z). Using this potential, we estimate the radial action as

$$J_R = \frac{1}{\pi}\int_{R_p}^{R_a} \mathrm{d}R\, v_R, \tag{2.35}$$

where R_p and R_a are the radii where the radial velocity, v_R, is zero.

2.5 Ellipsoidal Adiabatic Approximation

The adiabatic approximation of the previous section assumed the motion could be considered separable in cylindrical polar coordinates. We call this method the polar adiabatic approximation (PAA) to differentiate it from the method presented here. It is not obvious that the coordinates in which the motions are 'most separable' are the polar coordinates, (R, z). Prolate spheroidal coordinates (λ, ϕ, ν) provide an alternative set of coordinates for describing motion in an axisymmetric potential, which are important as they are linked to Stäckel potentials. Here we present an alternative method based on the polar adiabatic approximation, but instead considering the motion in the prolate spheroidal coordinates to be separable. We call this method the ellipsoidal adiabatic approximation (EAA).

For a general axisymmetric potential, $\Phi(R, z)$, we recall from Eq. (2.6) that the Hamiltonian in prolate spheroidal coordinates is given by

$$H = \frac{1}{2}\left(\frac{p_\lambda^2}{P_\lambda^2} + \frac{p_\nu^2}{P_\nu^2} + \frac{L_z^2}{R^2(\lambda, \nu)}\right) + \Phi(\lambda, \nu) \tag{2.36}$$

where we have chosen a specific coordinate system, $p_{\lambda,\nu}$ are the conjugate momenta to λ, ν and $P_\lambda^2 = \frac{\lambda-\nu}{(\lambda-a^2)(\lambda-c^2)}$ and $P_\nu^2 = \frac{\nu-\lambda}{(\nu-a^2)(\nu-c^2)}$.

Now we assume the 'vertical' motion follows an ellipse of constant $\lambda = \lambda_0$ such that the ν coordinate is determined by the potential

$$\Psi_\nu(\nu) = \frac{L_z^2}{2R^2(\lambda_0, \nu)} - \frac{L_z^2}{2R^2(\lambda_0, c^2)} + \Phi(\lambda_0, \nu) - \Phi(\lambda_0, c^2). \tag{2.37}$$

The energy of the ν oscillations is given by

$$E_\nu = \frac{1}{2}P_\nu^2\dot{\nu}^2 + \Psi_\nu(\nu) \tag{2.38}$$

such that the vertical action is found as

$$J_z = \frac{2}{\pi}\int_{c^2}^{\nu_+} \mathrm{d}\nu\, p_\nu = \frac{2}{\pi}\int_{c^2}^{\nu_+} \mathrm{d}\nu \sqrt{2P_\nu^2(\lambda, \nu)}\sqrt{E_\nu - \Psi_\nu(\lambda, \nu)}, \tag{2.39}$$

where ν_+ is the root of $(E_\nu - \Psi_\nu(\lambda, \nu))$. Therefore, given a 6D phase-space point $(\boldsymbol{x}, \boldsymbol{v})$ we may find the best prolate spheroidal coordinate system using Eq. (2.15), evaluate E_ν at this point via the coordinate transformation and then carry out the integration along the curve of constant λ that passes through the phase-space point.

Given a value of λ, L_z and the vertical energy E_ν we are able to find the vertical action. The vertical energy varies with λ whilst the vertical action should, by definition, remain constant. Therefore, this calculation may be reversed such that at a given λ we may determine the vertical energy $E_\nu = E_\nu(\lambda, L_z, J_z)$. This is done by tabulating E_ν for a range of values of λ, L_z and J_z. Armed with this tabulation, we calculate the radial action in much the same way as when using the polar adiabatic approximation. The radial motion is governed by the effective potential along the $z = 0$ axis given by

$$\Psi_\lambda(\lambda) = \Phi(\lambda, c^2) + \frac{L_z^2}{2R^2(\lambda, c^2)} + E_\nu(\lambda, L_z, J_z). \tag{2.40}$$

The total energy of the orbit is then given by

$$E_{\rm tot} = \frac{1}{2}P_\lambda^2\dot{\lambda}^2 + \Psi_\lambda(\lambda) \tag{2.41}$$

and the radial action is calculated as

$$J_R = \frac{1}{\pi}\int_{\lambda_-}^{\lambda_+} \mathrm{d}\lambda\, p_\lambda = \frac{1}{\pi}\int_{\lambda_-}^{\lambda_+} \mathrm{d}\lambda \sqrt{2P_\lambda^2(\lambda, c^2)}\sqrt{E_{\rm tot} - \Psi_\lambda(\lambda)} \tag{2.42}$$

where λ_+ and λ_- are the roots of $(E_{\rm tot} - \Psi_\lambda(\lambda))$.

An illustration of how this method differs from the polar adiabatic approximation of the previous section is shown in Fig. 2.12. The blue line shows the line along which the PAA vertical-action integration is performed. The red line shows the line of constant λ along which the EAA vertical-action integration is performed. The points at which they intersect gives the $(R, \pm|z|)$ coordinates of the input phase-space point. The EAA clearly captures the shape of the boundaries of the orbit.

As we will see in Sect. 2.8, the EAA offers a considerable improvement over the PAA but unfortunately, due to the energy correction now being a function of three

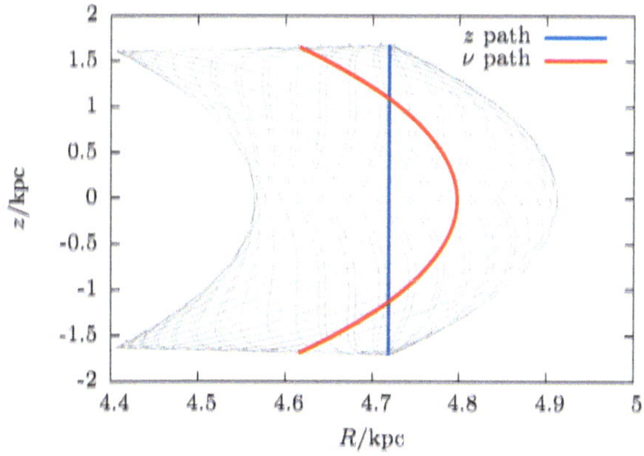

Fig. 2.12 Illustration of the difference between the PAA and EAA. An example orbit is shown by the *black line*. The *blue line* shows the line along which the PAA vertical-action integration is performed. The *red line* shows the line of constant λ along which the EAA vertical-action integration is performed. The points at which they intersect gives the $(R, \pm|z|)$ coordinates of the input phase-space point

variables (λ, L_z, J_z), this approach takes slightly longer. However, the E_ν tabulation requires a very small number of L_z values (here we use five) over the required L_z range ($0.5\,\mathrm{kpc^2Myr^{-1}} \leq L_z \leq 4.5\,\mathrm{kpc^2Myr^{-1}}$) for a sufficient level of accuracy. Note that the PAA is contained within the EAA. In the limit of very large focal distance, the surfaces of constant λ and ν tend to those of constant R and z. In this limit, the vertical action becomes independent of L_z and the EAA tends to the PAA.

2.6 Axisymmetric Stäckel Fudge

Binney (2012) presented a method for estimating the actions in a general axisymmetric potential by assuming it is close to a Stäckel potential. The presentation here differs from that in Binney (2012), but is chosen to match better to the triaxial extension presented in Chap. 4. For a general oblate axisymmetric potential, Φ, we define

$$\begin{aligned}\chi_\lambda(\lambda,\nu) &\equiv -(\lambda-\nu)\Phi,\\ \chi_\nu(\lambda,\nu) &\equiv (\lambda-\nu)\Phi.\end{aligned} \tag{2.43}$$

If Φ were a Stäckel potential, these quantities would be given by

$$\begin{aligned}\chi_\lambda(\lambda,\nu) &= f(\lambda)-f(\nu),\\ \chi_\nu(\lambda,\nu) &= f(\nu)-f(\lambda).\end{aligned} \tag{2.44}$$

Therefore, for a general potential we can write,

$$f(\tau) \approx \chi_\tau(\lambda,\nu) + D_\tau, \tag{2.45}$$

where D_τ are constants provided we evaluate χ_λ at constant ν and vice versa. We can write the equations of motion (Eq. 2.10) as

$$2(\tau-a^2)(\tau-c^2)p_\tau^2 = E(\tau-c^2) - \left(\frac{\tau-c^2}{\tau-a^2}\right)\frac{L_z^2}{2} - B_\tau + \chi_\tau(\lambda,\nu), \tag{2.46}$$

where we have defined the integrals of motion $B_\tau = I_3 - D_\tau$. Given an initial phase-space point, we use Eq. (2.15) to find a suitable coordinate system, calculate λ, ν, p_λ and p_ν, and use Eq. (2.46) to find the integrals B_τ. Equation (2.46) is then integrated over an oscillation in τ to find the actions as in Eq. (2.12). We note that for the λ integral we keep ν fixed at the input value, and vice versa. This is the procedure followed in Binney (2012).

2.7 Iterative Torus Machine

The methods presented so far suffer from the disadvantage that the error in the action is not controlled. As the methods are not convergent, we are unable to find the exact action regardless of how much computation we perform. McMillan and Binney (2008) used the torus machine iteratively to find the actions and angles of a given phase-space point. This typically involved around 20 torus fits per phase-space point and relied on a good initial guess for the actions and angles for fast convergence. Such a method is potentially the most accurate way to determine the angle-action variables but also the most costly. The authors report that it takes around 15 seconds to perform the iterative procedure on a single phase-space point. Improved methods for the initial estimate will naturally improve the usability of an iterative approach. We can use the results from the Stäckel-fitting approach presented in this chapter as an input to an iterative torus scheme. Here we detail such an approach.

1. Given an initial phase-space point $(\boldsymbol{x}_i, \boldsymbol{v}_i)$, we use the Stäckel-fitting method to find an approximate angle-actions $(\boldsymbol{\theta}_S, \boldsymbol{J}_S)$.
2. We construct a torus of action $\boldsymbol{J}_S$.
3. We find the closest phase-space point to $(\boldsymbol{x}_i, \boldsymbol{v}_i)$ on the torus by minimizing the tolerance
$$\eta = |\boldsymbol{\Omega}|^2|\boldsymbol{x}(\boldsymbol{\theta}) - \boldsymbol{x}_i|^2 + |\boldsymbol{v}(\boldsymbol{\theta}) - \boldsymbol{v}_i|^2 \tag{2.47}$$
with respect to the angle coordinates of the torus, $\boldsymbol{\theta}$, using $\boldsymbol{\theta}_S$ as an initial guess. Here $\boldsymbol{\Omega}$ is the frequency of the torus and is used to equate position and velocity separations. We perform the minimisation using the Nelder-Mead simplex algorithm (Nelder and Mead 1965). This minimisation procedure produces a new phase-space point $(\boldsymbol{x}_S, \boldsymbol{v}_S)$.
4. For the phase-space point $(\boldsymbol{x}_S, \boldsymbol{v}_S)$, we use the Stäckel-fitting method to find an estimate of the angle-actions $(\boldsymbol{\theta}_P, \boldsymbol{J}_P)$.
5. The error in the action reported by the Stäckel approximation is given by $\Delta\boldsymbol{J} = \boldsymbol{J}_P - \boldsymbol{J}_S$. Therefore, if we assume that the error made by the Stäckel approximation is approximately constant over this small region of phase space, we expect that our initial estimate, $\boldsymbol{J}_S$, is also in error by $\Delta\boldsymbol{J}$. A better approximation to the action is
$$\boldsymbol{J}_{\rm new} = \boldsymbol{J}_S - \Delta\boldsymbol{J} = \boldsymbol{J}_S + (\boldsymbol{J}_S - \boldsymbol{J}_P). \tag{2.48}$$
6. We can then construct a torus of action $\boldsymbol{J}_{\rm new}$ and repeat the procedure to refine our estimate of the mis-estimate $\Delta\boldsymbol{J}$.

This method should converge provided the errors in the Stäckel algorithm and the errors in the torus construction vary smoothly with position and velocity.

This procedure is preferable to the Stäckel-fitting method as the accuracy of the procedure can be assessed by evaluating the tolerance, η. However, this increased accuracy will necessarily come with an increase in computation time. To test the procedure, we produce a series of phase-space points on the torus with actions $(J_R, L_z, J_z) = (0.25, 3.5, 0.5)\,\mathrm{kpc^2 Myr^{-1}}$ using the torus machine with an accuracy $\Delta J/J = 10^{-7}$. In Fig. 2.13, we plot the action estimates for this series of phase-space points from the Stäckel-fitting algorithm along with the recovered actions after a single iteration of the above procedure, and the action estimates after the algorithm has reached a tolerance of $\eta = (0.1\,\mathrm{km\,s^{-1}})^2$, or the number of iterations exceeds 20. We find that the Stäckel algorithm initially produces a spread in the action estimates of approximately 4 %. After one iteration of the above routine, we yield actions accurate to ~0.1 %. Further iterations yield slightly more accurate actions (~0.05 %). It appears a single iteration is satisfactory and the benefit of further iterations is small. From the inset in Fig. 2.13, we see that the majority of initial conditions only took a few iterations to converge to the required tolerance. Note that some phase-space points took up to 20 iterations of the procedure and still did not converge to the required tolerance. This is not to say that the iterative procedure wandered away from the target actions, but just that the iterative procedure oscillated near the tar-

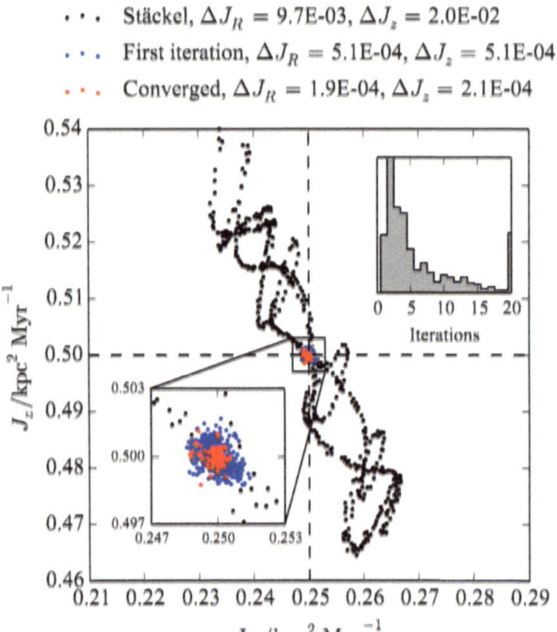

Fig. 2.13 Actions recovered using the iterative torus machine approach for a series of phase-space points on the torus with actions $(J_R, L_z, J_z) = (0.25, 3.5, 0.5)\,\mathrm{kpc^2 Myr^{-1}}$. The *black points* show the actions from the first estimate using the Stäckel-fitting method from this chapter. The *red points* show the recovered actions after one iteration of the procedure, and the blue points show the actions after the procedure was deemed to converge to a tolerance of $\eta = (0.1\,\mathrm{km\,s^{-1}})^2$. The standard deviations of the three sets of points are shown above the plot. The *bottom-left* inset shows a zoom-in of the highlighted rectangle in the main plot, and the *top-right* inset shows a histogram of the number of iterations performed

get actions without ever reaching the required tolerance. This is probably due to non-linearities in both the Stäckel-fitting method and the torus construction. If one chooses to construct more accurate tori i.e. lower $\Delta J/J$, the number that fail to converge is reduced. Finally, we note that the above scheme takes approximately 10 seconds to perform three iterations.

2.8 Method Comparison

When estimating actions, we are aiming for a balance between accuracy of the action recovery and speed. The most accurate method is the convergent iterative torus approach we presented in the previous section. However, this accuracy comes at the expense of additional computational time required to construct tori. Here we critically compare the methods presented in the chapter by investigating the accuracy of the action estimates and time taken to perform many action evaluations. We do not include the iterative torus approach in the following comparisons as the torus machinery is used to generate phase-space points of known actions. Therefore, we expect the approach to converge to the same accuracy as that used to generate the tori.

2.8.1 Total Angular Momentum

Before investigating the methods presented in this chapter, we show how the Stäckel-fitting method compares to the crudest estimate of the vertical action. Some authors have hunted for structure in the distribution of stars in spaces defined by phase space functions other than the actions. For example, Helmi and de Zeeuw (2000) use the set of variables (E, L_z, L), where $L = |\boldsymbol{x} \times \boldsymbol{v}|$ is the total angular momentum, to attempt to find substructure within numerical simulations of disrupted satellite galaxies, whereas Helmi et al. (2006) considered the 'APL space' of apocentre, pericentre and z-component of the angular momentum in order to identify signatures of past accretion events in the Geneva-Copenhagen Survey of the solar neighbourhood (Nordström et al. 2004). The total angular momentum is only conserved when we are considering spherical potentials. In a spherical potential, the vertical action is simply $J_z = L - |L_z|$. Here we investigate how much better we are doing when we estimate the vertical action using a Stäckel fit than if we simply use L. Figure 2.14 shows the absolute RMS error in the vertical action for the 100 low-action tori taken from the lower panel of Fig. 2.6 along with the RMS error in the spherical vertical action, $(L - |L_z|)$. The Stäckel-fitting method gives approximately two orders of magnitude improvement in the vertical-action error compared to simply using L.

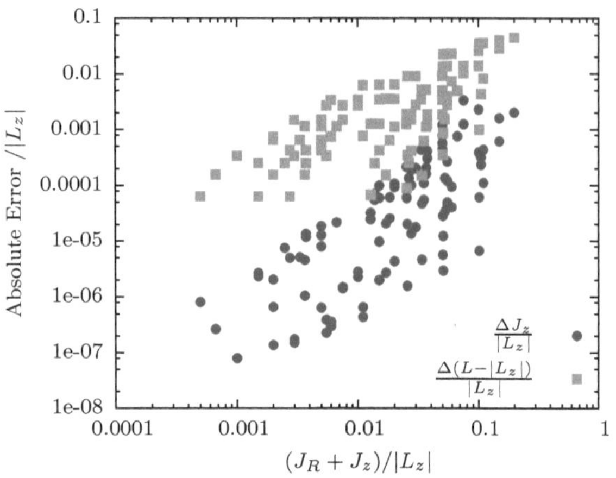

Fig. 2.14 RMS deviations of J_z for the Stäckel fitting method (*blue circles*) and the RMS deviations in the spherical vertical action $(L - |L_z|)$ (*red squares*) for the 100 low-action tori detailed in Sect. 2.3.3.2

2.8.2 Single Torus

We now compare the action recovery for two tori using four methods: the polar adiabatic approximation (PAA), the ellipsoidal adiabatic approximation (EAA), the Stäckel fudge (SF), and locally fitting Stäckel potentials (FIT). We do not compare the accuracy of the convergent iterative torus approach as this approach can be made arbitrarily accurate. We use the torus machine to generate 10,000 phase-space points on the tori $(J_R, L_z, J_z) = (0.001, 2.0, 0.001)\mathrm{kpc}^2\mathrm{Myr}^{-1}$, and $(J_R, L_z, J_z) = (0.1, 2.0, 0.1)\mathrm{kpc}^2\mathrm{Myr}^{-1}$ in the 'best' potential from McMillan (2011a), and estimate the actions using the four methods. In Fig. 2.15 we show the two tori in the meridional plane along with the action estimates for each method. The most accurate method for both tori is the FIT method, whilst the least accurate is the PAA. For the low-action torus, the PAA produces actions accurate to $\sim$1 %, whilst the methods based on Stäckel potentials all produce actions accurate to $\sim$0.01 %. The main difference between the three methods based on Stäckel potentials is the recovery of the radial action which is most accurate for the FIT method, and least accurate for the EAA. For the high-action torus, the PAA produces actions accurate to $\sim$10 %, whilst the other three methods produce actions accurate to $\sim$4 %. The SF method is the least accurate of these three and the FIT method is the most accurate, although the differences are very subtle.

2.8.3 Multiple Tori

We now inspect the accuracy of the action recovery for a selection of tori. We use the 'low-action' tori from Sect. 2.3.3.2. These 100 tori probe the region 3 kpc $< R <$ 22 kpc, $|z| <$ 5 kpc and are chosen to be representative of disc-type tori. For each

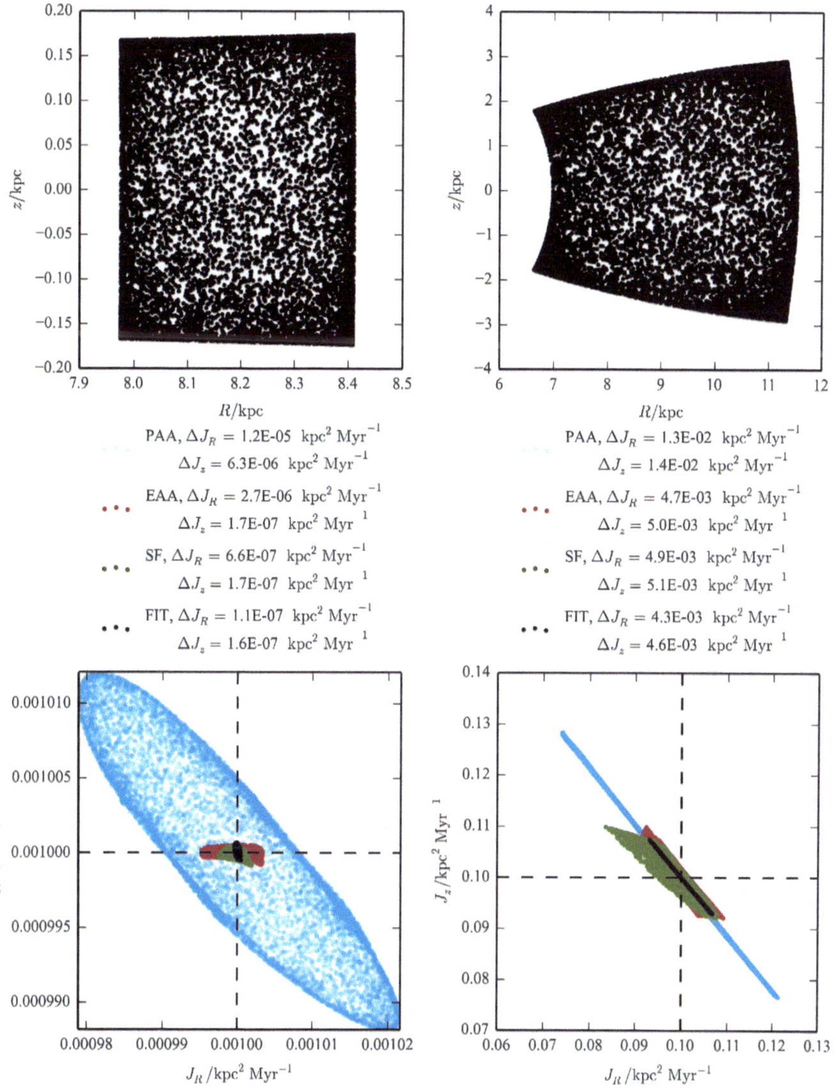

Fig. 2.15 Accuracy of actions using the polar adiabatic approximation (PAA), ellipsoidal adiabatic approximation (EAA), Stäckel fudge (SF) and locally fitting Stäckel potentials. The *top two panels* show 10,000 (R, z) points on two tori in McMillan (2011a) 'best' potential with actions $(J_R, L_z, J_z) = (0.001, 2.0, 0.001)\mathrm{kpc}^2\mathrm{Myr}^{-1}$ (*left*), and $(J_R, L_z, J_z) = (0.1, 2.0, 0.1)\mathrm{kpc}^2\mathrm{Myr}^{-1}$ (*right*). The *lower two panels* show the radial and vertical action estimates for these phase-space points using four methods: the PAA in *blue*, the EAA in *red*, the Stäckel fudge method (SF) from Binney (2012) in *green*, and locally fitting Stäckel potentials (FIT) in *black*. We show the standard deviations of the action estimates for each method between the plots

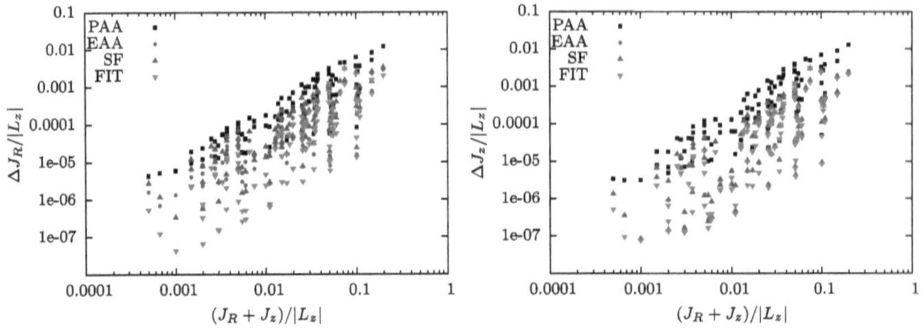

Fig. 2.16 Absolute RMS deviations of the radial (*left*) and vertical (*right*) actions for the polar adiabatic approximation (PAA, *black squares*), the ellipsoidal adiabatic approximation (EAA, *red diamonds*), the Stäckel fudge method (SF, *blue upwards-pointing triangles*), and the Stäckel fitting method (FIT, *green downwards-pointing triangles*). In the *right panel*, the *red diamonds* are not visible as the error in the vertical action reported by the EAA is near identical to that reported by the SF

Table 2.3 Comparison of the accuracy of the action recovery for three tori along with the time taken to evaluate the actions of 10,000 phase-space points for each of the four methods presented in this chapter

$J_R/\lvert L_z\rvert$	$J_z/\lvert L_z\rvert$	$\Delta J_R^{\mathrm{PAA}}/\Delta J_R^{\mathrm{FIT}}$	$\Delta J_R^{\mathrm{EAA}}/\Delta J_R^{\mathrm{FIT}}$		$\Delta J_R^{\mathrm{SF}}/\Delta J_R^{\mathrm{FIT}}$
0.001	0.001	91	7.4		6.5
0.01	0.01	46	7.0		8.4
0.1	0.1	6.5	1.5		1.7
$J_R/\lvert L_z\rvert$	$J_z/\lvert L_z\rvert$	$\mathrm{Time}_{\mathrm{PAA}}$/s	$\mathrm{Time}_{\mathrm{EAA}}$/s	$\mathrm{Time}_{\mathrm{SF}}$/s	$\mathrm{Time}_{\mathrm{FIT}}$/s
0.001	0.001	4.5	7.2	2.5	10.1
0.01	0.01	6.4	8.3	2.6	15.1
0.1	0.1	8.3	10.7	2.6	44.6

torus, we plot the standard deviation in the action estimates for 10,000 randomly selected phase-space points from each of the four methods in Fig. 2.16.

For all methods, we find that the error in the actions correlates approximately with $(J_R + J_z)/|L_z|$. The PAA is clearly the least accurate of the four methods in both the recovery of J_R and J_z. The SF and EAA give near identical errors in J_z such that we cannot see the EAA points in the plot as they are all behind the SF points. We are using identical choices of a^2 for the two methods so the accuracy of the J_z recovery seems limited by this choice. The error in the J_R recovery is, in general, larger for the SF method than the EAA method, but not significantly. The most accurate method for both J_R and J_z is the FIT method. In Table 2.3 we show the ratio of the RMS spreads in the J_R recovery for each method. We see for low-action tori, $J_i/|L_z| \lesssim 0.1$, where $i = R, z$, the FIT method is 2 orders of magnitude more accurate than the PAA, whilst for $J_i/|L_z| \approx 0.1$ this reduces to a single order of magnitude. For the low-action tori $J_i/|L_z| \lesssim 0.1$ the FIT method is 2 orders of magnitude more accurate than both the EAA and SF methods, but this reduces to only a 1.5 increase in accuracy for $J/|L_z| \lesssim 0.1$. We expect that the FIT method is the most accurate of the Stäckel methods as it uses the most information to calculate an appropriate value of a^2, whilst the EAA and SF use only the initial phase-space point to estimate a^2.

It is also worth noting that the relative errors for the PAA can be as much as order one for orbits with large J_z—the PAA performs best for orbits that stay near the plane. The accuracy of the actions estimated by the EAA are approximately an order of magnitude better than those determined using the PAA.

2.8.4 Computational Cost

In Table 2.3, we also give the time taken to evaluate 10,000 actions on a 2.5 GHz i5 processor. For all tori, the SF method is the fastest, followed by the PAA, then the EAA, and finally the FIT method takes the longest time. We can understand this hierarchy as the SF method requires no interpolation, the PAA uses a 2D interpolation, the EAA uses a 3D interpolation and finally the FIT method requires an orbit integration and interpolation of the resulting fitted potential. For $J_i/|L_z| = 0.001$ ($i = R, z$) the FIT method requires 4 times the computational expense for a factor of 6.5 increase in the accuracy over the SF method, whilst this increases to 20 times the computational expense for only a 1.7 increase in accuracy for $J_i/|L_z| = 0.1$. The EAA takes $\gtrsim 4$ times longer than the SF for only a marginal increase in the action accuracy.

In conclusion, the full Stäckel-fitting method presented in this chapter is only marginally more accurate than both the ellipsoidal adiabatic approximation and the Stäckel fudge method at the expense of more computation. Both the ellipsoidal adiabatic approximation and the Stäckel fudge method provide a clear route to tabulating the actions on a grid which then may be interpolated allowing significant speed-ups (see Binney (2012) and Chap. 8). However, the adiabatic approximations require an initial tabulation of the vertical energies and any call outside this grid is problematic. The Stäckel fudge does not suffer from this disadvantage as any action calls outside the grids may be found on the fly. For low-action applications, such as disc modelling, the ellipsoidal adiabatic approximation and the Stäckel fudge are superior methods, but for more accurate high-action applications, such as stream modelling, the Stäckel-fitting method is slightly superior.

2.9 Conclusions

We have detailed methods for estimating the angle-action variables in a general axisymmetric potential given a 6D phase-space point, $(\boldsymbol{x}, \boldsymbol{v})$. The first and most thoroughly explored method is based on locally fitting a Stäckel potential to the region of the potential an orbit probes and then estimating the true actions and angles as those in the fitted Stäckel potential. We have investigated the systematic errors of this method by producing phase-space points of known actions and angles using the torus machine (McMillan and Binney 2008) and then assessing how well the method can reproduce these variables. For a single representative disc-type torus, the errors in the angles are largest for phase-space points near apsis and the errors in the actions are of order a few percent. For a collection of tori chosen to be representative of both disc and halo-type tori, the absolute error in the actions is found to scale with the sum of the vertical and radial actions. The errors in the angles scale with the relative error in their corresponding action. We demonstrated the use of the method by application to the Geneva-Copenhagen Survey (GCS). As this survey is only local, the angle-action space does not reveal much more information than velocity space. However, we present angle-action coordinates for the peaks of the clumps and streams present in the survey and use them to study the relative impact on estimated angles and actions of observational errors and the known systematic errors of the method. We show that the observational errors are dominant. In Chap. 8, we will inspect the GCS data again in the context of extended distribution function modelling of the Galactic disc.

In addition to the Stäckel-fitting method, we have also presented two new methods for finding angles, actions and frequencies in an axisymmetric potential. The first builds on the adiabatic approximation presented in Schönrich and Binney (2012), but, instead of assuming the motion is separable in cylindrical polar coordinates, assumes the motion is separable in prolate spheroidal coordinates. We demonstrated the improvement in the accuracy of the actions calculated for a range of orbits, and showed that the accuracy rivalled the full Stäckel-fitting method for a much smaller computational cost. One shortcoming of the methods based on fitting Stäckel potentials is that they can not be made arbitrarily accurate. The accuracy is limited by how close the true potential is to a Stäckel potential over the region of interest. The torus machinery (McMillan and Binney 2008) constructs a series expansion for the generating function from a toy to target torus, and can be made arbitrarily accurate by using more terms in the generating function. Such a procedure is designed to find $(\boldsymbol{x}, \boldsymbol{v})$ given $(\boldsymbol{J}, \boldsymbol{\theta})$. We presented an iterative method for finding $(\boldsymbol{J}, \boldsymbol{\theta})$ from $(\boldsymbol{x}, \boldsymbol{v})$ by using the Stäckel-fitting procedure as a first estimate, and refining the estimate using the torus machinery. We showed that, for a single torus, only a single iteration was required to improve the error in the action by a factor of 20, and further iterations did little to improve this.

We closed the chapter by comparing the presented methods for estimating the actions in an axisymmetric potential. The Stäckel-fitting method gives results approximately two orders of magnitude more accurate than assuming the potential is spherical. The Stäckel fudge method from Binney (2012) provides the best compromise between speed and accuracy for both low- and high-action disc tori. However, it is not as accurate as the full Stäckel-fitting method, particularly at low actions. The polar adiabatic approximation is improved on by the ellipsoidal adiabatic approximation which offers an accuracy comparable to the Stäckel fudge method but at greater computational expense.

We have demonstrated that the methods presented here are suitable for many disc and halo-type orbits. The procedure will not work for resonant orbits or chaotic orbits. However, the occurrence of these orbits in realistic galaxy axisymmetric potentials is rare and the great majority of stars are on quasi-periodic non-resonant orbits (Ollongren 1962; Martinet and Mayer 1975).

It is hoped that these methods will lead to more widespread use of angle-action variables when analysing data. Whilst the GCS can be easily analysed in velocity space, angle-action variables should enable us to reveal structures, which are more dispersed in phase-space, in larger surveys.

2.9.1 Future Work

The potential used for testing the methods in this chapter consists of two stellar discs, bulge and halo. The gas disc is also a crucial component of any Galactic model (Dehnen and Binney 1998), and is confined to the plane with a scale-height of 40 pc. This component will affect the action recovery for low vertical action tori and some methods presented in this chapter may be more appropriate than others for dealing with this.

We have limited the discussion in this chapter to axisymmetric potentials. Whilst for our own spiral galaxy the axisymmetric approach may suffice, for analysing elliptical galaxies a triaxial approach must be developed. There are also triaxial Stäckel potentials (see de Zeeuw 1985 and Chap. 4), so it should be possible to expand the approach outlined in this chapter to triaxial potentials. The extension of the angle-action estimation is simple, but the fitting procedure is more complex when the potential is triaxial. de Zeeuw and Lynden-Bell (1985) discuss how a general triaxial potential may be fitted both locally and globally by a Stäckel potential. The method for global fitting is the three-dimensional generalisation of the method used in this chapter so involves multiple multi-dimensional integrals. Also, the best choice of coordinate system involves minimising the least-square difference with respect to two coordinate parameters, so a more computationally expensive procedure than the simple method used in this chapter may be required for finding the best coordinate system. In the next two chapters, we will go on to discuss other approaches to estimating the actions in a general triaxial potential.

References

Antoja T, Figueras F, Torra J, Valenzuela O, Pichardo B (2010) Lect Notes Essays Astrophys 4:13
Bacon R et al (2001) MNRAS 326:23
Bahcall JN, Schmidt M, Soneira RM (1982) ApJL 258:L23
Binney J (2010) MNRAS 401:2318
Binney J (2012) MNRAS 426:1324
Binney J, McMillan P (2011) MNRAS 413:1889

Binney J, Tremaine S (2008) Galactic dynamics, 2nd edn. Princeton University Press, Princeton
Cappellari M et al (2011) MNRAS 413:813
De Bruyne V, Leeuwin F, Dejonghe H (2000) MNRAS 311:297
De Simone R, Wu X, Tremaine S (2004) MNRAS 350:627
de Zeeuw PT, Lynden-Bell D (1985) MNRAS 215:713
de Zeeuw T (1984) Ph.D. thesis, Leiden University
de Zeeuw T (1985) MNRAS 216:273
de Zeeuw T, Peletier R, Franx M (1986) MNRAS 221:1001
Dehnen W (1998) AJ 115:2384
Dehnen W, Binney J (1998) MNRAS 294:429
Dejonghe H, de Zeeuw T (1988) ApJ 329:720
Famaey B, Jorissen A, Luri X, Mayor M, Udry S, Dejonghe H, Turon C (2005) A & A 430:165
Helmi A, de Zeeuw PT (2000) MNRAS 319:657
Helmi A, Navarro JF, Nordström B, Holmberg J, Abadi MG, Steinmetz M (2006) MNRAS 365:1309
Holmberg J, Nordström B, Andersen J (2009) A & A 501:941
Jasevicius V (1994) Baltic Astron 3:232
Kent SM, de Zeeuw T (1991) AJ 102:1994
Martinet L, Mayer F (1975) A & A 44:45
McMillan PJ (2011a) MNRAS 414:2446
McMillan PJ (2011b) MNRAS 418:1565
McMillan PJ, Binney JJ (2008) MNRAS 390:429
Navarro JF, Frenk CS, White SDM (1996) ApJ 462:563
Nelder JA, Mead R (1965) Comput J 7:308
Nordström B et al (2004) A & A 418:989
Ollongren A (1962) Bull Astron Inst Neth 16:241
Perryman MAC et al (2001) A & A 369:339
Sanders J (2012) MNRAS 426:128
Schönrich R, Binney J (2012) MNRAS 419:1546
Schönrich R, Binney J, Dehnen W (2010) MNRAS 403:1829
Schwarzschild M (1979) ApJ 232:236
Sellwood JA (2010) MNRAS 409:145
Zwitter T et al (2008) AJ 136:421

Chapter 3
Actions, Angles and Frequencies from Numerically Integrated Orbits

3.1 Introduction

In the previous chapter, we presented various methods for estimating the angle-action variables in axisymmetric potentials. Whilst calculating angles and actions in axisymmetric potentials is appropriate for constructing models of the Galactic disc, for other components of the Galaxy, such as the Galactic bulge or dark matter halo, more general modelling is required. The haloes that form in baryon-free cosmological simulations almost always have triaxial shapes (Jing and Suto 2002; Allgood et al. 2006; Vera-Ciro et al. 2011). When baryons are added to the simulations, many dark haloes become more spherical (Kazantzidis et al. 2004; Bailin et al. 2005; Valluri et al. 2010). Law and Majewski (2010) and Vera-Ciro and Helmi (2013) present evidence that the tidal tails of the Sagittarius dwarf galaxy can only be fitted if the Milky Way has a triaxial dark matter halo. Moreover, there is considerable observational evidence that the so-called "cored", slowly-rotating elliptical galaxies are generically triaxial (Cappellari et al. 2011). Dynamical models of triaxial stellar systems are of considerable astronomical interest. Hence we need to be able to determine angle-action coordinates for stars in triaxial potentials. In this chapter, we show how to evaluate the angles and actions of particles in a given triaxial potential. If the potential is axisymmetric, the actions can be evaluated using the methods presented in Chap. 2.

The methods of the previous chapter were divided into two classes: convergent and non-convergent. Torus construction (McMillan and Binney 2008) is a convergent procedure that yields $(\boldsymbol{x}, \boldsymbol{v})$ as a function of the angle-action variables. This method operates by constructing the generating function for the transformation from a toy potential, in which the angle-actions are readily calculable, to the target potential. In this chapter, we present a very similar convergent procedure to find the angle-action variables given $(\boldsymbol{x}, \boldsymbol{v})$. The algorithm operates by finding the components of the generating function that satisfy the canonical transformation from toy to target angle-actions at a series of $(\boldsymbol{x}, \boldsymbol{v})$ from an orbit integration.

J. Sanders, *Dynamics of the Milky Way*, Springer Theses,
DOI 10.1007/978-3-319-18772-3_3

In Sect. 3.2, we derive the equations that yield values of angles, frequencies and actions. In Sect. 3.3, we test our solutions of these equations by comparing the resulting angles, frequencies and actions for orbits in a Stäckel potential with analytic values. In Sect. 3.4, we use the equations to explore a constant-energy surface in the action space of the triaxial potential for our Galaxy that Law and Majewski (2010) fitted to the tidal stream of the Sagittarius dwarf. In Sect. 3.5, we relate our work to previous work in the field and discuss possible extensions. Section 3.6 sums up and looks to the future. The work of this chapter was published in Sanders and Binney (2014).

3.2 Formalism

Angles and actions can be assigned to orbits that are "regular" or quasiperiodic because such an orbit is confined to a torus labelled by the actions (Arnold 1978). We will work in three dimensions so will have three actions denoted as $\boldsymbol{J} = (J_1, J_2, J_3)$. Each action quantifies the magnitude of the oscillation in a suitable coordinate.

The transformation from ordinary phase-space coordinates $(\boldsymbol{x}, \boldsymbol{v})$ to angle-action coordinates $(\boldsymbol{\theta}, \boldsymbol{J})$ is possible analytically in only a few cases. McGill and Binney (1990) used one of these cases as a starting point for the numerical construction of more general transformations by "torus mapping". The key point about torus mapping is that it yields orbits with specified actions rather than orbits with specified initial conditions $(\boldsymbol{x}, \boldsymbol{v})$. When analysing an N-body model, we require actions given an initial condition and not vice versa. Here we adapt the approach of McGill and Binney (1990) into a procedure that finds the actions, angles and frequencies given a series of phase-space coordinates $(\boldsymbol{x}_i, \boldsymbol{v}_i)$ sampled along an orbit at times t_i, where $0 \leq t_i \leq T$. With this time series, we seek a generating function that will map a "toy torus" of a simple "toy potential" into the "target torus" to which the orbit is confined. The toy potential must have analytically tractable angles and actions and permit orbits that have the correct geometry.

In the absence of figure rotation, a general triaxial potential admits two basic classes of non-resonant orbit: loop orbits and box orbits (Schwarzschild 1979; de Zeeuw 1985). Loop orbits have a definite sense of rotation either around the long- or short-axis of the potential, whilst a box orbit has no sense of rotation and can reach down to the centre of the potential. Hence, the class of an orbit can be determined by inspection of components of the angular momentum along the orbit: if all components of the angular momentum change sign, the orbit has no sense of circulation and is a box orbit; when a component of the angular momentum retains its sign, the orbit is a loop orbit around the corresponding axis (Carpintero and Aguilar 1998). For each class of orbit, we use a toy potential that provides tori with the same geometrical structure as the tori of the given orbit class.

For a box orbit, the actions J_1, J_2 and J_3 quantify the oscillation in the x, y and z directions, respectively. For loop orbits, J_1 quantifies oscillation in a generalized radial coordinate. For a short-axis loop, J_2 quantifies the particle's circulation around the short axis, whilst J_3 quantifies oscillation parallel to this axis. For a long-axis loop orbit, J_3 quantifies circulation around the long axis, whilst J_2 quantifies oscillation parallel to this axis. We choose this definition such that our actions match J_λ, J_μ and J_ν for a Stäckel potential (de Zeeuw 1985), and each class of orbit occupies a distinct region of action space (see Sect. 3.4).

3.2.1 Toy Potentials

We now present the toy angle-action variables appropriate for each orbit class.

3.2.1.1 Triaxial Harmonic Oscillator

For box orbits, we use the potential of the triaxial harmonic oscillator,

$$\Phi_{\rm ho}(\boldsymbol{x}) = \tfrac{1}{2}\sum_{i=1}^{3}\omega_i^2 x_i^2, \tag{3.1}$$

which has three parameters, ω_i. Here we have chosen the principal axes of the potential to lie along the Cartesian x, y, z directions on the assumption that the time series has already been rotated into the coordinate system that is aligned with the principal axes of the true potential. The actions and angles in this potential are given by

$$\begin{aligned} J_i &= \tfrac{p_i^2+\omega_i^2 x_i^2}{2\omega_i}, \\ \theta_i &= \arctan\left(\tfrac{p_i}{\omega_i x_i}\right). \end{aligned} \tag{3.2}$$

3.2.1.2 Isochrone Sphere

For loop orbits, we use the isochrone potential,

$$\Phi_{\rm iso}(\boldsymbol{x}) = \frac{-GM}{b+\sqrt{b^2+r^2}}, \tag{3.3}$$

where r is the spherical radius. This potential has two free parameters: the mass M and the scale radius, b. The full expressions for the angle-action coordinates in the isochrone potential are given in Appendix C. The three actions in the isochrone potential are given by the radial action J_r, the z-component of the angular momentum L_z and the vertical action $J_z \equiv L - |L_z|$, where L is the total angular momentum. With this choice we must orient our coordinate system, such that the orbit circulates around the z-axis, before finding the actions.

3.2.1.3 Offsets

One might also include the offset of the centre of the potential from the coordinate centre as a free parameter, but we shall not do so here, presuming instead that the time samples x_i have already been adjusted to be relative to one's best estimate of the centre of the true potential.

3.2.1.4 Parameter Choice

Once a class of potential has been chosen, we set the parameters of the potential by minimizing (McGill and Binney 1990)

$$\chi^2 = \sum_i (H_i - \langle H \rangle)^2, \tag{3.4}$$

where the sum is over the times, H_i is the value of the toy Hamiltonian at $(\boldsymbol{x}_i, \boldsymbol{v}_i)$, and $\langle H \rangle$ is the mean of these values. The minimization of χ^2 is done using the Levenberg–Marquardt algorithm (Press et al. 2002).

The experiments described below suggest that this method for selecting the parameters is sub-optimal in that it leads to a rather centrally concentrated toy potential being selected. This central concentration then leads to high-order Fourier components being required in the generating function. However, our attempts to find a better procedure for selecting the toy potential have not met with success.

3.2.2 *Generating Function*

With a toy potential chosen, we construct the generating function to transform between the angle-actions $(\boldsymbol{\theta}, \boldsymbol{J})$ of the toy potential, and those $(\boldsymbol{\theta}', \boldsymbol{J}')$ of the target potential. The generating function for this transformation, $S(\boldsymbol{\theta}, \boldsymbol{J}')$, can be written

$$S(\boldsymbol{\theta}, \boldsymbol{J}') = \boldsymbol{\theta} \cdot \boldsymbol{J}' - \mathrm{i} \sum_{\boldsymbol{n} \neq \boldsymbol{0}} S_{\boldsymbol{n}}(\boldsymbol{J}') \mathrm{e}^{\mathrm{i}\boldsymbol{n} \cdot \boldsymbol{\theta}}, \tag{3.5}$$

where the vector $\boldsymbol{n}$ has integer components. The first term on the right generates the identity transformation, whilst the structure of the second part is required by the periodicity of the angle variables.

McGill and Binney (1990) show that if the Hamiltonian is time-reversible, the $S_{\boldsymbol{n}}$ must satisfy

$$S_{\boldsymbol{n}} = -S_{-\boldsymbol{n}}. \tag{3.6}$$

For this condition to be satisfied, there must exist a point on the toy torus at which $\dot{\boldsymbol{J}} = \mathbf{0}$—in Appendix D we demonstrate that this is true for the toy potentials of the previous section. Additionally, the requirement that the actions are real coupled with the constraint from Eq. (3.6) requires the $S_{\boldsymbol{n}}$ to be real. With these constraints, the generating function can be written

$$S(\boldsymbol{J}', \boldsymbol{\theta}) = \boldsymbol{\theta} \cdot \boldsymbol{J}' + 2 \sum_{\boldsymbol{n} \in \mathbf{N}} S_{\boldsymbol{n}}(\boldsymbol{J}') \sin \boldsymbol{n} \cdot \boldsymbol{\theta}, \tag{3.7}$$

where the integer vectors $\boldsymbol{n}$ are now restricted to just half of a three-dimensional lattice. We take this half to be the set $\mathbf{N} = \{(i, j, k)\}$, where either $(k > 0)$, $(k = 0, j > 0)$ or $(k = 0, j = 0, i > 0)$. Symmetries of the target potential require some of the $S_{\boldsymbol{n}}$ to be zero. This is discussed further in Appendix D.

From the generating function (3.7), we find that the toy actions are

$$\boldsymbol{J} = \frac{\partial S}{\partial \boldsymbol{\theta}} = \boldsymbol{J}' + 2 \sum_{\boldsymbol{n} \in \mathbf{N}} \boldsymbol{n} S_{\boldsymbol{n}}(\boldsymbol{J}') \cos \boldsymbol{n} \cdot \boldsymbol{\theta}, \tag{3.8}$$

and the target angles are

$$\boldsymbol{\theta}' = \frac{\partial S}{\partial \boldsymbol{J}'} = \boldsymbol{\theta} + 2 \sum_{\boldsymbol{n} \in \mathbf{N}} \frac{\partial S_{\boldsymbol{n}}}{\partial \boldsymbol{J}'}(\boldsymbol{J}') \sin \boldsymbol{n} \cdot \boldsymbol{\theta}. \tag{3.9}$$

Note that by the choice of our generating function, the target angle zero-point coincides with the toy-angle zero-point.

Given the choice of a toy Hamiltonian, we find the toy actions and angles $(\boldsymbol{J}(t_i), \boldsymbol{\theta}(t_i))$. Each time then produces a separate equation (3.8) with common unknowns: the target actions and the Fourier components of the generating function, $S_{\boldsymbol{n}}$. We cannot solve these equations exactly because we are dealing with equations in an infinite number of unknowns. Because we can include only a finite number of terms on the right-hand side of each equation, the right-hand sides should not agree exactly with the left-hand sides, and the correct procedure is to minimize the sum of the squares of the residuals of individual equations. This sum is

$$F = \sum_i \sum_k \left(J_k(t_i) - J_k' - 2 \sum_{\boldsymbol{n} \in \mathbf{N}} n_k S_{\boldsymbol{n}}(\boldsymbol{J}') \cos \boldsymbol{n} \cdot \boldsymbol{\theta}(t_i) \right)^2, \tag{3.10}$$

where the inner sum is over the dimension of the action space and the set $\mathbf{N}$ is limited to a finite number of vectors $\boldsymbol{n}$. We take this set to be the N vectors that satisfy the condition $|\boldsymbol{n}| \leq N_{\max}$, where $N_{\max} \simeq 6$.

We minimize F by setting to zero its derivatives with respect to the unknowns:

$$0 = \frac{\partial F}{\partial J'_k} = -2\sum_i \Big(J_k(t_i) - J'_k - 2\sum_{\boldsymbol{n}\in\mathbf{N}} n_k S_{\boldsymbol{n}}(\boldsymbol{J}') \cos \boldsymbol{n}\cdot\boldsymbol{\theta}(t_i)\Big)$$
$$0 = \frac{\partial F}{\partial S_{\boldsymbol{m}}} = -2\sum_i\sum_k 2m_k \cos \boldsymbol{m}\cdot\boldsymbol{\theta}(t_i)\Big(J_k(t_i) - J'_k - 2\sum_{\boldsymbol{n}\in\mathbf{N}} n_k S_{\boldsymbol{n}}(\boldsymbol{J}') \cos \boldsymbol{n}\cdot\boldsymbol{\theta}(t_i)\Big). \tag{3.11}$$

To solve these equations, we define a matrix $\mathbf{c}_{\boldsymbol{n}k}$ that has as subscripts the vector $\boldsymbol{n}$ and the integer $k = (1, 2, 3)$ that selects a particular spatial dimension. This N-by-3 matrix is

$$\mathbf{c}_{\boldsymbol{n}k}(t_i) \equiv 2n_k \cos\big(\boldsymbol{n}\cdot\boldsymbol{\theta}(t_i)\big), \ \text{(no sum over } \boldsymbol{n}\text{)}. \tag{3.12}$$

We further define two $(3+N)$-vectors

$$\boldsymbol{x_J} \equiv (\boldsymbol{J}', S_{\boldsymbol{n}}), \quad \boldsymbol{b_J} \equiv \sum_i (\boldsymbol{J}(t_i), \mathbf{c}_{\boldsymbol{n}}(t_i)\cdot\boldsymbol{J}(t_i)), \tag{3.13}$$

and the symmetric matrix

$$\mathbf{A}_{\boldsymbol{J}} \equiv \sum_i \begin{pmatrix} \mathbf{I}_3 & \mathbf{c}^{\mathrm{T}}(t_i) \\ \mathbf{c}(t_i) & \mathbf{c}(t_i)\cdot\mathbf{c}^{\mathrm{T}}(t_i) \end{pmatrix}. \tag{3.14}$$

Here $\mathbf{I}_3$ is the 3-by-3 identity matrix. With these definitions, the Eqs. (3.11) to be solved can be written as

$$\mathbf{A}_{\boldsymbol{J}}\cdot\boldsymbol{x_J} = \boldsymbol{b_J}. \tag{3.15}$$

We solve these equations for $\boldsymbol{x_J}$ by LU decomposition (Press et al. 2002).

A similar procedure yields the target angles from Eq. (3.9). We note that, at time t_i, the orbit has $\boldsymbol{\theta}'(t_i) = \boldsymbol{\theta}'(0) + \boldsymbol{\Omega}' t_i$, where $\boldsymbol{\Omega}'$ is the target frequency, and $\boldsymbol{\theta}'(0)$ is the angle corresponding to the initial point in the orbit integration. The relevant sum of squared residuals is

$$G = \sum_i\sum_k \Big(\theta'_k(0) + \Omega'_k t_i - \theta_k(t_i) - 2\sum_{\boldsymbol{n}\in N} \frac{\partial S_{\boldsymbol{n}}}{\partial J'_k}(\boldsymbol{J}') \sin \boldsymbol{n}\cdot\boldsymbol{\theta}\Big)^2. \tag{3.16}$$

The unknowns are $\boldsymbol{\theta}'(0)$, $\boldsymbol{\Omega}'$ and the set of $\partial S_{\boldsymbol{n}}/\partial \boldsymbol{J}'$, which we denote as $(\partial_1 S_{\boldsymbol{n}}, \partial_2 S_{\boldsymbol{n}}, \partial_3 S_{\boldsymbol{n}})$. For each time, we define the N-vector

$$\mathbf{s}_{\boldsymbol{n}}(t_i) = -2\sin\big(\boldsymbol{n}\cdot\boldsymbol{\theta}(t_i)\big). \tag{3.17}$$

We also define the $3(2+N)$-vectors

$$\begin{aligned} \boldsymbol{x}_\theta &\equiv (\boldsymbol{\theta}'(0), \boldsymbol{\Omega}', \partial_1 S_{\boldsymbol{n}}, \partial_2 S_{\boldsymbol{n}}, \partial_3 S_{\boldsymbol{n}}), \\ \boldsymbol{b}_\theta &\equiv \textstyle\sum_i (\boldsymbol{\theta}(t_i), t_i\boldsymbol{\theta}(t_i), \theta_1(t_i)\mathbf{s}(t_i), \theta_2(t_i)\mathbf{s}(t_i), \theta_3(t_i)\mathbf{s}(t_i)), \end{aligned} \tag{3.18}$$

and the symmetric matrix

$$\mathbf{A}_\theta \equiv \sum_i \begin{pmatrix} \mathbf{I}_3 & t_i\mathbf{I}_3 & \mathbf{s}^{1T} & \mathbf{s}^{2T} & \mathbf{s}^{3T} \\ t_i\mathbf{I}_3 & t_i^2\mathbf{I}_3 & t_i\mathbf{s}^{1T} & t_i\mathbf{s}^{2T} & t_i\mathbf{s}^{3T} \\ \mathbf{s}^1 & t_i\mathbf{s}^1 & \mathbf{s}\cdot\mathbf{s}^T & 0 & 0 \\ \mathbf{s}^2 & t_i\mathbf{s}^2 & 0 & \mathbf{s}\cdot\mathbf{s}^T & 0 \\ \mathbf{s}^3 & t_i\mathbf{s}^3 & 0 & 0 & \mathbf{s}\cdot\mathbf{s}^T \end{pmatrix} \tag{3.19}$$

where each $\mathbf{s}^m$ is an N-by-3 matrix with the N-vector $\mathbf{s}$ in the mth column, and each $\mathbf{s}$ is evaluated at the ith time. Setting the partial derivatives of G with respect to the unknowns to zero yields the matrix equation,

$$\mathbf{A}_\theta \cdot \boldsymbol{x}_\theta = \boldsymbol{b}_\theta. \tag{3.20}$$

The toy angles will be 2π-periodic, and we require the same for the target angles $\boldsymbol{\theta}'(0) + \boldsymbol{\Omega}' t_i$. However, in order to solve the matrix equation we must first make the $\boldsymbol{\theta}(t_i)$ from the orbit integration continuously increase, and then we solve for the target angles and take the 2π-modulus.

3.2.2.1 Axisymmetric Case

In axisymmetric potentials, the angular momentum about the z-axis is one of the actions. This action is independent of the potential such that $J_2(t_i) = J_2' = L_z$. From Eq. (3.11), we see that this implies $S_{\boldsymbol{n}} = 0$ for $n_2 > 0$, so we need only consider the Fourier components with $n_2 = 0$.

3.2.3 Choice of N_T, $N_{\rm max}$ and T

Given the scheme presented above, the only questions that remain are how to select the orbit integration time T, the number of time samples, N_T, to use, and what value to use for $N_{\rm max}$, which determines the number N of Fourier components we solve for. Here we discuss how we can automatically choose these parameters such that we have good recovery of the unknowns.

A necessary condition is that the number of unknowns must be less than the number of time samples, N_T. For the action calculation, the number of unknowns is approximately $N_{\max}^3/2$, whilst for the angle-frequency calculation we have $\sim 3N_{\max}^3/2$ unknowns. We also expect our ability to recover the unknowns to depend upon the sampling of the toy-angle space.

Let us first consider an idealised 1D case. If we were able to sample uniformly in the toy angle of a 1D system, we would select N_T points in a single period separated in toy angle by $\Delta = 2\pi/N_T$. With this sampling rate, we would be able to constrain all modes $\mathrm{e}^{in\theta}$ with $n\Delta \leq \pi$. We can choose to constrain only the $N_{\max}$ modes with $n < \pi/\Delta$ as then we would be super-sampling the highest considered modes. Here we are using a time series that is a product of an orbit integration, so is not uniformly spaced in toy angles—the toy-angle distribution depends on the target Hamiltonian, the toy potential and the distribution of sampling times. The recovery of Fourier components from non-uniform samples is discussed in Marvasti (2001). To constrain modes from a 1D non-uniform sampling, we must sample on average at or above the Nyquist frequency. If we have toy-angle samples, θ_i we require

$$\frac{n}{N_T - 1}\sum_{i=1}^{i=N_T-1}(\theta_{i+1} - \theta_i) \leq \pi, \tag{3.21}$$

to constrain mode n.

Here we are attempting to recover components from samples, $\boldsymbol{\theta}_i$, in 3D toy-angle space. As we are restricted to using samples generated from an orbit integration, our sampling is limited to some sub-space of the full 3D toy-angle space. The 3D sampling can be considered as a series of 1D samples in $\boldsymbol{n} \cdot \boldsymbol{\theta}_i$ (we first unroll the angles such that they increase continuously). In order to recover the $S_{\boldsymbol{n}}$ from this toy-angle sampling, we need to satisfy two conditions:

1. As in the 1D case, we need to sample on average at or above the Nyquist frequency such that

$$\frac{1}{N_T - 1}\sum_{i=1}^{i=N_T-1}\boldsymbol{n} \cdot (\boldsymbol{\theta}_{i+1} - \boldsymbol{\theta}_i) \leq \pi. \tag{3.22}$$

2. For every included mode, $\boldsymbol{n}$, we would also like a good total coverage in $\boldsymbol{n} \cdot \boldsymbol{\theta}$. We choose to require that the $\boldsymbol{n} \cdot \boldsymbol{\theta}$ samples cover the full range from 0 to 2π:

$$\max(\boldsymbol{n} \cdot \boldsymbol{\theta}) - \min(\boldsymbol{n} \cdot \boldsymbol{\theta}) > 2\pi. \tag{3.23}$$

If this condition is not satisfied, we are including a mode which will not be well constrained by the toy-angle sampling i.e. the average of $\cos \boldsymbol{n} \cdot \boldsymbol{\theta}$ will not be near zero. We therefore expect that the corresponding $S_{\boldsymbol{n}}$ will not be well recovered from this sampling. It could be that this $S_{\boldsymbol{n}}$ is not significant so will not affect the recovered actions and frequencies significantly. However, a conservative approach would ensure that Eq. (3.23) is satisfied for all included modes.

The second of these conditions is the stricter. To ensure that the toy-angle sampling satisfies Eq. (3.23) when an orbit is near-resonant, we require time samples which span a very large number of periods. This is an inevitable drawback of the approach taken here because we have very little control over the sampling in the toy angle space.

Having identified a mode that will not be well constrained, one possibility is to set $S_n = 0$ for this mode. However, by doing this, we risk throwing out a mode which is significant, and the recovery of the actions and frequencies will deteriorate, so we opt not to do this.

Another requirement is that the $S_{\boldsymbol{n}}$ (and $\partial S_{\boldsymbol{n}}/\partial \boldsymbol{J}'$) decrease as we go to larger $\boldsymbol{n}$ such that the truncation at $N_{\max}$ is valid. If the S_n do not decrease with $\boldsymbol{n}$, this is evidence of aliasing such that these higher $\boldsymbol{n}$ modes are not well recovered and we expect the actions, angles and frequencies will also not be well recovered.

3.2.3.1 Procedure

We will now summarize the above discussion into a procedure that can be implemented:

- We first select a reasonable $N_{\max}$, for instance $N_{\max} = 6$ is used in the later examples.
- We then integrate for some time T recording at least $N_T = 3N + 6$ time samples (or $N_T = N + 3$ if we only need the actions) such that we have as many equations as unknowns. This is always satisfied if we choose

$$N_T = \max(200, \frac{9N_{\max}^3}{4}). \tag{3.24}$$

- For each time sample, we find the toy angles and check that Eqs. (3.22) and (3.23) are satisfied for each mode. If Eq. (3.22) is not satisfied, T is much longer than the fundamental orbital periods, so we require a finer time sampling from the orbit integration. If Eq. (3.23) is not satisfied, we continue integrating the orbit until this equation *is* satisfied for all the modes.
- We then perform the procedure outlined in Sect. 3.2.2 to find the $S_{\boldsymbol{n}}$. We require the S_n to be decreasing with $\boldsymbol{n}$ such that on the boundaries the values of the S_n are small. If we find that the boundary values of S_n are large, we have not included a sufficient number of modes in the generating function so we must increase $N_{\max}$ and repeat the above procedure until we are satisfied that all dominant modes are included.

As we will see below this procedure is very conservative but should ensure that the recovery of the actions, angles, frequencies and components of the generating function are accurate.

3.3 Example

As a test of the above, let us look at an example. The most general separable triaxial potential is the triaxial Stäckel potential (see de Zeeuw 1985 and Chap. 4 for more details). We choose to work with the perfect ellipsoid, which has density profile

$$\rho(x, y, z) = \frac{\rho_0}{(1+m^2)^2}, \tag{3.25}$$

where

$$m^2 \equiv \frac{x^2}{x_P^2} + \frac{y^2}{y_P^2} + \frac{z^2}{z_P^2}, \quad x_P \geq y_P \geq z_P \geq 0. \tag{3.26}$$

The associated coordinates are confocal ellipsoidal coordinates in which the actions can be expressed as one-dimensional integrals. These may be calculated numerically using Gauss-Legendre quadrature. Similarly, the frequencies can also be determined from one-dimensional integrals. Equations for these quantities are given in de Zeeuw (1985). Here we work with the potential with parameters $\rho_0 = 7.2 \times 10^8 M_\odot\,\mathrm{kpc}^{-3}$, $x_P = 5.5\,\mathrm{kpc}$, $y_P = 4.5\,\mathrm{kpc}$ and $z_P = 1\,\mathrm{kpc}$.

In this potential we examine three orbits—a short-axis loop orbit with initial condition $(x, y, z) = (10, 1, 8)$ kpc, $(v_x, v_y, v_z) = (40, 152, 63)\ \mathrm{km\,s^{-1}}$, a box orbit with initial condition $(x, y, z) = (0.1, 0.1, 0.1)$ kpc, $(v_x, v_y, v_z) = (142, 140, 251)\ \mathrm{km\,s^{-1}}$, and a long-axis loop orbit with initial condition $(x, y, z) = (-0.5, 18, 0.5)$ kpc and $(v_x, v_y, v_z) = (25.0, 20.0, -133.1)\ \mathrm{km\,s^{-1}}$. Each orbit was integrated for 8 periods of the lowest frequency, T_F. We set $N_{\max} = 6$ and calculated the corresponding number of uniformly-spaced time samples required from Eq. (3.24). We ensured that Eqs. (3.22) and (3.23) were satisfied for all the included modes. In Figs. 3.1 and 3.2 we show the orbits in the (x, y) and (x, z) planes, the sampling of the toy-angle space and the resultant actions. We also show, in faint red, the result of integrating in the best-fitting toy potential. This gives us an idea of the work that the generating function has to do to deform the toy torus into the target torus.

For the short-axis loop orbit, the true and recovered actions are

$$\begin{aligned}
\boldsymbol{J}_{\mathrm{true}} &= (212.09, 1307.54, 708.15)\ \mathrm{kpc\ \ km\,s^{-1}} \\
\boldsymbol{J}_{\mathrm{recov}} &= (213.33, 1307.29, 709.16)\ \mathrm{kpc\ \ km\,s^{-1}},
\end{aligned}$$

and the true and recovered frequencies are

$$\begin{aligned}
\boldsymbol{\Omega}_{\mathrm{true}} &= (21.28048, 15.30346, 18.90759)\ \mathrm{Gyr}^{-1} \\
\boldsymbol{\Omega}_{\mathrm{recov}} &= (21.28081, 15.30360, 18.90754)\ \mathrm{Gyr}^{-1}.
\end{aligned}$$

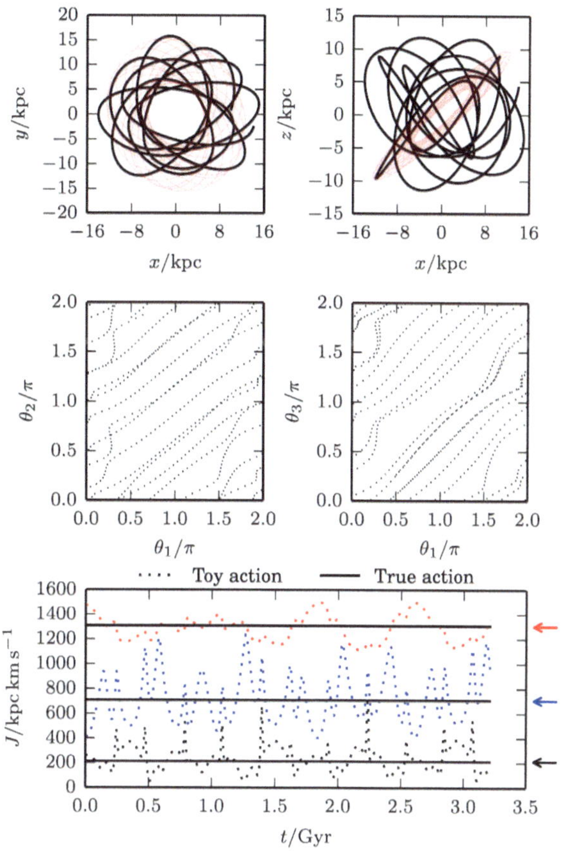

Fig. 3.1 Example short-axis loop orbit in the triaxial Stäckel potential—the *top panels* show the orbit integrated in the test potential in *black*. This is a short-axis loop orbit so circulates about the axis $z = 0$. In *faint red*, we show the initial point integrated in the best-fitting isochrone potential. In the *middle panels*, we show the toy angles calculated at each time sample. In the *bottom panel*, we show the toy actions at each time-step as a *dotted line* (*black* for J_1, *blue* for J_2 and *red* for J_3). The *solid lines* show the true actions and the *arrows* mark the estimated actions

In the left panel of Fig. 3.3, we depict two cross-sections of $\boldsymbol{n}$-space showing the absolute value of the components of the generating function (for the isochrone potential, we use the convention that subscript 1 refers to J_r, subscript 2 refers to L_z and subscript 3 refers to $J_z \equiv L - |L_z|$.) We see that the two most significant modes are $\boldsymbol{n} = (-1, 2, 0)$, which causes a mixing between the radial motion and azimuthal motion, and $\boldsymbol{n} = (0, 0, 2)$. Note that the $S_{\boldsymbol{n}}$ decrease towards the boundary, so we are content that we have included the relevant modes.

For the box orbit, the true and recovered actions are

$$\begin{aligned} \boldsymbol{J}_{\text{true}} &= (336.39, 137.78, 237.96)\,\text{kpc km s}^{-1} \\ \boldsymbol{J}_{\text{recov}} &= (336.85, 137.26, 238.17)\,\text{kpc km s}^{-1}, \end{aligned}$$

and the true and recovered frequencies are

$$\begin{aligned} \boldsymbol{\Omega}_{\text{true}} &= (38.8673, 45.3762, 72.1720)\,\text{Gyr}^{-1} \\ \boldsymbol{\Omega}_{\text{recov}} &= (38.8656, 45.3737, 72.1689)\,\text{Gyr}^{-1}. \end{aligned}$$

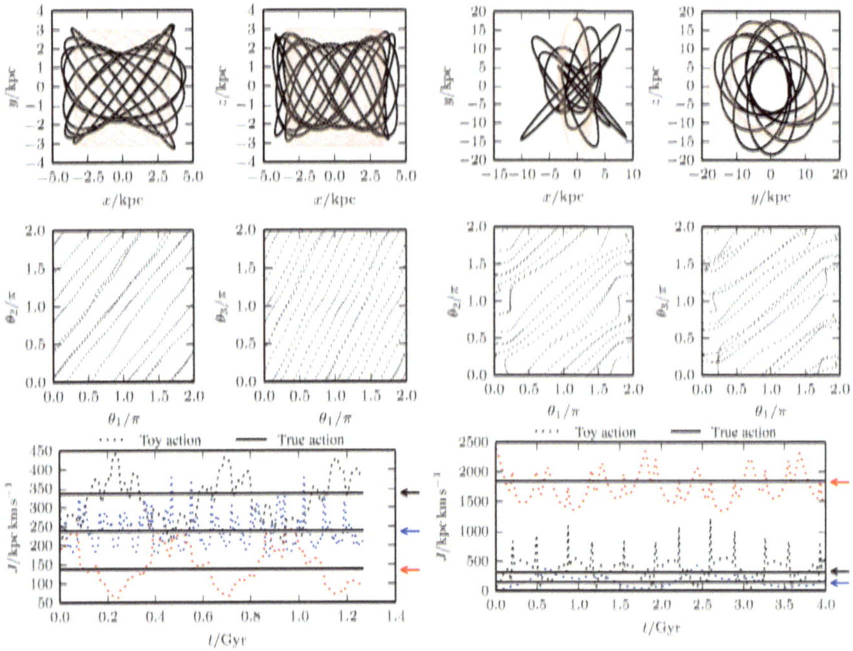

Fig. 3.2 Example box orbit (*left panels*) and long-axis loop orbit (*right panels*) in the triaxial Stäckel potential—the top panels show the orbit integrated in the test potential in *black*. In *faint red*, we show the initial point integrated in the best-fitting toy potential. In the *middle panels*, we show the toy angles calculated at each time sample. In the *bottom panel*, we show the toy actions at each time-step as a *dotted line* (*black* for J_1, *blue* for J_2 and *red* for J_3). The *solid lines* show the true actions and the *arrows* mark the estimated actions

In the right panel of Fig. 3.3, we depict two cross-sections of $\boldsymbol{n}$-space showing the absolute value of the components of the generating function. The two most significant modes are $\boldsymbol{n} = (2, -2, 0)$, which causes a mixing between the x motion and the y motion, and $\boldsymbol{n} = (2, 0, -2)$, which mixes the x and z motions. These modes are required to distort the rectangular orbits of the triaxial harmonic oscillator into those bounded by surfaces of constant confocal ellipsoidal coordinate. Note that the $S_{\boldsymbol{n}}$ decrease at the boundaries as required. The structure of the $S_{\boldsymbol{n}}$ for the loop orbit is much richer than the $S_{\boldsymbol{n}}$ for the box orbits as the generating function has many more significant terms.

For the long-axis loop orbit, the true and recovered actions are

$$\begin{aligned}\boldsymbol{J}_{\rm true} &= (312.89, 146.48, 1839.2)\,\text{kpc km s}^{-1}\\ \boldsymbol{J}_{\rm recov} &= (323.12, 146.03, 1842.0)\,\text{kpc km s}^{-1},\end{aligned}$$

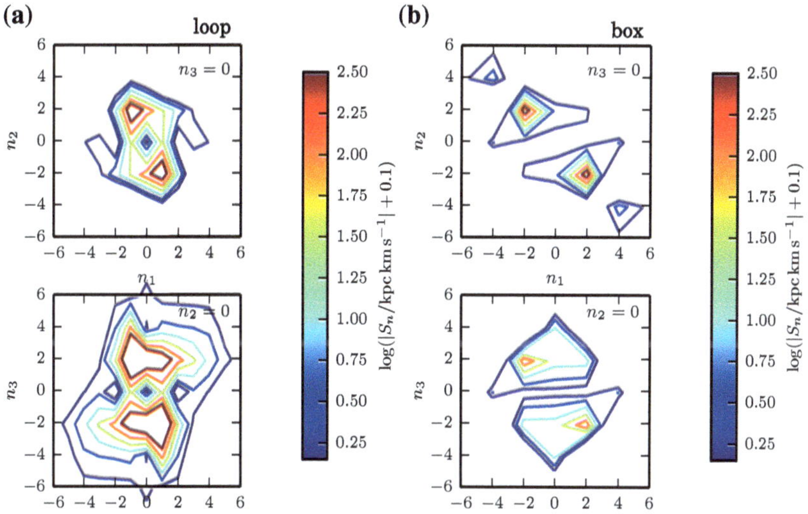

Fig. 3.3 Cross-sections of the $S_{\boldsymbol{n}}$ as a function of $\boldsymbol{n}$ for the example loop orbit (*left*) and the example box orbit (*right*). In the *top panels* we show the cross-section $n_3 = 0$. The most significant mode in this plane for the loop orbit is $(-1, 2, 0)$, which causes a mixing between the radial motion and azimuthal motion. For the box orbit the most significant mode in the plane $n_3 = 0$ (*top*) is $(2, -2, 0)$, which causes a mixing between the x motion and the y motion. In the *lower panels* we show the cross-section $n_2 = 0$, in which the mode $(0, 0, 2)$ is the most significant for the loop orbit, and the most significant mode for the box orbit is $(2, 0, -2)$. **a** Example loop orbit, **b** Example box orbit

and the true and recovered frequencies are

$$\begin{aligned}\boldsymbol{\Omega}_{\text{true}} &= (18.0665, 12.29, 14.0504)\,\text{Gyr}^{-1}\\ \boldsymbol{\Omega}_{\text{recov}} &= (18.0749, 12.3008, 14.0497)\,\text{Gyr}^{-1}.\end{aligned}$$

3.3.1 Accuracy of the Method

Figure 3.4 shows errors in J_3' and Ω_3' for the box orbit as a function of N_{max} for various choices of the total integration time T. We have linked N_{max}–N_T via Eq. (3.24). However, we have not ensured that Eqs. (3.22) and (3.23) are satisfied for each case. The weight of each point is proportional to the largest gap in coverage for the N modes. We see that, in general, a longer integration time provides a more accurate estimate of the action and particularly the frequency. We can understand this as a longer line segment provides a better measurement of the gradient for noisy data. From Fig. 3.4, we see that when working with high N_{max} it is not sufficient to satisfy Eq. (3.24). We must also satisfy equation (3.23) such that we have a sufficient sampling in toy-angle space to constrain these higher modes.

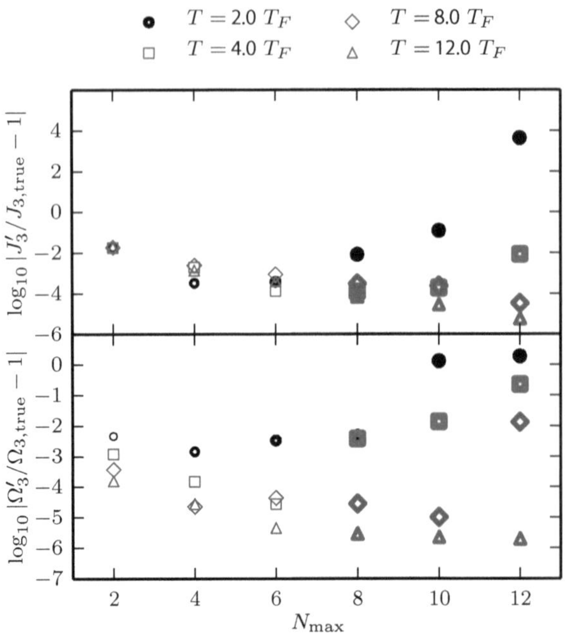

Fig. 3.4 Error in the recovered values of J_3' and Ω_3' for the box orbit as a function of N_{max} and the total integration time. We work with even multiples of the period, T_F, corresponding to the lowest frequency. The size of the points is proportional to the largest gap in coverage for the N modes. In general, a longer integration time provides more accurate actions and frequencies. When attempting to constrain higher modes, it is necessary to integrate the orbit for a longer period to ensure that the sampling in toy angle is sufficient

For $T = 2T_F$, Eq. (3.23) is not satisfied for $N_{\mathrm{max}} \geq 4$. For large N_{max} and $T = 2T_F$, many modes have insufficient coverage and the results are very poor. For the other three integration times, Eq. (3.23) is not satisfied for $N_{\mathrm{max}} \geq 8$. For $T = 4T_F$, this results in an immediate deterioration of the frequency recovery as we have included a mode with $\max(\boldsymbol{n} \cdot \boldsymbol{\theta}) - \min(\boldsymbol{n} \cdot \boldsymbol{\theta}) \approx \pi/2$. For $T = 8T_F$ and $T = 12T_F$, a lack of coverage has not affected the results apart from for $T = 8T_F$ and $N_{\mathrm{max}} = 12$ where the frequency recovery is poorer. The mode that is not well covered is also not well covered for $N_{\mathrm{max}} = 8$, but we only see the effects of this lack of coverage when we try to include more modes. For $T = 12T_F$, both the action and frequency recovery are very good despite equation (3.23) not being satisfied when $N_{\mathrm{max}} \geq 8$. In particular, there is one mode for which $\max(\boldsymbol{n} \cdot \boldsymbol{\theta}) - \min(\boldsymbol{n} \cdot \boldsymbol{\theta}) \approx 4.3$. It seems that this coverage is sufficient to not degrade the results. In conclusion, when Eq. (3.23) is satisfied we recover the frequencies and actions well, whilst when it is not satisfied the recovery deteriorates in some cases, particularly that of the frequency.

Finally, we find that, when we double the number of time samples used for the examples shown in Fig. 3.4, the results change significantly only when Eq. (3.23) is not satisfied. Therefore, we conclude that, provided we have more equations than unknowns and have satisfied Eqs. (3.22) and (3.23), the actions and frequency recovery will be satisfactory.

3.3.2 Near-Resonant Orbit

To illustrate some of the points discussed, we show results for a near-resonant orbit. This orbit is a box orbit with the initial conditions $(x, y, z) = (0.1, 0.1, 0.1)$ kpc, $(v_x, v_y, v_z) = (142, 150, 216.5)$ km s^{-1}. Again we integrate for time $T = 8T_F$ and set $N_{\max} = 6$. The results are shown in Fig. 3.5. The frequency vector of this orbit is nearly parallel to $\boldsymbol{n} = (-4, 0, 2)$ so the coverage of this mode is very poor and $\max(\boldsymbol{n} \cdot \boldsymbol{\theta}) - \min(\boldsymbol{n} \cdot \boldsymbol{\theta}) \approx 1.11$ for $\boldsymbol{n} = (-4, 0, 2)$. However, the true and recovered actions are

$$\begin{aligned} \boldsymbol{J}_{\text{true}} &= (301.74, 147.63, 165.36)\,\text{kpc km s}^{-1} \\ \boldsymbol{J}_{\text{recov}} &= (300.69, 147.66, 165.89)\,\text{kpc km s}^{-1}, \end{aligned}$$

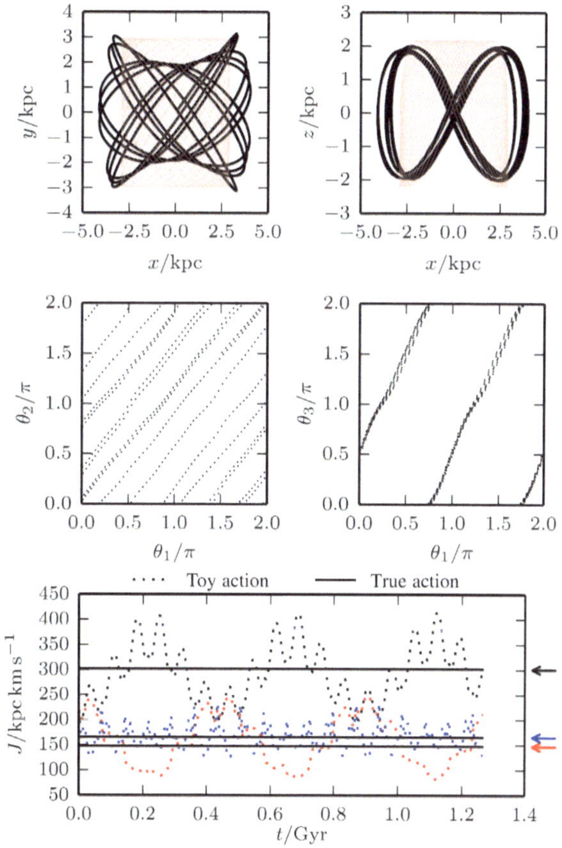

Fig. 3.5 Example near-resonant orbit in the triaxial Stäckel potential—the *top panels* show the orbit integrated in the test potential in *black*. In *faint red*, we show the initial point integrated in the best-fitting triaxial harmonic oscillator potential. In the *middle panels*, we show the toy angles calculated at each time sample. In the *bottom panel*, we show the toy actions at each time-step as a *dotted line* (*black* for J_1, *blue* for J_2 and *red* for J_3). The *solid lines* show the true actions and the *arrows* mark the estimated actions

and the true and recovered frequencies are

$$\begin{aligned}\boldsymbol{\Omega}_{\rm true} &= (43.318, 50.369, 86.724)\,{\rm Gyr}^{-1}\\ \boldsymbol{\Omega}_{\rm recov} &= (43.386, 50.371, 86.777)\,{\rm Gyr}^{-1}.\end{aligned}$$

As seen before, poor coverage in one of the modes is not detrimental to the action and frequency recovery.

3.4 Application

As a brief application of the method outlined in this chapter, we will inspect the action diagram for a realistic triaxial Galactic potential. We take the potential from Law and Majewski (2010). This potential was found to produce the best fit to the Sagittarius stream data. This potential has three components: a disc defined by the Miyamoto-Nagai potential

$$\Phi_{\rm disc}(x,y,z) = \frac{-GM_{\rm disc}}{\sqrt{x^2+y^2+(a+\sqrt{z^2+b^2})^2}}, \tag{3.27}$$

with $M_{\rm disc} = 10^{11}\,M_\odot$, $a = 6.5\,{\rm kpc}$ and $b = 0.26\,{\rm kpc}$; a spherical bulge described by the Hernquist profile

$$\Phi_{\rm bulge}(r) = \frac{-GM_{\rm bulge}}{r+c}, \tag{3.28}$$

with $M_{\rm bulge} = 3.4\times10^{10}\,M_\odot$ and $c = 0.7\,{\rm kpc}$; and the triaxial logarithmic halo

$$\Phi_{\rm halo}(x,y,z) = v_{\rm halo}^2 \ln\left(C_1x^2 + C_2y^2 + C_3xy + \frac{z^2}{q_z^2} + r_{\rm halo}^2\right) \tag{3.29}$$

with $v_{\rm halo} = 121.7\,{\rm km\,s^{-1}}$, $C_1 = 0.99\,{\rm kpc}^{-2}$, $C_2 = 0.53\,{\rm kpc}^{-2}$, $C_3 = 0.11\,{\rm kpc}^{-2}$, $q_z = 1.36$ and $r_{\rm halo} = 12\,{\rm kpc}$.

3.4.1 An Example Orbit

We inspect a single orbit in this potential in Fig. 3.6. The chosen orbit is a short-axis loop orbit with initial condition $(x, y, z) = (14.69, 1.80, 0.12)$ kpc, $(v_x, v_y, v_z) = (15.97, -128.90, 44.68)$ km s^{-1}. We use different, but overlapping, $8T_F$ long segments of the orbit with $N_T = 500$ to calculate the actions, angles and frequencies

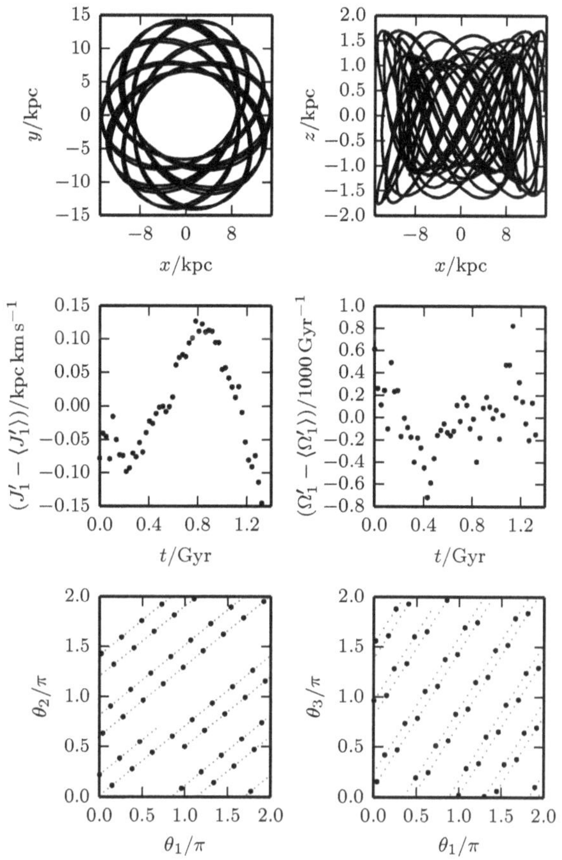

Fig. 3.6 An example orbit in the Law and Majewski (2010) potential. It is a short-axis loop orbit with actions $\boldsymbol{J}' \approx (160, 2186, 36)$ kpc km s^{-1}. In the *top panel*, we show a $16T_F$ long orbit segment in the (x, y) and (x, z) planes. In the *central two panels*, we show the spread in J_1' and Ω_1' calculated using 500 time-samples from an $8T_F$ orbit segment labelled by its initial time sample. In the *bottom panel*, we show the calculated angles at these times with *black dots*. We also show the angles found using $\boldsymbol{\theta}'(0) + \boldsymbol{\Omega}' t_i$ with one of the calculated frequencies and initial angles in smaller *blue dots*

using $N_{\max} = 6$. We ensure Eq. (3.23) is satisfied for these time samplings. This orbit lies in the surface of constant energy explored in the next section. We find that the action and frequency are

$$\begin{aligned}\boldsymbol{J}' &\approx (160.18, 2186.16, 36.09)\,\text{kpc km s}^{-1}\\ \boldsymbol{\Omega}' &\approx (27.26, 19.12, 37.01)\,\text{Gyr}^{-1}.\end{aligned}$$

The error in the actions and frequencies can be estimated by the spread of the estimates from each segment. We find

$$\begin{aligned}\Delta\boldsymbol{J}' &\approx (0.07, 0.08, 0.03)\,\text{kpc km s}^{-1},\\ \Delta\boldsymbol{\Omega}' &\approx (3\times10^{-4}, 6\times10^{-5}, 2\times10^{-3})\,\text{Gyr}^{-1}.\end{aligned}$$

For each orbit segment, we find $\boldsymbol{\theta}'(0)$ and these different values should all lie along straight lines with gradients given by the derived frequencies. In Fig. 3.6, we show that the condition is well satisfied.

Using different orbit segments is perhaps the only way to estimate the error in an action or frequency found using the present method. It is simplest to use consecutive orbit segments as we have here. However, a better method is to use orbit segments separated by a large time interval. This can be achieved most effectively by utilizing the estimated generating function to find an initial condition for a second orbit integration. A simple choice is to increase one of the derived angle coordinates by $\pi/2$.

3.4.2 A Typical Constant Energy Surface

Now we turn to constructing the action diagram for the chosen potential. For a given energy (that of a particle dropped at 18 kpc on the intermediate axis), we launched particles at a series of points linearly spaced between 0.2 and 18 kpc along the potential's intermediate axis with the velocity vector perpendicular to the axis and inclined at linearly spaced angles to the z-axis between 0 and $\pi/2$.[1] We integrated each initial condition for $\sim$10 Gyr saving $N_T = 1000$ samples. For all orbits, the energy was conserved to one part in 10^6. We set $N_{\max} = 6$ and ensured that Eqs. (3.22) and (3.23) were satisfied. If Eq. (3.22) was not satisfied, we had undersampled the orbit, so we took a finer sampling. If Eq. (3.23) was not satisfied, we did not have sufficient coverage, so we continued integrating for another 10 Gyr, taking another 1000 samples. We then calculated the actions from the time series. Figure 3.7 shows each orbit as a point in 3D action-space.[2] We see that the surface of constant energy is a triangle-shaped plane in action-space. The points are coloured based on their orbit classification. An equivalent figure for a Stäckel potential can be found in de Zeeuw (1985).

In a triaxial potential, the loop orbits can be divided into two classes: the short-axis loops that loop around the short axis (in our case the z-axis) and the long-axis loops that loop around the long axis (the x-axis). Along with the box orbits, these three classes of orbit occupy distinct regions on the action-space plane of constant energy. At each corner of the plane, only one action is non-zero and the corresponding orbit is the parent orbit of each of the three classes: the $J_2 = 0$, $J_3 = 0$ orbit is a radial orbit along the long axis, the $J_1 = 0$, $J_3 = 0$ orbit is a closed orbit in the (x, y) plane and the $J_1 = 0$, $J_2 = 0$ orbit is a closed orbit in the (y, z) plane. We note that near the interface between the different orbit classes some regions of the plane are depleted of points (our choice of initial sampling causes an increased density of points near the edges of the plane). Also, there is some overlap between the different

---

[1] Note that the intermediate axis of the halo model proposed by Law and Majewski is actually the z-axis. However, at small radii ($\lesssim$18 kpc), the intermediate axis of the full potential is in the (x, y) plane due to the disc contribution, and the z-axis is the short axis.

[2] To produce a continuous plane in action-space we must scale the 'radial' actions of the loop orbits, J_1, by a factor of 2. J_1 for a loop orbit corresponds to a single oscillation from minimum to maximum coordinate and back, whilst for a box orbit a single oscillation covers the interval 0 to maximum coordinate four times.

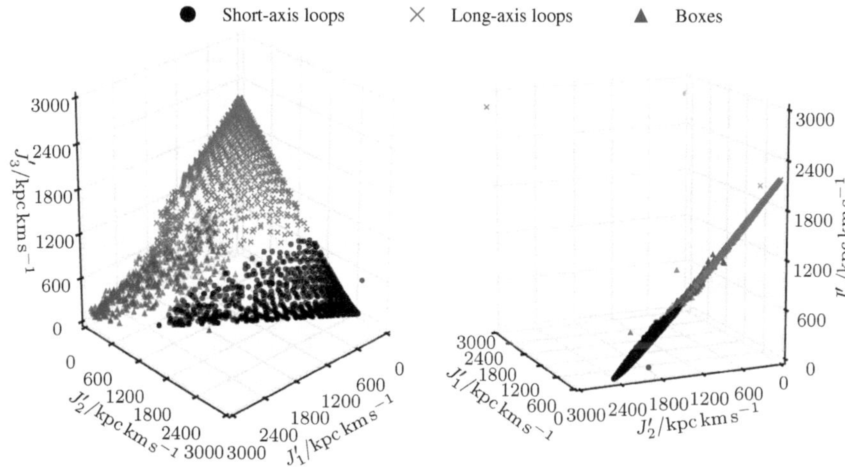

Fig. 3.7 Two projections of a surface of constant energy in the 3D action space of the potential proposed by Law and Majewski (2010). *Black circles* show short-axis loop orbits, *red crosses* show long-axis loop orbits and *blue triangles* show box orbits

orbit classes in the action space. These features are due to the presence of resonant islands with surrounding chaotic orbits at the interface of the regular orbit regions (see Sect. 3.5.3). Additionally, for orbits near the box/loop boundary, it can take many orbital periods to correctly identify the orbit class (Carpintero and Aguilar 1998), and some may be misclassified.

3.5 Discussion

3.5.1 Relation to Previous Work

The problem addressed here goes back to Binney and Spergel (1982, 1984), who Fourier transformed the time series $x(t_i)$ of individual coordinates and assigned to each line in the resulting spectrum appropriate integers n_j so that ωt could be identified with $\sum_j n_j \Omega_j t$. Once this identification had been successfully accomplished, $\Omega_j t$ could be replaced with θ_j to yield the orbit's angle representation. This approach is inferior to that introduced here in several respects: (i) Whereas the generating function is a scalar, a star's location is described by a vector, so it is wasteful to construct the angle representations of all three coordinates rather than the angle representation of the generating function: Binney and Spergel (1984) failed to take advantage of the strong restrictions on tori that arise from angle-action coordinates being canonical. (ii) It is not straightforward to measure correctly the complex amplitudes A from the discrete Fourier transform of a time series such as $x(t_i)$ because the required amplitude will, in general, not lie at one of the discrete frequencies sampled. (iii) When an orbit is near-resonant, there is often dangerous ambiguity in the integers n_j that should be assigned to a particular line. With the present technique, we work from the outset with periodic functions and their Fourier series so the issue of how frequencies fall on a discrete grid does not arise. Moreover, the assignment of integers n_j to Fourier terms is unambiguous.

The method described here has significant overlap with the work of Warnock (1991) on the construction of magnetic coordinates and the related method of Kaasalainen and Binney (1994) for the construction of angle coordinates. In both these studies, angle-action variables were evaluated along numerically computed orbits. The coordinates evaluated were not those of a toy potential but of a trial torus that had been previously constructed: Warnock (1991) was refining the Fourier coefficients S_n while Kaasalainen and Binney (1994) were solving for the $\partial_i S_n$ given the S_n. In both these studies, several initial conditions for orbit integration were chosen on each torus to overcome the problem that with a single short integration a resonant orbit yields a highly non-uniform distribution of sample points on the torus. Since we do not have a good representation of the target torus until the equations have been set up and solved, we cannot take advantage of this possibility.

Warnock (1991) solved for the discrete Fourier transforms of the $\boldsymbol{n} S_{\boldsymbol{n}}$ rather than for the $S_{\boldsymbol{n}}$ because the matrix that then has to be inverted is nearly diagonal when the toy and target tori are close to one another and the sample points provide a nearly regular grid in the space of toy angles. Since our toy and target tori can be quite different, and it is hard to achieve a uniform sampling of toy-angle space, we have not used Warnock's technique.

3.5.2 Possibility of Using Stäckel Tori

We have used completely different toy potentials for each class of orbit, and it is natural to ask whether it would not be advantageous to use always a Stäckel potential since such a potential has tori of every type. We have not pursued this option for two reasons. First, the actions and angles of Stäckel potentials require the evaluation of integrals whereas the potentials we have used yield algebraic expressions for angles and actions. Secondly, and more fundamentally, when integrating an orbit that lies close to the box/loop interface, it would be non-trivial to ensure that the toy torus with the actions of the target orbit had the same geometry as the target torus. By using potentials that support only one type of torus, we are assured from the outset that this condition is satisfied. However, this rests on our correct identification of the orbit type from the time series. As we saw with the Law and Majewski potential, in some marginal cases it may take many orbital periods to correctly identify the orbit.

Another option is to use the approximate methods, such as those detailed in the previous chapter and the next, to find the toy angle-actions. However, this approach suffers from the same disadvantages as using Stäckel potentials.

3.5.3 Resonances and Chaos

We have focused here on orbits that are non-resonant members of the major orbital families. In real galactic potentials, one encounters orbits that are either resonantly trapped or chaotic (e.g. Sect. 3.7, Binney and Tremaine 2008). Chaotic orbits can be thought of as sequences of sections of resonantly trapped orbits, so these two types of orbit raise similar issues.

In a generic integrable potential, the frequencies Ω_i depend on the actions, so on some tori a resonant condition $\boldsymbol{n} \cdot \boldsymbol{\Omega} = 0$ is satisfied. Consequently, individual orbits on these resonant tori do not cover the entire torus since the condition $\boldsymbol{n} \cdot \boldsymbol{\theta} =$ constant constrains the angle variables. This lack of coverage makes it impossible to determine some of the Fourier coefficients $S_{\boldsymbol{n}}$.

When the potential is strictly integrable, orbits on tori that are adjacent to a resonant torus completely cover their tori although they take a long time to do so. In a generic potential, however, such orbits move over a series of tori without covering any of them, as they librate around the strictly resonant orbit. Consequently, these orbits have some of the characteristics of a strictly resonant torus. When the present technique is used on a resonantly trapped orbit, the generating function will map the toy torus into a close approximation to the strictly resonant torus, so, in an N-body model, the density of stars on this torus will seem to be larger than it really is. Hence with the present technique, resonantly trapped orbits will give rise to apparent crowding in action space that is analogous to the signature of resonances when particles are mapped into frequency space by determining orbital frequencies by Fourier decomposition of coordinates (Dumas and Laskar 1993): when the ratios Ω_2/Ω_1 and Ω_3/Ω_1 are used to place orbits in frequency-ratio space, the existence of resonantly trapped orbits leads to a crowding of points along the straight lines associated with certain resonance conditions $\boldsymbol{n} \cdot \boldsymbol{\theta} =$ constant (Binney and Tremaine 2008, Sect. 3.7.3(b)).

Chaotic orbits can be considered as moving through a series of quasi-periodic orbits. Therefore the recovered actions and frequencies from our method will be a function of the total integration time. We see that the region of the constant energy surface occupied by the box orbits in Fig. 3.7 has considerable crowding and the regular grid of initial conditions is not visible. This is indicative of chaotic orbits which have been allocated very different actions from one initial condition to the next.

In Fig. 3.8 we perform the same procedure as outlined in Sect. 3.7.3(b) of Binney and Tremaine (2008) to inspect the ratio of frequencies plane of a logarithmic potential. We use the potential

$$\Phi(\boldsymbol{x}) = \frac{1}{2} \ln \left(x^2 + \frac{y^2}{q_y^2} + \frac{z^2}{q_z^2} + r_c^2 \right), \tag{3.30}$$

with $q_y = 0.9$, $q_z = 0.7$ and $r_c^2 = 0.1$. We drop a series of test particles on the surface $\Phi(\boldsymbol{x}) = 0.5$ regularly spaced in the spherical polar coordinates ϕ and $\cos\theta$, and integrate each initial condition for a time $T = 200$ extracting $N_T = 2048$

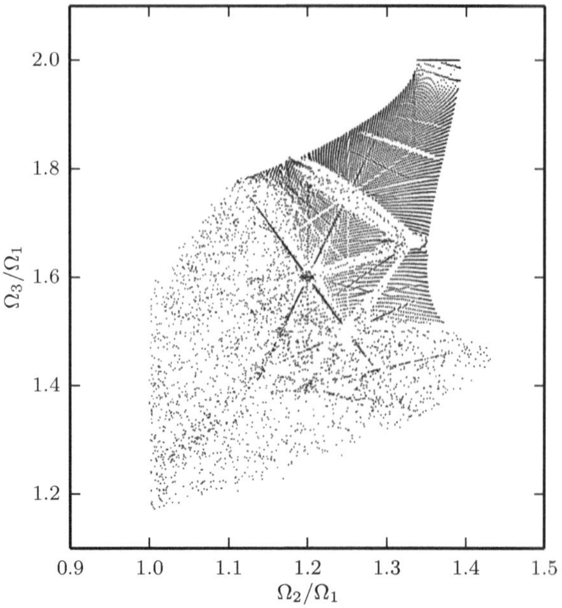

Fig. 3.8 Frequency ratios in the triaxial logarithmic potential extracted from orbital time series using the method presented in this chapter. Each point corresponds to an initial condition for a particle dropped on the surface $\Phi(\boldsymbol{x}) = 0.5$

samples. We then use our method to find the corresponding orbital frequencies and plot their ratios in Fig. 3.8. As noted in Binney and Tremaine (2008), the top-right corner of this plane shows the regular spacing of the initial conditions whilst the lower-left corner shows a more irregular distribution with no evidence of the regular grid of initial conditions used to produce it. Also, we find that there are overdensities along lines corresponding to resonances. Our plot is very similar to that shown in Binney and Tremaine (2008). However, the structure of the irregular bottom-left region differs. This is to be expected as it is these orbits which are irregular, and how one assigns regular properties to them depends on the method employed.

To illustrate this further we inspect two orbits in the surface of constant energy of the triaxial logarithmic potential. One orbit is chaotic with initial condition $\theta = 1.066\,\mathrm{rad}$, $\phi = 1.419\,\mathrm{rad}$, whilst the other is regular with initial condition $\theta = 1.441\,\mathrm{rad}$, $\phi = 0.7601\,\mathrm{rad}$. The orbits and spectra are shown in Fig. 3.9. The spectra are simply the Fourier transforms of each coordinate (Binney and Spergel 1982). We see that the regular orbit has a very simple spectrum of single peaks broadened by the finite integration time, whilst the chaotic orbit has a raggedy spectrum of many lines. In Fig. 3.10, we show the toy angles for each orbit. We see for the regular orbit that the toy angles increase at a near constant rate with small fluctuations around this constant rate. However, the chaotic orbit has a trend in the toy angles with time. Two straight line segments have been fitted to the points to indicate the change with time. It appears that the orbit is regular for small times, but over larger times the

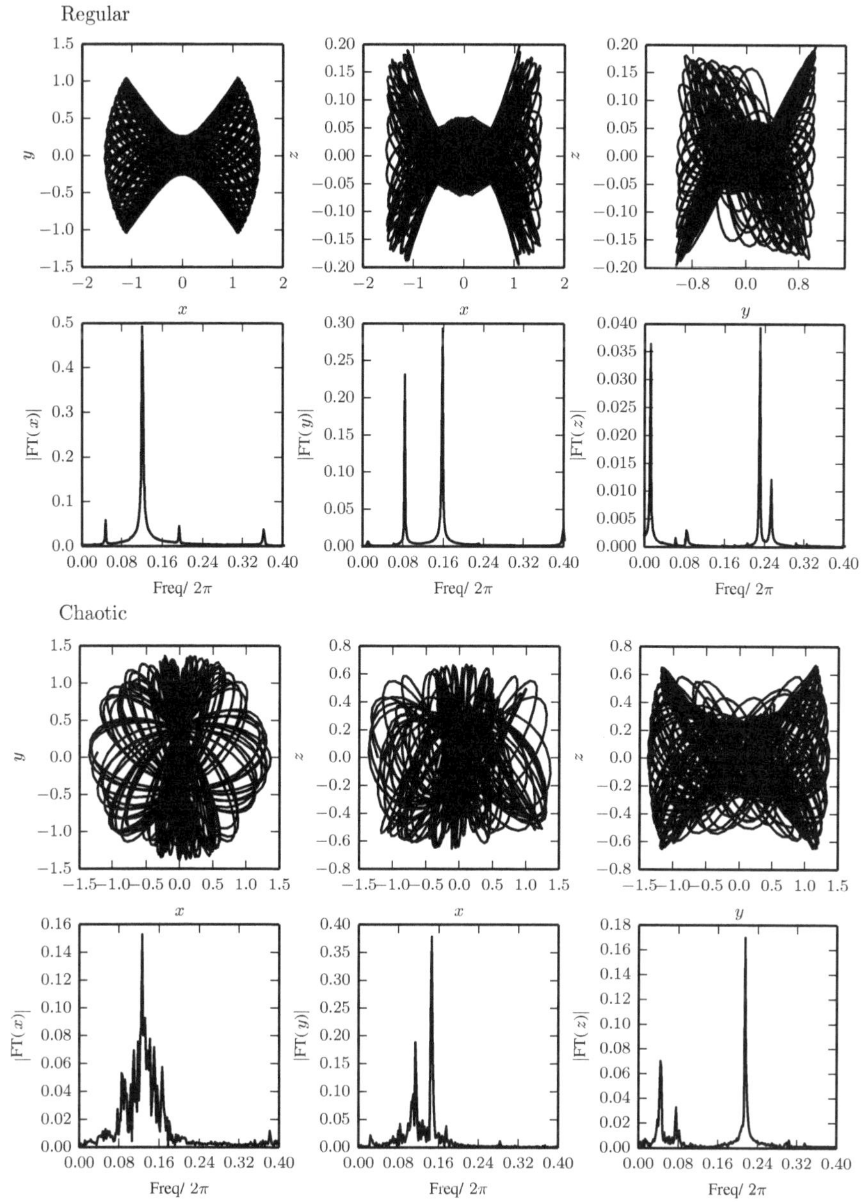

Fig. 3.9 Two orbits dropped on the surface $\Phi(\boldsymbol{x}) = 0.5$ in the triaxial logarithmic potential. The *top six panels* correspond to a regular orbit, whilst the *bottom six* correspond to a chaotic orbit. The *top three panels* of each section show three spatial projections of the orbit and the *lower three* show the Fourier transforms of each Cartesian coordinate

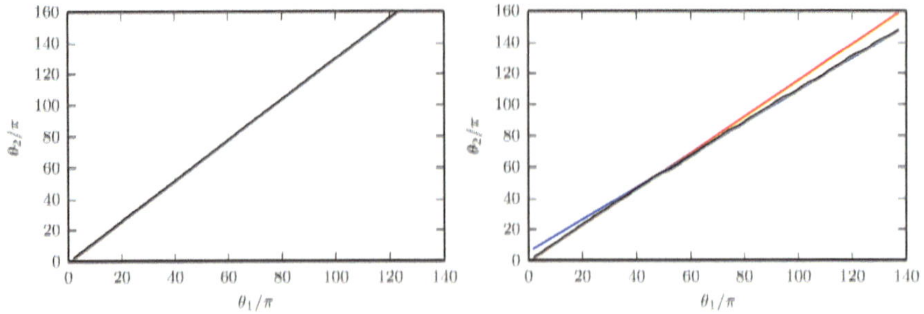

Fig. 3.10 Toy angles for the two orbits from Fig. 3.9. The *left panel* shows the toy angles for the regular orbit, whilst the *right panel* corresponds to the chaotic orbit. *Two straight lines* have been fitted to the first 500 (*red*) and the final 500 toy angles (*blue*). They are barely visible for the regular orbit as they near coincide, whilst for the chaotic orbit there is a clear change in the gradient of the toy angles with time

gradient of the toy angles shifts such that at large times the orbit also appears regular. The regular frequencies we assign to this orbit are a function of the integration time. For the chosen integration time, the assigned frequency is some average of the initial and final frequencies such that the initial condition is scattered randomly in Fig. 3.8.

3.6 Conclusions

We have presented a method for finding actions, frequencies and angles from numerically integrated orbits in a general potential. The method relies on estimating the Fourier components of the generating function that maps a toy torus into the torus on which the computed orbit lies by solving systems of linear algebraic equations. This method enables one to determine the angle-action coordinates $(\boldsymbol{\theta}, \boldsymbol{J})$ of a given phase-space point $(\boldsymbol{x}, \boldsymbol{v})$ and has numerous possible applications in astronomy.

We inspected the method by recovering the actions of different orbit classes in a triaxial Stäckel potential, including a near-resonant orbit. The method has several parameters that affect the accuracy of the recovered actions. We showed how these parameters can be set automatically, and how the error in the actions varies with the parameters. As an application of the method, we constructed the 3D action diagram for a surface of constant energy in the Law and Majewski (2010) potential. Finally, we closed with a discussion of the effects of chaotic and resonant orbits on the method.

Ours is the first method presented in the literature for finding the actions in a general triaxial potential. Triaxiality is an essential ingredient of dark-matter distributions, and a realistic Galactic model which should include non-axisymmetric features such as the bar, and the potentially triaxial halo. This method is a necessary first step towards constructing distribution functions, $f(\boldsymbol{J})$, for these more complex Galactic components. However, the method lacks the speed to evaluate many actions required for numerically integrating the moments of the distribution function. Binney

(2012a) introduced an algorithm for rapidly estimating the actions in an axisymmetric potential by assuming the potential is close to a Stäckel potential. In the next chapter, we present an extension of this approach to triaxial potentials that is more appropriate for constructing $f(\boldsymbol{J})$ models.

3.6.1 Future Work

Here we discussed time-independent triaxial potentials. In this case, we can determine a priori the phases of the terms in the generating function. Rotation of the figure of the potential destroys the time-reversibility of the Hamiltonian and we lose the ability to set the phases a priori. In the worst case, the $S_{\boldsymbol{n}}$ in Eq. (3.5) become complex numbers that are only limited by the condition $S_{-\boldsymbol{n}} = S^*_{\boldsymbol{n}}$ required to make the generating function real. Extending the current framework to this case, approximately doubles the dimensionality of the matrices we must solve for given $N_{\max}$.

Action-based distribution functions have proved important for the understanding and analysis of the Galactic disc (Binney et al. 2014). These simple axisymmetric distribution functions make use of the speed of the Stäckel approximation methods for finding angles and actions (Binney 2012a). Much of the dynamics of our Galaxy is governed by the central bar (Fux 2004), and, as surveys probe deeper through the Galactic disc, it is imperative that this structure is included in the models. This triaxial time-dependent feature needs dynamically consistent models for its description and the method presented here provides a route for such an approach. One direction for future work is to construct a simple distribution function for the bar and inspect the velocity histograms at various locations in the bar for comparison with results from ARGOS (Freeman et al. 2013; Ness et al. 2013a, b) and APOGEE (Ahn et al. 2014). One inhibiting feature of this course of action is the speed of the method presented in this chapter. In the next chapter, we present a faster, non-convergent method for estimating actions in a general triaxial potential that may be more appropriate for this application.

An important application of the presented method is to the analysis of N-body simulations. A single N-body snapshot consists of 3D positions and velocities for $\sim 10^9$ particles. Letting the simulation evolve for a few time steps produces another snapshot with a completely different set of 10^9 positions and velocities. Thus the particles' phase-space coordinates constitute a highly degenerate and non-compact representation of the simulation. Effective analysis of the simulation should start by condensing the coordinates into a smaller set of numbers. This can be done by replacing the $6N_T$ numbers $(\boldsymbol{x}_i, \boldsymbol{v}_i)$ with just three numbers J_i and plotting each particle as a point in 3D action space. The simulation then becomes a density of particles in a 3D space. This representation will greatly facilitate the comparison of different N-body models. Also, it may prove possible to find good fits to the stellar action-space density in terms of analytic functions, as Pontzen and Governato (2013) have done for numerical dark-matter haloes and appears to be possible for the Galactic discs (Binney 2012b; Binney et al. 2014).

It should be noted that it is not advisable to take the N_T time samples of a given orbit directly from the simulation. Rather, at some time t, the potential should be computed on a spatial grid (e.g. Magorrian 2007), and the equations of motion in this potential should be integrated for N_T time-steps starting from the phase-space location of each particle at time t. These integrations in a fixed potential lend themselves to massive parallelization, for example on a Graphical Processor Unit (GPU), so it should be possible to compute angle-action coordinates for very large numbers of particles.

Ueda et al. (2014) present a similar method to that outlined in this chapter to find Fourier coefficients of the generating function for a set of representative tori. These coefficients may then be interpolated to construct tori of any action. Such a scheme is possible here for speeding up the calculation of actions for many particles.

References

Ahn CP et al (2014) ApJS 211:17
Allgood B, Flores RA, Primack JR, Kravtsov AV, Wechsler RH, Faltenbacher A, Bullock JS (2006) MNRAS 367:1781
Arnold VI (1978) Mathematical methods of classical mechanics. Springer, New York
Bailin J et al (2005) ApJL 627:L17
Binney J (2012a) MNRAS 426:1324
Binney J (2012b) MNRAS 426:1328
Binney J et al (2014) MNRAS 439:1231
Binney J, Spergel D (1982) ApJ 252:308
Binney J, Spergel D (1984) MNRAS 206:159
Binney J, Tremaine S (2008) Galactic dynamics, 2nd edn. Princeton University Press, New Jersey
Cappellari M et al (2011) MNRAS 413:813
Carpintero DD, Aguilar LA (1998) MNRAS 298:1
de Zeeuw T (1985) MNRAS 216:273
Dumas HS, Laskar J (1993) Phys Rev Lett 70:2975
Freeman K et al (2013) MNRAS 428:3660
Fux R (2004) In: Alfaro EJ, Pérez E, Franco J (eds) How does the galaxy work?, Astrophysics and space science library. Kluwer, Dordrecht, p 213
Jing YP, Suto Y (2002) ApJ 574:538
Kaasalainen M, Binney J (1994) MNRAS 268:1033
Kazantzidis S, Kravtsov AV, Zentner AR, Allgood B, Nagai D, Moore B (2004) ApJL 611:L73
Law DR, Majewski SR (2010) ApJ 714:229
Magorrian J (2007) MNRAS 381:1663
Marvasti F (ed) (2001) Nonuniform sampling. Theory and practice. Kluwer Academic/Plenum Publishers, Dordrecht, pp xxvi + 924
McGill C, Binney J (1990) MNRAS 244:634
McMillan PJ, Binney JJ (2008) MNRAS 390:429
Ness M et al (2013a) MNRAS 430:836
Ness M et al (2013b) MNRAS 432:2092
Pontzen A, Governato F (2013) MNRAS 430:121
Press WH, Teukolsky SA, Vetterling WT, Flannery BP (2002) Numerical recipes in C++: the art of scientific computing. Cambridge University Press, Cambridge
Sanders JL, Binney J (2014) MNRAS 441:3284

Schwarzschild M (1979) ApJ 232:236
Ueda H, Hara T, Gouda N, Yano T (2014) ArXiv e-prints
Valluri M, Debattista VP, Quinn T, Moore B (2010) MNRAS 403:525
Vera-Ciro C, Helmi A (2013) ApJL 773:L4
Vera-Ciro CA, Sales LV, Helmi A, Frenk CS, Navarro JF, Springel V, Vogelsberger M, White SDM (2011) MNRAS 416:1377
Warnock RL (1991) Phys Rev Lett 66:1803

Chapter 4
Action Estimation Using a Triaxial Stäckel Approximation

4.1 Introduction

Triaxial models are of considerable importance for modelling of slowly-rotating elliptical galaxies, as well as potentially triaxial components of our own Galaxy such as the bulge, stellar halo and dark halo. The first triaxial models were made by violent relaxation of an N-body model (Aarseth and Binney 1978), and these models prompted Schwarzschild (1979) to develop the technique of orbit superposition so triaxial models with prescribed density profiles could be constructed. Schwarzschild's work gave significant insight into how triaxial systems work for the first time, and this insight was enhanced by de Zeeuw (1985), who showed that Stäckel potentials provided analytic models of orbits in a very interesting class of triaxial systems.

Work on axisymmetric models in the context of our Galaxy has increased awareness of the value in stellar dynamics of the intimately related concepts of the Jeans' theorem and action integrals. If the system is axisymmetric, the energy E and component of angular momentum L_z are integrals that are defined for any axisymmetric potential $\Phi(R, z)$, and equilibrium models of axisymmetric systems have been constructed from distribution functions (DFs) of the form $f(E, L_z)$ (Prendergast and Tomer 1970; Wilson 1975; Rowley 1988). However, these two-integral models are not generic, and they are much harder to construct than generic models when the DF is specified as a function $f(\boldsymbol{J})$ of the actions (Binney 2014). Moreover, knowledge of the DF as a function of the actions is the key to Hamiltonian perturbation theory, and the ability to perturb models is crucial if we are to really understand how galaxies work, and evolve over time.

The self-consistent axisymmetric action-based DFs of Binney (2014) relied on a fast routine for estimating the actions in a general axisymmetric potential. Binney (2014) used the 'Stäckel fudge' method detailed in Chap. 2. The target potential is assumed to be close to a Stäckel potential. This method has been used to construct axisymmetric distribution functions for the Galactic disc and we will use this method for constructing extended distribution functions in Chap. 8.

J. Sanders, *Dynamics of the Milky Way*, Springer Theses,
DOI 10.1007/978-3-319-18772-3_4

In order to construct triaxial action-based DFs for the type of modelling detailed above we require an efficient algorithm for estimating the actions. The routine presented in the previous chapter finds the actions, angles and frequency to arbitrary accuracy in a general triaxial potential. In Chap. 2, we presented several methods for estimating the actions in axisymmetric potentials that were not convergent. We had to make approximations to estimate the actions, and, regardless of the amount of computation, we could not get arbitrarily close to the true actions. Convergent methods are preferable for finding accurate action estimates, but this is at the expense of additional computational time. For instance, the method for finding the Fourier components of the generating function of the previous chapter involves time-consuming orbit integration and construction of large matrices. For working with distribution functions for triaxial features in the Galaxy, we need a fast routine for finding $\boldsymbol{J}(\boldsymbol{x}, \boldsymbol{v})$ to enable fast evaluation of moments of the distribution function. In this chapter, we present an extension to the method from Binney (2012) to the triaxial case, and show how it can be used to construct triaxial DFs in general triaxial potentials. The method relies on insights gained from the triaxial Stäckel case.

We begin in Sect. 4.2 by showing how to find the actions in a triaxial Stäckel potential. In Sect. 4.3, we extend the Stäckel fudge to general triaxial potentials. In Sect. 4.4 we apply this algorithm to a series of orbits in a triaxial NFW potential, and in Sect. 4.6 we construct the first triaxial stellar systems with specified DFs $f(\boldsymbol{J})$, and demonstrate that, notwithstanding the uncontrolled nature of the fudge as an approximation, the models satisfy the Jeans equations to good accuracy. Finally, we conclude in Sect. 4.7.

4.2 Triaxial Stäckel Potentials

In this section, we show how actions can be found in a triaxial Stäckel potential. Triaxial Stäckel potentials are associated with ellipsoidal coordinates (λ, μ, ν). The presentation here follows that given by de Zeeuw (1985).

4.2.1 Ellipsoidal Coordinates

Triaxial Stäckel potentials are expressed in terms of ellipsoidal coordinates. These coordinates are related to the Cartesian coordinates (x, y, z) as the three roots of the cubic in τ

$$\frac{x^2}{(\tau+\alpha)} + \frac{y^2}{(\tau+\beta)} + \frac{z^2}{(\tau+\gamma)} = 1, \tag{4.1}$$

where α, β and γ are constants defining the coordinate system. For the potential explored later, we choose to set x as the major axis, y as the intermediate axis and z as the minor axis, such that $-\gamma \leq \nu \leq -\beta \leq \mu \leq -\alpha \leq \lambda$. Surfaces of constant λ are ellipsoids, surfaces of constant μ are hyperboloids of one sheet (flared tubes of elliptical cross section that surround the x axis), and surfaces of constant ν are hyperboloids of two sheets that have their extremal point on the z axis. In the plane $z = 0$ lines of constant λ are ellipses with foci at $y = \pm\Delta_1 \equiv \pm\sqrt{\beta - \alpha}$, whilst in the plane $x = 0$ lines of constant μ are ellipses with foci at $z = \pm\Delta_2 \equiv \pm\sqrt{\gamma - \beta}$. The expressions for the Cartesian coordinates as a function of the ellipsoidal coordinates are

$$x^2 = \frac{(\lambda+\alpha)(\mu+\alpha)(\nu+\alpha)}{(\alpha-\beta)(\alpha-\gamma)}, \quad y^2 = \frac{(\lambda+\beta)(\mu+\beta)(\nu+\beta)}{(\beta-\alpha)(\beta-\gamma)},$$
$$z^2 = \frac{(\lambda+\gamma)(\mu+\gamma)(\nu+\gamma)}{(\gamma-\beta)(\gamma-\alpha)}. \tag{4.2}$$

Note that a Cartesian coordinate (x, y, z) gives a unique (λ, μ, ν), whilst the point (λ, μ, ν) corresponds to eight points in (x, y, z). Therefore, we will only consider potentials with this symmetry i.e. triaxial potentials with axes aligned with the Cartesian axes.

The generating function, S, to take us between Cartesian, (x, y, z, p_x, p_y, p_z), and ellipsoidal coordinates, $(\lambda, \mu, \nu, p_\lambda, p_\mu, p_\nu)$, is

$$S(p_x, p_y, p_z, \lambda, \mu, \nu) = p_x x(\lambda, \mu, \nu) + p_y y(\lambda, \mu, \nu) + p_z z(\lambda, \mu, \nu). \tag{4.3}$$

Using $p_\tau = \partial S/\partial\tau$, we find, for instance,

$$p_\lambda = \frac{p_x}{2}\sqrt{\frac{(\mu+\alpha)(\nu+\alpha)}{(\alpha-\beta)(\alpha-\gamma)(\lambda+\alpha)}} + \frac{p_y}{2}\sqrt{\frac{(\mu+\beta)(\nu+\beta)}{(\beta-\alpha)(\beta-\gamma)(\lambda+\beta)}}$$
$$+ \frac{p_z}{2}\sqrt{\frac{(\mu+\gamma)(\nu+\gamma)}{(\gamma-\alpha)(\gamma-\beta)(\lambda+\gamma)}}. \tag{4.4}$$

There are similar equations for p_μ and p_ν. Inversion of these three equations gives us expressions for p_x, p_y and p_z as functions of p_τ and τ. For a general triaxial potential, Φ, we can express the Hamiltonian, H, in terms of the ellipsoidal coordinates as

$$H = \tfrac{1}{2}\left(p_x^2 + p_y^2 + p_z^2\right) + \Phi(x, y, z) = \tfrac{1}{2}\left(\frac{p_\lambda^2}{P_\lambda^2} + \frac{p_\mu^2}{P_\mu^2} + \frac{p_\nu^2}{P_\nu^2}\right) + \Phi(\lambda, \mu, \nu). \tag{4.5}$$

where

$$P_\lambda^2 = \tfrac{(\lambda-\mu)(\lambda-\nu)}{4(\lambda+\alpha)(\lambda+\beta)(\lambda+\gamma)}, \quad P_\mu^2 = \tfrac{(\mu-\nu)(\mu-\lambda)}{4(\mu+\alpha)(\mu+\beta)(\mu+\gamma)}, \quad P_\nu^2 = \tfrac{(\nu-\mu)(\nu-\lambda)}{4(\nu+\alpha)(\nu+\beta)(\nu+\gamma)}. \tag{4.6}$$

4.2.2 Stäckel Potentials

The most general triaxial Stäckel potential, Φ_S, can be written as

$$\Phi_S(\lambda, \mu, \nu) = \frac{f(\lambda)}{(\lambda-\mu)(\nu-\lambda)} + \frac{f(\mu)}{(\mu-\nu)(\lambda-\mu)} + \frac{f(\nu)}{(\nu-\lambda)(\mu-\nu)}. \tag{4.7}$$

Φ_S is composed of three functions of one variable. Here we denote the three functions with the same letter, f, as their domains are distinct. Additionally, for Φ_S to be finite at $\lambda = \mu = \alpha$ and $\mu = \nu = \beta$, $f(\tau)$ must be continuous at $\tau = \alpha$ and $\tau = \beta$. With this form for the potential, we can solve the Hamilton-Jacobi equation (de Zeeuw 1985). Let $W(\lambda, \mu, \nu, \boldsymbol{J})$ be the unknown generating function for the transformation between (τ, p_τ) and the as yet unknown actions $\boldsymbol{J}$ and their corresponding angle coordinates. Then we have that $p_\tau = \partial W/\partial\tau$ and at fixed $\boldsymbol{J}$ we can equate the Hamiltonian to the total energy, E, in Eq. (4.5). We then multiply through by $(\lambda-\mu)(\mu-\nu)(\nu-\lambda)$ to find

$$\begin{aligned}
&(\nu-\mu)\Big(2(\lambda+\alpha)(\lambda+\beta)(\lambda+\gamma)\Big(\tfrac{\partial W}{\partial\lambda}\Big)^2 - f(\lambda) - \lambda^2 E\Big)\\
&+(\lambda-\nu)\Big(2(\mu+\alpha)(\mu+\beta)(\mu+\gamma)\Big(\tfrac{\partial W}{\partial\mu}\Big)^2 - f(\mu) - \mu^2 E\Big)\\
&+(\mu-\lambda)\Big(2(\nu+\alpha)(\nu+\beta)(\nu+\gamma)\Big(\tfrac{\partial W}{\partial\nu}\Big)^2 - f(\nu) - \nu^2 E\Big)\\
&=0.
\end{aligned} \tag{4.8}$$

We make the Ansatz $W = \sum_\tau W_\tau(\tau)$ and define

$$U(\tau) = 2(\tau+\alpha)(\tau+\beta)(\tau+\gamma)\Big(\frac{\partial W}{\partial\tau}\Big)^2 - f(\tau) - \tau^2 E, \tag{4.9}$$

such that the Hamilton-Jacobi equation becomes

$$(\nu-\mu)U(\lambda) + (\lambda-\nu)U(\mu) + (\mu-\lambda)U(\nu) = 0. \tag{4.10}$$

Taking the second derivative of this expression with respect to $\tau = \{\lambda, \mu, \nu\}$, we find that

$$U(\tau) = a\tau - b, \tag{4.11}$$

where a and b are constants. Therefore, the equations for the momenta can be written as

$$2(\tau+\alpha)(\tau+\beta)(\tau+\gamma)p_\tau^2 = \tau^2 E - \tau a + b + f(\tau). \tag{4.12}$$

For an initial phase-space point, $(\boldsymbol{x}_0, \boldsymbol{v}_0)$, we find $\tau_0(\boldsymbol{x}_0, \boldsymbol{v}_0)$ and $p_{\tau 0}(\boldsymbol{x}_0, \boldsymbol{v}_0)$ using the coordinate transformations and can then find the integrals a and b by solving Eq. (4.12) (see de Zeeuw 1985, for more details). These integrals are related to the classical integrals I_2 and I_3 in a simple way. As p_τ is only a function of τ, the actions are then given by the 1D integrals

$$J_\tau = \frac{2}{\pi}\int_{\tau_-}^{\tau_+} \mathrm{d}\tau\,|p_\tau(\tau)|. \tag{4.13}$$

where (τ_-, τ_+) are the roots of $p_\tau(\tau) = 0$, which we find by using Brent's method to find points where the right side of Eq. (4.12) vanishes. Note that for loop orbits we must divide the 'radial' action by two (J_λ for the short-axis loops and outer long-axis loops, J_μ for the inner long-axis loops). The approach to finding the actions presented here requires an explicit form for f. In the next section, we will show how we can circumnavigate the need for this explicit form, which allows us to use the same equations for a general potential.

4.3 Triaxial Stäckel Fudge

We now show how we can use the insights from Stäckel potentials to estimate actions in a more general potential. For a general triaxial potential, Φ, we can attempt to find the actions by assuming that the general potential is close to a Stäckel potential. Given a general potential, we define the quantities

$$\begin{aligned}\chi_\lambda(\lambda,\mu,\nu) &\equiv (\lambda-\mu)(\nu-\lambda)\Phi(\lambda,\mu,\nu),\\ \chi_\mu(\lambda,\mu,\nu) &\equiv (\mu-\nu)(\lambda-\mu)\Phi(\lambda,\mu,\nu),\\ \chi_\nu(\lambda,\mu,\nu) &\equiv (\nu-\lambda)(\mu-\nu)\Phi(\lambda,\mu,\nu).\end{aligned} \tag{4.14}$$

where we have chosen a particular coordinate system, (α, β, γ) (see Sect. 4.4.1). If Φ were a Stäckel potential, these quantities would be given by, for instance,

$$\chi_\lambda(\lambda,\mu,\nu) = f(\lambda) - \lambda\frac{f(\mu)-f(\nu)}{\mu-\nu} + \frac{\nu f(\mu)-\mu f(\nu)}{\mu-\nu}. \tag{4.15}$$

Therefore, for a general potential, we can write

$$f(\tau) \approx \chi_\tau(\lambda,\mu,\nu) + C_\tau\tau + D_\tau, \tag{4.16}$$

where C_τ and D_τ are constants provided we always evaluate χ_τ with two of the ellipsoidal coordinates fixed. For instance, we always evaluate χ_λ at fixed μ and ν.

When we substitute these expressions into Eq. (4.12), we find

$$2(\tau+\alpha)(\tau+\beta)(\tau+\gamma)p_\tau^2 = \tau^2E - \tau A_\tau + B_\tau + \chi_\tau(\lambda,\mu,\nu). \tag{4.17}$$

For each τ coordinate, there are two new integrals of motion given by $A_\tau = a - C_\tau$ and $B_\tau = b + D_\tau$.

Given an initial phase-space point, $(\boldsymbol{x}_0, \boldsymbol{v}_0)$, and a coordinate system, (α, β, γ), we can calculate the ellipsoidal coordinates $(\lambda_0, \mu_0, \nu_0, p_{\lambda 0}, p_{\mu 0}, p_{\nu 0})$. Inserting this initial phase-space point into Eq. (4.17) gives us an expression for B_τ as

$$B_\tau = 2(\tau_0 + \alpha)(\tau_0 + \beta)(\tau_0 + \gamma)p_{\tau 0}^2 - \tau_0^2 E + \tau_0 A_\tau - \chi_\tau(\lambda_0, \mu_0, \nu_0). \quad (4.18)$$

It remains to find an expression for A_τ as a function of the initial phase-space point. To proceed we consider the derivative of the Hamiltonian with respect to τ. In a Stäckel potential we can stay on the orbit while changing τ and $p_\tau(\tau)$ with all the other phase-space variables held constant. Therefore, in a Stäckel potential $\partial H/\partial\tau = 0$. Here we consider $\partial H/\partial\lambda$ and will give the results for μ and ν afterwards. Using Eq. (4.5) we write

$$0 = \left(\frac{\partial H}{\partial \lambda}\right)_{\mu,\nu} = \frac{1}{2}\frac{\partial}{\partial\lambda}\left[\frac{p_\lambda^2}{P_\lambda^2}\right] + \frac{1}{2}\frac{p_\mu^2}{(\mu-\lambda)P_\mu^2} + \frac{1}{2}\frac{p_\nu^2}{(\nu-\lambda)P_\nu^2} + \frac{\partial\Phi}{\partial\lambda}. \quad (4.19)$$

To evaluate $\partial[p_\lambda^2/P_\lambda^2]/\partial\lambda$ we use Eq. (4.18) to write

$$\begin{aligned} 2(\lambda+\alpha)(\lambda+\beta)(\lambda+\gamma)p_\lambda^2 &= 2(\lambda_0+\alpha)(\lambda_0+\beta)(\lambda_0+\gamma)p_{\lambda 0}^2 \\ &+(\lambda^2-\lambda_0^2)E - (\lambda-\lambda_0)A_\lambda - \chi_\lambda(\lambda, \mu_0, \nu_0) + \chi_\lambda(\lambda_0, \mu_0, \nu_0), \end{aligned} \quad (4.20)$$

such that

$$\frac{1}{2}\frac{p_\lambda^2}{P_\lambda^2} = \frac{Q + (\lambda^2-\lambda_0^2)E - (\lambda-\lambda_0)A_\lambda}{(\lambda-\mu)(\lambda-\nu)} - \Phi(\lambda, \mu_0, \nu_0), \quad (4.21)$$

where

$$Q = 2(\lambda_0+\alpha)(\lambda_0+\beta)(\lambda_0+\gamma)p_{\lambda 0}^2 + \chi_\lambda(\lambda_0, \mu_0, \nu_0). \quad (4.22)$$

Upon substitution into Eq. (4.19), we note that the derivatives of Φ cancel. Therefore, evaluating $\partial H/\partial\lambda$ at the initial phase-space point we find

$$\begin{aligned} A_\lambda = 2\lambda_0 E - (2\lambda_0 - \mu_0 - \nu_0)\left(\Phi(\lambda_0, \mu_0, \nu_0) + \frac{1}{2}\frac{p_{\lambda 0}^2}{P_{\lambda 0}^2}\right) - \frac{1}{2}\frac{p_{\mu 0}^2(\lambda_0-\nu_0)}{P_{\mu 0}^2} \\ - \frac{1}{2}\frac{p_{\nu 0}^2(\lambda_0-\mu_0)}{P_{\nu 0}^2}. \end{aligned} \quad (4.23)$$

This can be simplified further to

$$A_\lambda = (\mu_0+\nu_0)E + \frac{1}{2}\frac{p_{\mu 0}^2(\lambda_0-\mu_0)}{P_{\mu 0}^2} + \frac{1}{2}\frac{p_{\nu 0}^2(\lambda_0-\nu_0)}{P_{\nu 0}^2}. \quad (4.24)$$

Note that A_λ is independent of λ_0 and $p_{\lambda 0}$ (except implicitly in the energy, E). Similarly

$$\begin{aligned} A_\mu &= (\lambda_0 + \nu_0)E + \tfrac{1}{2}\frac{p_{\lambda 0}^2(\mu_0-\lambda_0)}{P_{\lambda 0}^2} + \tfrac{1}{2}\frac{p_{\nu 0}^2(\mu_0-\nu_0)}{P_{\nu 0}^2}, \\ A_\nu &= (\lambda_0 + \mu_0)E + \tfrac{1}{2}\frac{p_{\lambda 0}^2(\nu_0-\lambda_0)}{P_{\lambda 0}^2} + \tfrac{1}{2}\frac{p_{\mu 0}^2(\nu_0-\mu_0)}{P_{\mu 0}^2}. \end{aligned} \tag{4.25}$$

For a true Stäckel potential, given an initial phase-space point we can find 6 integrals of motion, $(A_\lambda, A_\mu, A_\nu, B_\lambda, B_\mu, B_\nu)$ from Eqs. (4.18), (4.24) and (4.25). Note that a general Stäckel potential only admits three integrals of motion so the 6 derived integrals of motion are not independent. This procedure gives identical results to evaluating the integrals as in de Zeeuw (1985). Note that the expressions for these integrals do not explicitly involve the function $f(\tau)$—they only involve the potential, Φ. With the integrals of motion calculated, we are in a position to find $p_\tau(\tau)$ and hence the actions from Eq. (4.13).

For a general potential, we may find six approximate integrals of motion using the same equations, and hence estimate the actions. In this case, although the potential may admit only three true integrals of motion, the 6 approximate integrals of motion are independent estimates of true integrals of motion. Again, as the expressions do not require $f(\tau)$ they can be evaluated for a general potential. In Appendix E, we show how the angles and frequencies can be estimated using the same approach.

4.3.1 Relation to Axisymmetric Case

The above procedure extends the work of Binney (2012). Binney (2012) constructed the "Stäckel fudge" algorithm for estimating actions in a general axisymmetric potential $\Phi(R, z)$, where R and z are the usual cylindrical polar coordinates. We now relate the procedure to that of Binney (2012) to develop further understanding. The following is a repeat of the method outlined in Sect. 2.6.

Oblate axisymmetric Stäckel potentials are associated with prolate elliptic coordinates (λ, ν) given by the roots for τ of

$$\frac{R^2}{\tau+\alpha} + \frac{z^2}{\tau+\gamma} = 1, \tag{4.26}$$

where $-\gamma \leq \nu \leq -\alpha \leq \lambda$. Binney (2012) uses the coordinates (u, v) which are related to (λ, ν) via

$$\begin{aligned} \sinh^2 u &= \tfrac{\lambda+\alpha}{\gamma-\alpha}, \\ \cos^2 v &= \tfrac{\nu+\gamma}{\gamma-\alpha}, \end{aligned} \tag{4.27}$$

such that

$$\begin{aligned} R &= \sqrt{\gamma - \alpha}\sinh u \sin v, \\ z &= \sqrt{\gamma - \alpha}\cosh u \cos v. \end{aligned} \tag{4.28}$$

An oblate axisymmetric Stäckel potential can be written as

$$\Phi_S(\lambda, \nu) = -\frac{f(\lambda) - f(\nu)}{\lambda - \nu}, \tag{4.29}$$

and the equations for the momenta are given by (de Zeeuw 1985)

$$2(\tau + \alpha)(\tau + \gamma)p_\tau^2 = E(\tau + \gamma) - \left(\frac{\tau + \gamma}{\tau + \alpha}\right) I_2 - I_3 + f(\tau). \tag{4.30}$$

For axisymmetric potentials $I_2 = \frac{1}{2}L_z^2$, where L_z is the z-component of the angular momentum. For a general oblate axisymmetric potential, Φ, we define

$$\begin{aligned} \chi_\lambda(\lambda, \nu) &\equiv -(\lambda - \nu)\Phi, \\ \chi_\nu(\lambda, \nu) &\equiv -(\nu - \lambda)\Phi. \end{aligned} \tag{4.31}$$

If Φ were a Stäckel potential, these quantities would be given by

$$\begin{aligned} \chi_\lambda(\lambda, \nu) &= f(\lambda) - f(\nu), \\ \chi_\nu(\lambda, \nu) &= f(\nu) - f(\lambda). \end{aligned} \tag{4.32}$$

Therefore, for a general potential, we can write,

$$f(\tau) \approx \chi_\tau(\lambda, \nu) + D_\tau, \tag{4.33}$$

where D_τ are constants provided we evaluate χ_λ at constant ν and vice versa. We can write the equations for the momenta as

$$2(\tau + \alpha)(\tau + \gamma)p_\tau^2 = E(\tau + \gamma) - \left(\frac{\tau + \gamma}{\tau + \alpha}\right) I_2 - B_\tau + \chi_\tau(\lambda, \nu), \tag{4.34}$$

where we have defined the integral of motion $B_\tau = I_3 - D_\tau$. B_τ may be found given an initial phase-space point and we then integrate the equations for the momenta to find the actions. Note that, in this case, only two integrals of the motion, B_τ, need to be found, as, in the axisymmetric case, we can find two exact integrals of motion, E and L_z. This is the procedure followed in Binney (2012) and, despite the differing conventions and presentation, this method gives identical results to that of Binney (2012).

4.4 Tests

For the purposes of testing the triaxial Stäckel fudge, we use a triaxial NFW halo (Navarro et al. 1996; Jing and Suto 2002):

$$\Phi(x, y, z) = \Phi(m) = \frac{-GM_s}{m} \ln\left(1 + \frac{m}{m_0}\right) \text{ where } m = \sqrt{x^2 + \frac{y^2}{y_s^2} + \frac{z^2}{z_s^2}}. \quad (4.35)$$

We set $y_s = 0.95$, $z_s = 0.85$, $m_0 = 10\,\text{kpc}$ and $GM_s = (1109\,\text{km}\,\text{s}^{-1})^2\,\text{kpc}$. In Fig. 4.1 we show the equipotential contours in the $z = 0$ and $y = 0$ planes. It is perhaps more conventional to include the triaxiality in the density (e.g. Jing and Suto 2002), but, for simplicity, we have chosen to include triaxiality in the potential. For our choice of parameters this does not lead to negative densities anywhere.

4.4.1 Selection of Coordinate System

The accuracy of the above routine for a general potential will depend upon our choice of coordinate system, (α, β, γ). We can freely set $\gamma = -1\,\text{kpc}^2$ as the coordinate system only depends on Δ_1 and Δ_2.

In Chap. 2 we used the mixed derivative $\partial_\lambda \partial_\nu[(\lambda - \nu)\Phi]$ to select an appropriate coordinate system. For the triaxial case, we could construct a similar quantity: $\partial_\lambda \partial_\mu \partial_\nu[(\lambda - \mu)(\mu - \nu)(\nu - \lambda)\Phi]$. However, this expression would involve third derivatives of the potential so is undesirable. Binney (2014) selected a coordinate

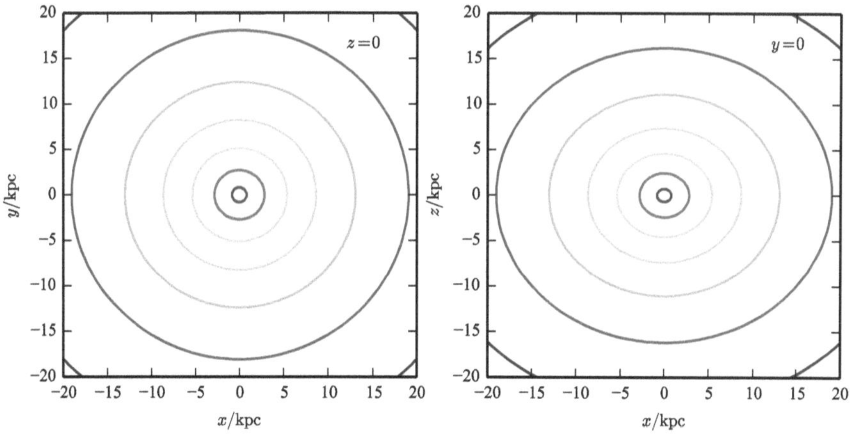

Fig. 4.1 Equipotential contours for the triaxial NFW potential in the two planes $z = 0$ (*left*) and $y = 0$ (*right*). The central contour shows $\Phi/GM_sm_0 = -0.0096$ and the contours increase linearly by $\Delta(\Phi/GM_sm_0) = 0.0008$ outwards

system by fitting ellipses to shell orbits at each energy, E. We follow a similar procedure: we assume that the best choice of coordinate system is solely a function of E. Given a value for E, we find the two closed loop orbits—one around the short axis and one around the long axis. We select a point along the intermediate axis, $y = y_i$, and launch an orbit with speed $v = \sqrt{2(E - \Phi(0, y_i, 0))}$ in either the x (for the short-axis loop) or z direction (for the long-axis loop). The next time the orbit crosses the y-axis we note the y-intercept, $y = y_f$, change y_i and repeat till we have minimised $|-y_f - y_i|$ using Brent's method. We only integrate half of the orbit to avoid misidentifying fish-tail resonant orbits as closed loop orbits.

Fitting an ellipse to the short-axis closed loop will give an estimate of $\Delta_1 = \sqrt{\alpha - \beta}$, whilst fitting an ellipse to the long-axis closed loop will give an estimate of $\Delta_2 = \sqrt{\gamma - \beta}$. Additionally, for these closed loop orbits only one of the actions is non-zero. We therefore integrate the closed long-axis loop orbit and minimise the standard deviation of the J_ν estimates from each time-step with respect to β using Brent's method. This procedure is not sensitive to the choice of α. Once we have found β, we perform a similar procedure for the short-axis loop: vary α until we have minimised the standard deviation of J_μ. We perform this procedure for a range of energies from $E_{\rm min} = \Phi(0, y_{\rm min}, 0)$ to $E_{\rm max} = \Phi(0, y_{\rm max}, 0)$, tabulating the found values of α and β for interpolation. For the NFW potential, we adopt $y_{\rm min} = 0.05$ kpc and $y_{\rm max} = 60$ kpc.

In Fig. 4.2, we plot the standard deviation for the actions of the closed orbits against Δ_2 and Δ_1 for the constant energy surface with $E = \Phi(0, m_0, 0) = -(290\,{\rm km\,s^{-1}})^2$. In both cases, there is a clear minimum in the standard deviation. In Fig. 4.2 we show the standard deviation in J_ν as a function of $\Delta_2 = \sqrt{\gamma - \beta}$ using two different values for α. The results are indistinguishable. Provided we initially choose a sufficiently negative value of α that the optimal β satisfies $\beta > \alpha$, we are free to first set Δ_2 and then choose Δ_1.

In the left panel of Fig. 4.3, we have plotted our choice of Δ_1 and Δ_2 as a function of the energy. We see that for low energies (very centrally confined orbits) Δ_i tends to zero. Due to the cusp at the centre of the NFW potential, loop orbits exist right down to the centre of the potential. The foci must lie within these loop orbits so Δ_i must decrease as we go to lower energy. As we increase the energy, Δ_i increases with $\Delta_1 < \Delta_2$.

To check this procedure, we launch a series of orbits of constant energy ($E = \Phi(0, m_0, 0) = -(290\,{\rm km\,s^{-1}})^2$) at linearly-spaced intervals along the y-axis with velocity vectors in the (x, z) plane oriented at differing linearly-spaced angles, θ, to the x axis. Each orbit is integrated and then we minimise the sum of the variances of the actions with respect to α and β. The results of this procedure are shown in the right panel of Fig. 4.3. We see that the majority of orbits yield optimal Δ_i similar to that found from inspecting the closed loop orbits. At the extremes of y, Δ_i deviates from this choice. These are the box orbits and they seem to favour lower Δ_i. At fixed y, the choice of Δ_i is not so sensitive to θ.

We could improve our choice of Δ_1 and Δ_2 by making the choice a function of an additional variable. For instance, we could make the choice a function of the total

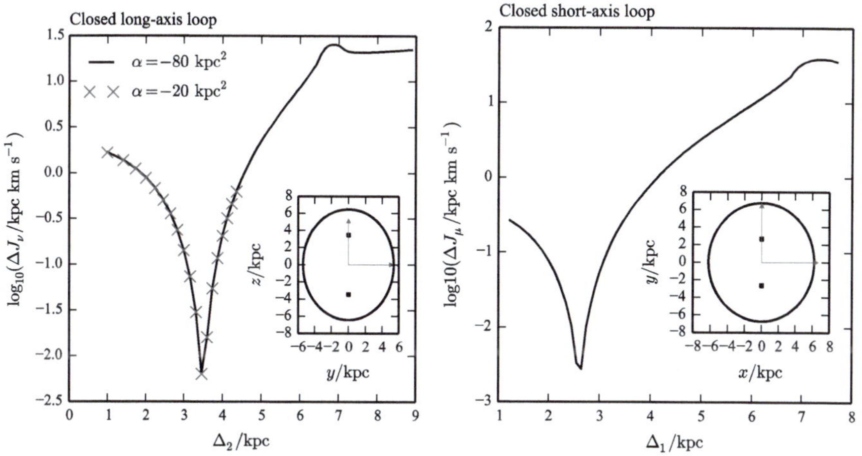

Fig. 4.2 Standard deviation in actions as a function of the focal lengths for the closed loop orbits: The *left panel* shows the standard deviation in J_ν as a function of Δ_2 for the closed long-axis loop orbit shown in the inset. The *solid line* shows the results if we set $\alpha = -80$ kpc^2, whilst the red crosses show the results if we set $\alpha = -20$ kpc^2. The choice of Δ_2 is insensitive to α. In the inset, the *two red arrows* show the initial position vector for the orbit and that position vector rotated by 90° anticlockwise, and the black squares show the best choice of the location of the foci, $z = \pm\Delta_2$. The *right panel* shows the standard deviation in J_μ as a function of Δ_1 for the closed short-axis loop orbit shown in the inset. In the inset, the *two red arrows* show the initial position vector for the orbit and that position vector rotated by 90° clockwise, and the *black squares* show the best choice of the location of the foci, $y = \pm\Delta_1$

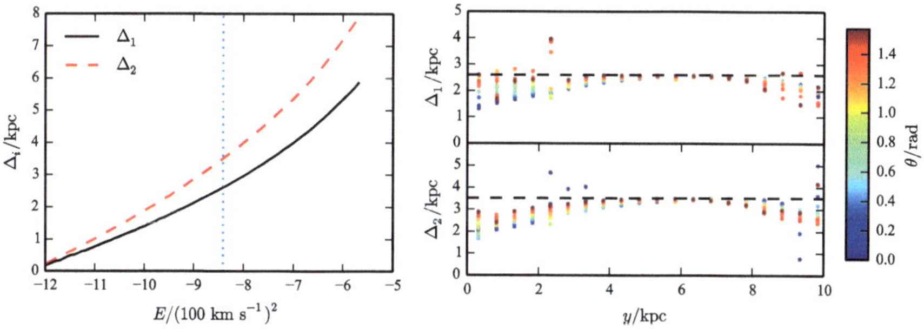

Fig. 4.3 Choice of Δ_1 and Δ_2: in the *left panel*, we show our choice of Δ_1 (*solid black*) and Δ_2 (*dotted red*) as a function of energy, E, for the NFW potential described in Sect. 6.3. The range of energies covered corresponds to the energies of particles dropped from 0.5 *pc* to 30 *pc* along the intermediate axis. The *vertical blue dotted line* gives the energy of the surface explored in Sect. 4.5. In the right panel, we show the choice of Δ_1 and Δ_2 that minimises the variation in the actions for a range of orbits confined to a constant energy surface. Each orbit was launched at y on the intermediate axis with angle θ from the long axis. The *dashed black line* gives the values chosen by inspecting the closed loop orbits as specified in Sect. 4.4.1

angular momentum, which is not an integral of motion. However, we will see that we cannot significantly improve the action recovery with a better choice of Δ_i.

4.5 Accuracy

We now briefly inspect the accuracy of the action recovery using the above procedure. We take three orbits from the surface of constant energy explored in the previous section. The three orbits are a box orbit with $y = 1.8234$ kpc, $\theta = 0.6$ rad (shown in Fig. 4.4), a short-axis loop orbit with $y = 4.8234$ kpc, $\theta = 0.4$ rad (shown in Fig. 4.5), and a long-axis loop orbit with $y = 3.8234$ kpc, $\theta = 1.2$ rad (shown in Fig. 4.6). In each figure, we show the orbit along with the action estimates calculated at each point along the orbit using our 'best-choice' Δ_i, and the choice of Δ_i that minimises the spread in the action estimates. Additionally, we show the 'true' actions calculated by using the method of Chap. 3 to determine the generating function for

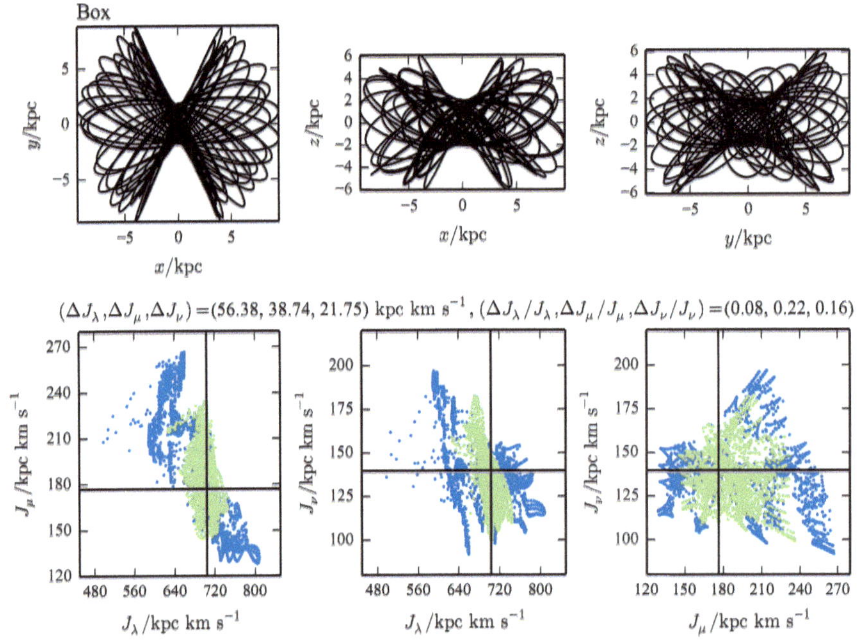

Fig. 4.4 Action estimates for example box orbit using triaxial Stäckel approximation: the *top three panels* show three projections of the orbit, and the *bottom three panels* show the action estimates for points along the orbit. The *dark blue points* show the action estimates calculated using our 'best-choice' Δ_i based on the energy, the *light green points* show the choice of Δ_i that minimises the spread in the action estimates, and the *black lines* show the 'true' actions found using the method presented in Chap. 3. Note the origin is not included in the plots. Between the *top* and *bottom* plots, we give the absolute and relative error in the actions

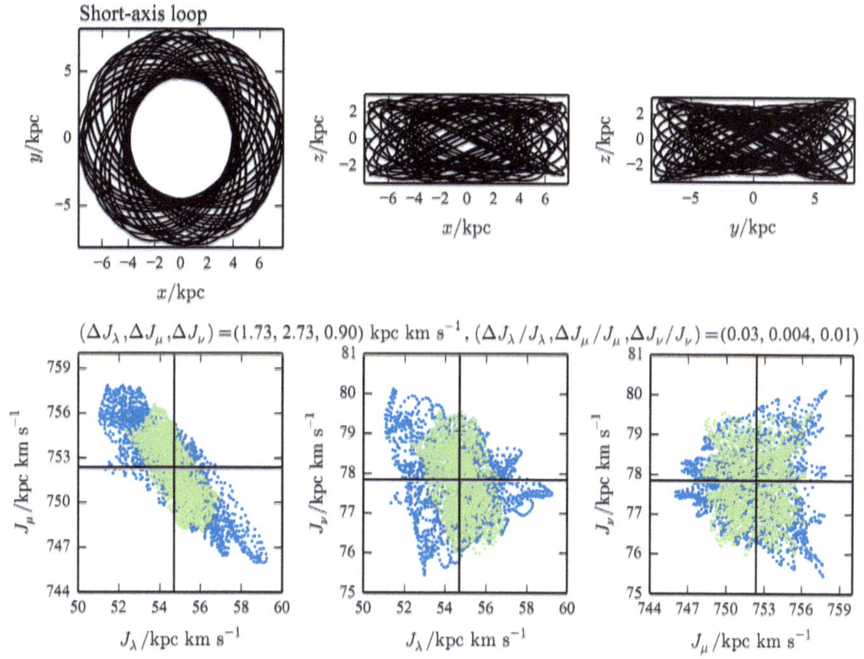

Fig. 4.5 Action estimates for example short-axis loop orbit using triaxial Stäckel approximation. See Fig. 4.4 for information on each panel

the transformation from toy to true angle-action variables. In Appendix E, we show how well the angle coordinates are recovered for these orbits.

The actions of the box orbit are $(J_\lambda, J_\mu, J_\nu) = (686, 192, 137)\,\mathrm{kpc\,km\,s^{-1}}$ and our method yields errors of $(\Delta J_\lambda, \Delta J_\mu, \Delta J_\nu) = (56, 39, 22)\,\mathrm{kpc\,km\,s^{-1}}$ so approximately 10–20 %. With the choice of Δ_i that minimises the spread in the action estimates, we find errors of $(\Delta J_\lambda, \Delta J_\mu, \Delta J_\nu) = (17, 19, 16)\,\mathrm{kpc\,km\,s^{-1}}$ so approximately $\lesssim$10 %. We can achieve a factor of two improvement for J_λ and J_μ.

The actions of the short-axis loop orbit are $(J_\lambda, J_\mu, J_\nu) = (55, 752, 78)$ kpc km s^{-1} and our method yields errors of $(\Delta J_\lambda, \Delta J_\mu, \Delta J_\nu) = (2, 3, 1)\,\mathrm{kpc\,km\,s^{-1}}$ so $\lesssim$4 %. With the choice of Δ_i that minimises the spread in the action estimates, we find errors of $(\Delta J_\lambda, \Delta J_\mu, \Delta J_\nu) = (0.8, 2.0, 0.9)\,\mathrm{kpc\,km\,s^{-1}}$.

The actions of the outer long-axis loop orbit are $(J_\lambda, J_\mu, J_\nu) = (50, 102, 680)$ kpc km s^{-1} and our method yields errors of $(\Delta J_\lambda, \Delta J_\mu, \Delta J_\nu) = (4, 5, 6)\,\mathrm{kpc\,km\,s^{-1}}$ so $\lesssim$8 %. With the choice of Δ_i that minimises the spread in the action estimates, we yield errors of $(\Delta J_\lambda, \Delta J_\mu, \Delta J_\nu) = (2.0, 2.5, 4.2)\,\mathrm{kpc\,km\,s^{-1}}$.

For all the orbits shown in the right panel of Fig. 4.3 (sampled from the constant energy surface $E = \Phi(0, m_0, 0) = -(290\,\mathrm{km\,s^{-1}})^2$), we have plotted the logarithm of the fractional error in the actions in Fig. 4.7. We find the most accurate action recovery occurs for the orbits with the initial condition $y \approx m_0/2$, where we have mostly loop orbits. For these loop orbits, J_μ and J_ν are accurate to $\lesssim$1 % but the

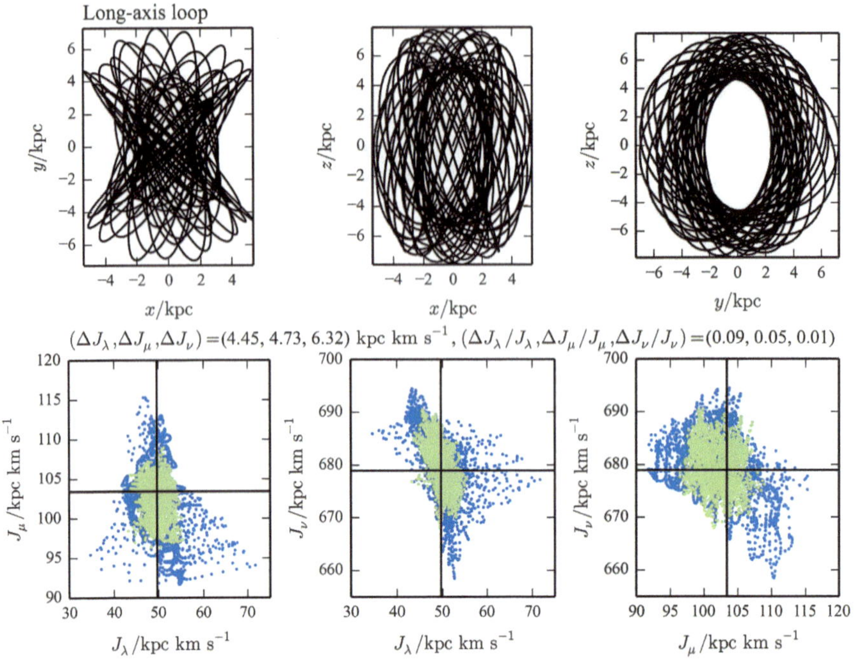

Fig. 4.6 Action estimates for example long-axis loop orbit using triaxial Stäckel approximation. See Fig. 4.4 for information on each panel

'radial' action J_λ is small for these orbits so the relative accuracy can be large. For the box orbits at the extremes of y, the relative error increases to $\sim$10 % but can be as large as order one in J_μ for low y.

In Fig. 4.8, we show the absolute errors in the actions as a function of action for the constant energy surface along with the orbit classification. Each phase-space point along the orbit is allocated a classification number based on the limits of τ found in the Stäckel approximation: $\lambda_- = -\alpha$, $\mu_- = -\beta$ and $\nu_- = -\gamma$ correspond to a box orbit (classification number 0), $\mu_- = \beta$, $\mu_+ = -\alpha$ to a short-axis loop orbit (1), $\lambda_- = -\alpha$, $\nu_+ = -\beta$ to an inner long-axis loop (2), and $\mu_+ = -\alpha$, $\nu_+ = -\beta$ to an outer long-axis loop (3). The orbit classification number is calculated as an average of these classifications along the orbit. With this scheme, orbits near the boundaries of the orbit classes that are chaotic or resonant are allocated non-integer orbit classification numbers. We see that the largest action errors occur at the interfaces between the orbit classes. In particular, ΔJ_λ and ΔJ_μ are largest along the box-short-axis-loop interface, whilst ΔJ_ν is largest at the box-inner-long-axis-loop interface. It is at these boundaries that the orbits pass close to the foci so clearly our choice of foci affects the action recovery for these orbits. Additionally, we find that the absolute action errors, particular for J_ν, are larger for the inner long-axis loops that the outer long-axis loops. This is because the inner long-axis loops probe more central regions of the potential and also lie close to the chaotic box-long-axis-loop interface.

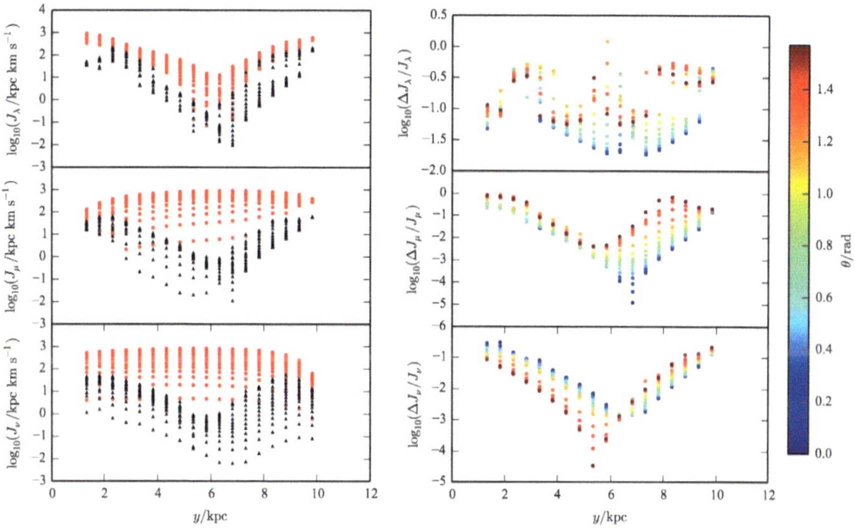

Fig. 4.7 Error in the actions for a selection of orbits in the constant energy surface $E = \Phi(0, m_0, 0) = -(290\,s)^2$ for the triaxial NFW potential. The *left panel* shows the logarithm of the mean of the action estimates for each orbit in *red*, and the standard deviation in *black*. The *right panel* shows the logarithm of the fractional error. The x-axis shows the position along the intermediate axis at which the orbits were launched (y), and the colour-coding in the *right panel* shows the angle, θ, in the $x-z$ plane at which the orbits were launched

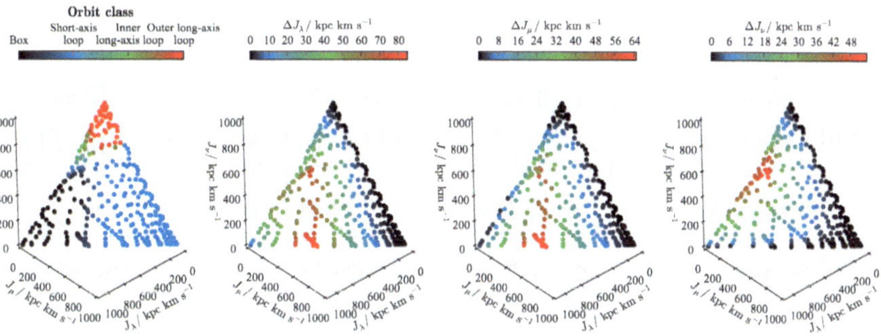

Fig. 4.8 Absolute errors in the actions as a function of action in the constant energy surface $E = \Phi(0, m_0, 0) = -(290\,kms)^2$ for the NFW potential. The *leftmost panel* shows the constant energy surface coloured by orbit class: boxes in *black*, short-axis loops in *blue*, inner long-axis loops in *green*, and outer long-axis loops in *red*. Note the classification is a continuum as it is calculated from an average of classifications along an orbit. The *second*, *third* and *fourth panels* show the absolute error in the three actions, J_λ, J_μ and J_ν respectively

In general, we find that the action recovery for loop orbits is good as these orbits probe a small radial range of the potential. For box orbits the recovery deteriorates as these orbits probe a larger central region of the potential. Additionally, we have seen that by altering Δ_i we can achieve up to a factor of two improvement in the accuracy of the actions for both the loop and box orbits.

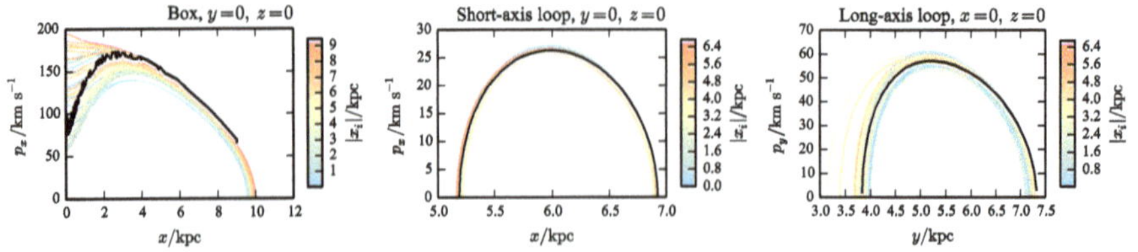

Fig. 4.9 Surfaces of section for the three test orbits in the triaxial NFW potential. In the *left panel* we show the box orbit, the central panel shows the short-axis loop orbit and the right panel shows the long-axis loop orbit. In each panel, the *solid black line* gives the true curve of consequents found from orbit integration. The narrower coloured lines give the consequents from the Stäckel approximation coloured by $|x_i|$ of the initial phase-space point, where $x_i = x$ for the short-axis loop and box orbit, and $x_i = y$ for the long-axis loop orbit. The text above each plot gives the plane that defines the surface of section

4.5.1 Surfaces of Section

For understanding the behaviour of dynamical systems, Poincaré (1892) introduced the concept of a surface of section. These diagrams simplify the motion of a high-dimensional dynamical system. A regular orbit in an integrable triaxial potential permits three constants of the motion, thus confining the motion to a 3-torus. If we choose to only plot the series of points where the orbit passes through a 4-surface in phase-space, e.g. defined by $y = 0$ and $z = 0$, the phase-space points will be confined to a line, or a consequent, which may be visualized in a 2D plot.

We can test the Stäckel approach outlined here by seeing how well it reproduces the surfaces of section. To produce the true surface of section, we integrate the orbit in the true potential and find the phase-space points where the orbit crosses our chosen 4-surface. Here we use 4-surfaces defined by one of the spatial axes. To produce the corresponding surface of section from the Stäckel method, we determine τ along our chosen spatial axis using Eq. (4.2) between the determined limits in τ, and use Eq. (4.17) to find the corresponding p_τ. From p_τ, we can use expressions such as Eq. (4.4) to calculate p_x, p_y and p_z: if we wish to draw the consequent defined by $y = 0$, $z = 0$ we have that $\mu = -\beta$ and $\nu = -\gamma$ such that $x = \sqrt{\lambda+\alpha}$, and $p_x = \sqrt{4(\lambda+\alpha)}p_\lambda$. If we wish to draw the consequent defined by $x = 0$, $z = 0$ we have that for $|y| > \Delta_1$, $\mu = -\alpha$ and $\nu = -\gamma$ such that $y = \sqrt{\lambda+\beta}$, and $p_y = \sqrt{4(\lambda+\beta)}p_\lambda$, whilst for $|y| < \Delta_1$, $\lambda = -\alpha$ and $\nu = -\gamma$ such that $y = \sqrt{\mu+\beta}$ and $p_y = \sqrt{4(\mu+\beta)}p_\mu$.[1]

In Fig. 4.9 we plot the consequents of the three orbits inspected in Sect. 4.5. We see that the Stäckel approximation consequents for the short-axis loop lie close to the true consequent. Those of long-axis loop are slightly worse. The box orbit seems problematic. For the phase-space points that lie close to the centre of the potential, the consequents turn over in the centre as required. However, they underestimate p_x

[1]If we wish to draw the consequent defined by $x = 0$, $y = 0$ we have for $|z| < \Delta_2$, $\lambda = -\alpha$ and $\mu = -\beta$ so $z = \sqrt{\nu+\gamma}$ and $p_z = \sqrt{4(\nu+\gamma)}p_\nu$, whilst for $|z| > \Delta_2$, $\mu = -\alpha$ and $\nu = -\beta$ so $z = \sqrt{\lambda+\gamma}$ and $p_z = \sqrt{4(\lambda+\gamma)}p_\lambda$.

at a given x. The phase-space points which lie further out fail to turn over at low x. The Stäckel tori for these orbits are near radial such that p_x is maximum for $x = 0$. However, we see from Fig. 4.4 that the orbit crosses through $x = 0$, $y = 0$ at an angle such that p_x is smaller than its maximum value. This behaviour is only captured for the initial phase-space points at low x.

4.6 A Triaxial Model with Specified DF

The main purpose of the algorithm presented here is to calculate efficiently the moments of triaxial distribution functions. We have seen that the errors in the actions reported by the scheme can be large. However, when calculating moments of a distribution function, many action evaluations are required and there is scope for errors to substantially cancel, leaving the final value of the moment quite accurate. In this section, we demonstrate this phenomenon by constructing triaxial models from an analytic DF $f(\boldsymbol{J})$.

We adopt a simple distribution function (Posti et al. 2015)

$$f(\boldsymbol{x}, \boldsymbol{v}) = f(\boldsymbol{J}(\boldsymbol{x}, \boldsymbol{v})) = (J_0 + |J_\lambda| + \zeta|J_\mu| + \eta|J_\nu|)^p, \tag{4.36}$$

where $J_0 = 10\,\mathrm{km\,s^{-1}\,kpc}$ is a scale action, ζ controls whether the model is tangentially/radially biased and η controls the flattening in z of the model. We set $p = -3$, which causes the density to go as r^0 in the centre and fall off as r^{-3} for large r. Note that the mass of this model diverges logarithmically. We set $\eta = 1.88$ and explore two values of $\zeta = 0.7$ (tangential bias) and $\zeta = 3.28$ (radial bias). Note that for the orbit classes to fill action space seamlessly, we must scale the radial action of the loop orbits by a factor of two (Binney and Spergel 1984). We proceed by calculating the moments of this distribution function in the test triaxial NFW potential at given spatial points, $\boldsymbol{x}$. These non-zero moments are

$$\begin{aligned} \rho(\boldsymbol{x}) &= \int \mathrm{d}^3\boldsymbol{v}\, f(\boldsymbol{x}, \boldsymbol{v}), \\ \sigma^2_{ij}(\boldsymbol{x}) &= \frac{1}{\rho}\int \mathrm{d}^3\boldsymbol{v}\, v_i v_j f(\boldsymbol{x}, \boldsymbol{v}). \end{aligned} \tag{4.37}$$

Note that, as the potential is time-independent, the Hamiltonian is time-reversible and we need only integrate over half the velocity space. We integrate up to the maximum velocity at $\boldsymbol{x}$, given by $\sqrt{2\Phi(\boldsymbol{x})}$. We will calculate these moments extensively later to demonstrate that the action-based distribution functions obey the Jeans equations. In Fig. 4.10, we plot the density of the radially-biased ($\zeta = 3.28$) and tangential-biased ($\zeta = 0.7$) models. We display contours of constant density in two planes along with the density along a line parallel to the x-axis decomposed into its contributions from each orbit class. The density is calculated using the adaptive Monte-Carlo Divonne routine in the CUBA package of Hahn (2005). The class of each orbit is determined by the limits of the motion in τ: $\lambda_- = -\alpha$, $\mu_- = -\beta$ and $\nu_- = -\gamma$ correspond to

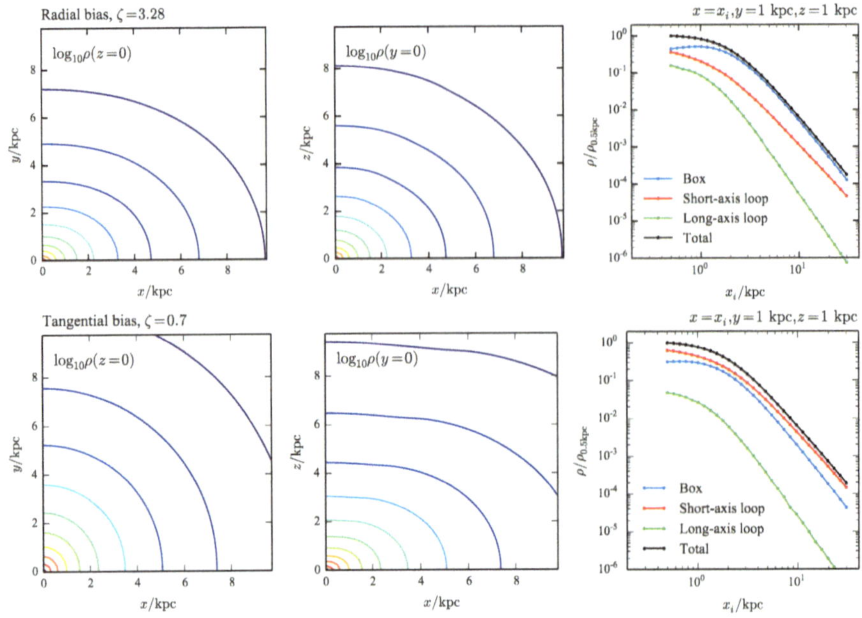

Fig. 4.10 Density for the radially-biased model ($\zeta = 3.28$, *top panels*) and for the tangentially-biased model ($\zeta = 0.7$, *bottom panels*). The *left panels* show equally-spaced contours of the logarithm of the density in the (x, y) plane and, similarly, in the (x, z) plane in the central panels. The outermost contour corresponds to $\log_{10}(\rho/\,\mathrm{kpc}^{-3}) = -5$ for the radially-biased model and $\log_{10}(\rho/kpc^{-3}) = -5.5$ for the tangentially-biased model, and the contours increase by 0.5 inwards. The *right panels* show the total density in *black* along the line $y = 1$ kpc, $z = 1$ kpc as well as the contributions from the box orbits in *blue*, the short-axis loop orbits in *red* and the long-axis loop orbits in *green*

a box orbit, $\mu_- = \beta$, $\mu_+ = -\alpha$ to a short-axis loop orbit, and $\nu_- = -\gamma$, $\nu_+ = -\beta$ to a long-axis loop. As we are calculating the density close to the x-axis, the long-axis loop orbits, which loop the x-axis, do not contribute significantly to the density integral. We see that for the radially-biased model the box orbits are the dominant contributors whilst for the tangentially-biased model the short-axis loop orbits are the major contributors.

We will now perform some checks to see whether our distribution functions are accurate.

4.6.1 Normalization

One check of our action estimation scheme is how accurately it recovers the normalization. To keep the normalization finite, we set $p = -3.5$ for this section. We are able to calculate the normalization of our DF in two distinct ways. Firstly, we calculate the normalization analytically from the DF as

$$M_{\text{true}} = (2\pi)^3 \int \mathrm{d}^3\boldsymbol{J}\, f(\boldsymbol{J}) = (2\pi)^3 \int_0^\infty \mathrm{d}J_\lambda \int_0^\infty \mathrm{d}J_\mu \int_0^\infty \mathrm{d}J_\nu\, f(\boldsymbol{J}). \quad (4.38)$$

Note that, for each $\boldsymbol{J}$ in the appropriate range, there are two loop orbits—one circulating clockwise and one anti-clockwise. Therefore, we must multiply the normalization by two for these orbits. However, we have defined the 'radial' action to be four times the integral from τ_- to τ_+ for these orbits so these factors cancel (Binney and Spergel 1984; de Zeeuw 1985). Additionally, we calculate the integral as

$$M_{\text{est}} = 8 \int_{(x,y,z)>0} \mathrm{d}^3\boldsymbol{x} \int \mathrm{d}^3\boldsymbol{v}\, f(\boldsymbol{J}(\boldsymbol{x}, \boldsymbol{v})). \quad (4.39)$$

For each spatial coordinate, we make the transformation $u_i = 1/(1+x_i)$ to make the integrand flatter. The limits of the integral are now $u_i = \{0, 1\}$. To reduce numerical noise, we split the integral such that we calculate the contribution near the axes separately. We perform the integral using the Monte Carlo Divonne routine. For the tangentially-biased model ($\zeta = 0.7$), we find $M_{\text{est}} \approx 1.006 M_{\text{true}}$.

4.6.2 The Jeans Equation

Our distribution function must satisfy the collisionless Boltzmann equation

$$\frac{\mathrm{d}f}{\mathrm{d}t} = 0. \quad (4.40)$$

In turn, this means the distribution function must satisfy the Jeans equations (see equation (4.209) of Binney and Tremaine 2008)

$$\frac{\partial(\rho\sigma_{ij}^2)}{\partial x_i} = -\rho\frac{\partial\Phi}{\partial x_j}. \quad (4.41)$$

A simple test of our action-based distribution functions is checking whether they satisfy these equations. The right hand side is calculated from analytic differentiation of the potential and multiplying by the density. The left hand side is found by numerically differentiating the three-dimensional integrals $\rho\sigma_{ij}^2$ and summing the appropriate contributions. Numerical differentiation of an integral leads to significant noise. To combat this, we use an adaptive vectorised integration-rule cubature scheme implemented in the CUBATURE package from Steven Johnson (http://ab-initio.mit.edu/wiki/index.php/Cubature). Using a fixed-rule adaptive routine means the noise in the integrals is controlled such that the numerical derivatives are less noisy.

In Fig. 4.11, we show how accurately the Jeans equations are satisfied along several lines through the potential for our two models. We plot each side of each Jeans equation for a choice of j along a range of lines, along with the percentage error

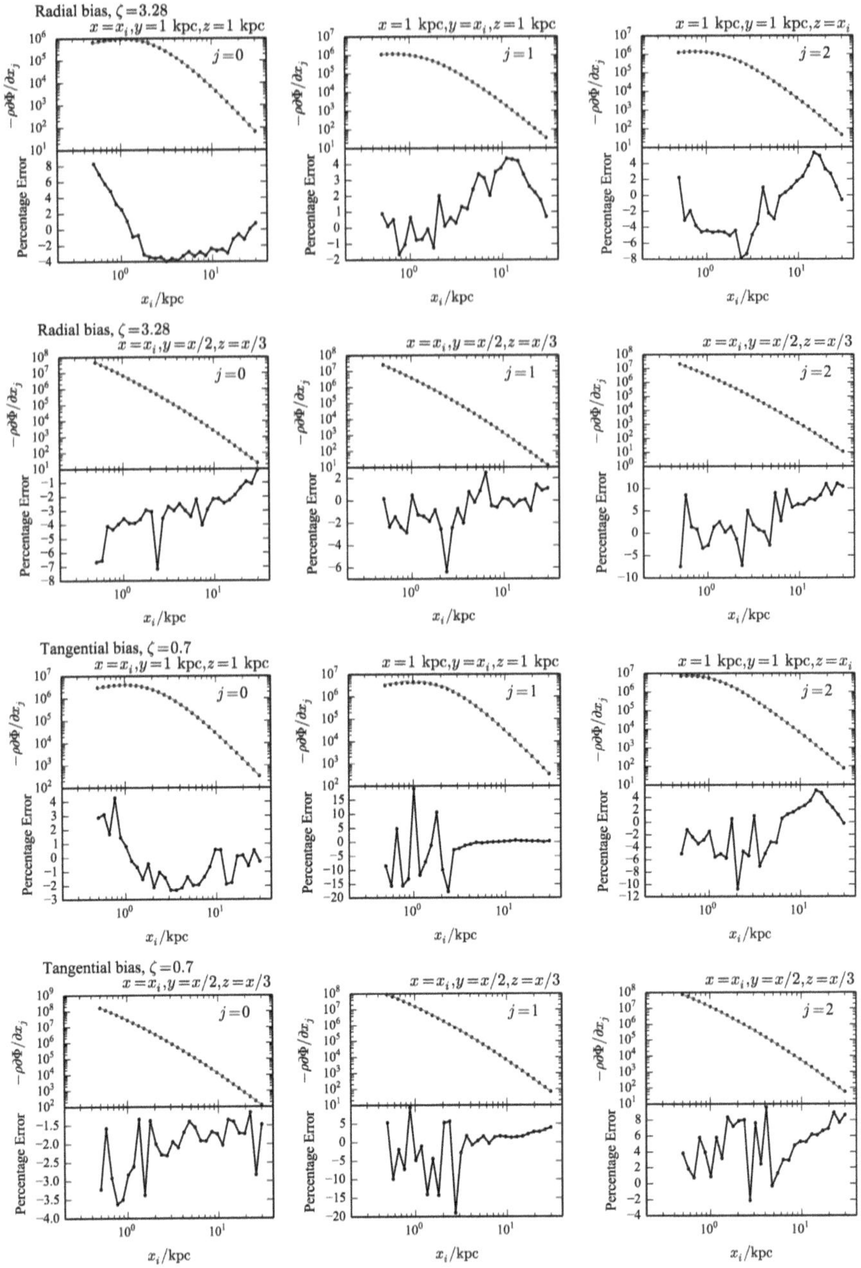

Fig. 4.11 Accuracy of Jeans' equation calculation for the radially-biased model ($\zeta = 3.28$, *top six plots*) and the tangentially-biased model ($\zeta = 0.7$, *bottom six plots*). In the *top half* of each panel, we show $-\rho\,\partial\Phi/\partial x_j$ as a series of *black dots* and $\partial(\rho\sigma_{ij}^2)/\partial x_i$ as a *red line*. In the *bottom half*, we show the percentage error difference between these quantities. Each panel shows a single component, i.e. a single j, along the line given above the *top-right corner* of each panel. The *second row* and *fourth row* of three panels all correspond to the same line

difference between the two sides of the equation. We avoid calculating the derivatives of the moments along the axes as the numerical differentiation is awkward there. In general, we find $\lesssim 10\,\%$ error for nearly all tested points with the majority having $\lesssim 4\,\%$ over a range of ~ 8 orders of magnitude.

Despite the large errors introduced by the action estimation scheme, we have produced a distribution function that satisfies the Jeans equations to reasonable accuracy. Even for the heavily radially-biased model, which has large contributions from the box orbits, the Jeans equations are well satisfied. This gives us confidence that models based on triaxial distribution functions can be constructed using the scheme we have presented.

4.7 Conclusions

We have presented a method for estimating the actions in a general triaxial potential using a Stäckel approximation. The method is an extension of the Stäckel fudge introduced by Binney (2012) for the axisymmetric case. We have investigated the accuracy of the method for a range of orbits in an astrophysically-relevant triaxial potential. We have seen that the recovery of the actions is poorest for the box orbits, which probe a large radial range of the potential, and much better for the loop orbits, which are confined to a more limited radial range. The only parameters in the method are the choice of the focal positions Δ_i. We have detailed a procedure for selecting these based on the energy of the input phase-space point. This choice is not optimal but, by adjusting Δ_i, we can, at best, increase the accuracy of the actions of a factor of two.

The advantage of this method over other methods for estimating the actions in a triaxial potential is speed. Unlike the convergent method introduced in Chap. 3, we obtain the actions without integrating an orbit—we only use the initial phase-space point. We have only to evaluate several algebraic expressions, find the limits of the orbits in the τ coordinate and perform Gaussian quadrature. These are all fast calculations. However, this speed comes at the expense of sometimes disappointing accuracy. If accurate results are required, the Stäckel fudge can be combined with torus construction to form a rapidly convergent scheme for the determination of $\boldsymbol{J}(\boldsymbol{x},\boldsymbol{v})$, as demonstrated in Chap. 2.

We went on to construct triaxial stellar systems from a specified DFs $f(\boldsymbol{J})$ in Sect. 4.6. We demonstrated the mass of these models is well recovered using the Stäckel fudge. Notwithstanding the errors in individual actions, both a radially-biased model and a tangentially-biased model satisfy the Jeans equations to good accuracy. This is because individual errors largely cancel out during integration over velocities when computing moments such as the density $\rho(\boldsymbol{x})$ and the pressure tensor $\rho\sigma^2_{ij}(\boldsymbol{x})$.

Whilst the scheme presented here does not give accurate enough actions for working with streams (see Chaps. 5 and 6) we have shown that it is an appropriate and powerful tool for constructing models from specified DFs $f(\boldsymbol{J})$. A key property of DFs of the form $f(\boldsymbol{J})$ is that they can be trivially added to build up a multi-component system. Hence the ability to extract observables from DFs of the form $f(\boldsymbol{J})$ is likely to prove extremely useful for interpreting data on both external galaxies (Cappellari et al. 2011) and our Galaxy, in which components such as the stellar and dark haloes may be triaxial, and the bulge certainly is. It is believed that $\sim$15 % of all galaxies are slow rotators with a generic triaxial shape (Houghton et al. 2013). The models introduced in this chapter are applicable to these structures.

4.7.1 Future Work

As discussed in the conclusions of the previous chapter, one purpose for designing algorithms to find actions is that they form the basis for the construction of distribution functions for triaxial components of the Galaxy, such as the bulge or possibly the dark halo, as well as for triaxial external galaxies. We have shown that the algorithm introduced in this chapter is appropriate for this purpose. Therefore, the next step is to begin building more realistic models with the intention of fully modelling the available data.

Binney (2014) has shown how one can use an iterative scheme for constructing a self-consistent triaxial distribution function. Such a scheme should be possible to implement with the triaxial DFs introduced here. Posti et al. (2015) has experimented with double power-law action-based DFs of the form

$$f(\boldsymbol{J}) \propto (J_\lambda + \zeta|J_\mu| + \eta|J_\nu|)^p (J_0 + J_\lambda + \zeta|J_\mu| + \eta|J_\nu|)^q, \tag{4.42}$$

where J_0 is a scale action and p and q control the inner and outer slopes of the density. The parameters can be chosen such that these models provide a good match to Hernquist profiles (Hernquist 1990), as well as other profiles of interest. One future step for building a realistic model would be to adopt a DF of this form and iteratively adjust the potential to construct a self-consistent DF. These models should then be compared to the available data on slow rotators from Cappellari et al. (2011). Additionally, as shown by Binney (2014), models with a sense of rotation can be constructed using odd functions of the actions. Certainly, many more experiments should be performed to understand the range of possible models one can construct with only a simply parametrised action-based DF.

Finally, the stellar halo and bulge of our own Galaxy are prime candidates for action-based DF modelling of the form discussed in this chapter. Deason et al. (2012) adopted a distribution-function-based modelling approach for extraction the large-scale structure of the Galactic potential and the velocity anisotropy of the stellar halo from a sample of BHB stars. Such an analysis could potentially be improved by the adoption of the models presented here. Similarly, the quantity of data for the bulge is increasing and the route to successfully modelling this structure must be through the construction of triaxial DFs.

References

Aarseth SJ, Binney J (1978) MNRAS 185:227
Binney J (2012) MNRAS 426:1324
Binney J (2014) MNRAS 440:787
Binney J, Spergel D (1984) MNRAS 206:159
Binney J, Tremaine S (2008) Galactic dynamics, 2nd edn. Princeton University Press, Princeton
Cappellari M et al (2011) MNRAS 413:813
de Zeeuw T (1985) MNRAS 216:273
Deason AJ, Belokurov V, Evans NW, An J (2012) MNRAS 424:L44
Hahn T (2005) Comput Phys Commun 168:78
Hernquist L (1990) ApJ 356:359
Houghton RCW et al (2013) MNRAS 436:19
Jing YP, Suto Y (2002) ApJ 574:538
Navarro JF, Frenk CS, White SDM (1996) ApJ 462:563
Poincaré H (1892) Les methodes nouvelles de la mecanique celeste
Posti L, Binney J, Nipoti C, Ciotti L (2015) MNRAS 447:3060
Prendergast KH, Tomer E (1970) AJ 75:674
Rowley G (1988) ApJ 331:124
Schwarzschild M (1979) ApJ 232:236
Wilson CP (1975) AJ 80:175

Chapter 5
Stream-Orbit Misalignment

5.1 Introduction

We will now move away from methods for action estimation and discuss tidal streams and how they may be used as probes of the Galactic potential. The action estimation schemes described in the first part of this thesis will form an important part of this work.

The halo of the Milky Way is rich with substructure. Large optical surveys have revealed enhancements in the density of stars in the halo, which trace out filaments across the sky. It is believed that each such filament is generated by stars being tidally stripped from a progenitor that has entered the influence of the Milky Way and hence these structures are called tidal streams.

Tidal streams probe the outer parts of the Galactic potential, where the potential is expected to be dark-matter dominated. By understanding their formation, it should be possible to constrain properties of the Galactic potential (McGlynn 1990; Johnston et al. 1996, 1999). One way of approaching this problem has been to assume the members of the stream delineate an orbit (Jin and Lynden-Bell 2007; Binney 2008). If we are given phase-space coordinates for objects that lie at different phases of a single orbit, then the path of the orbit, and hence the underlying potential, may be recovered. Even if the observables are not well known, the orbit and underlying potential can be recovered with reasonable accuracy. If the data lie along an orbit then full six-dimensional phase-space information is redundant: Eyre and Binney (2009b) showed that the orbit and potential could be recovered with positions, distances and line of sight velocities and Eyre and Binney (2009a) did the same with proper motions instead of line of sight velocities. The technique of orbit fitting to stream data has been utilised most successfully by Koposov et al. (2010), who used data for the stream GD-1 (Grillmair and Dionatos 2006b) to constrain a simple two-parameter logarithmic potential for the Milky Way as well as a more complex multi-component Milky Way potential.

J. Sanders, *Dynamics of the Milky Way*, Springer Theses,
DOI 10.1007/978-3-319-18772-3_5

The study of tidal streams has a natural expression in angle-action variables (Helmi and White 1999; Tremaine 1999; Eyre and Binney 2011) and in the correct potential a stream should reveal a clear signature in angle-action space. Eyre and Binney (2011) discussed and demonstrated the formation of streams in angle-action space through N-body simulations. The authors investigated the degree to which orbits delineate streams and found that assuming the stream lies along a single orbit can lead to systematic biases in estimates of the potential parameters. However, these authors were limited to quantifying the degree of misalignment for potentials in which the angle-action coordinates are analytically tractable (spherical and Stäckel potentials). The effect of the misalignment in more realistic Galactic potentials remained an open question. In particular, is the misalignment for known streams in the Galaxy expected to significantly bias orbit-fitting algorithms?

In the next section, we review the known tidal streams of the Milky Way and summarise the data from the literature that will be of use in answering this question. In Sect. 5.3, we present the angle-action formalism and discuss the formation of tidal streams in this framework. In Sect. 5.4, we motivate the need for an improvement on orbit-fitting algorithms by investigating the degree to which streams delineate orbits in realistic Galactic potentials, specifically focussing on which of the known streams can be reliably analysed using orbit-fitting algorithms. In Sects. 5.5 and 5.6, we discuss the validity of the presented formalism using N-body simulations and demonstrate the results obtained are independent of the progenitor's mass. We close by investigating the anticipated errors introduced by orbit-fitting when attempting to constrain the parameters of a two-parameter family of realistic Galactic potentials from stream data. The work of this chapter is based on that published in Sanders and Binney (2013).

5.2 Known Streams

Before we discuss the theory of tidal streams and how they may be used to constrain the Galactic potential we give a short description of known long streams. It is important that we understand the available data before we concern ourselves with the details of analysing stream data. For each stream, we have summarised the information from the literature that is useful for the following discussion. There are other streams, which we have not included. These streams are closely associated with globular clusters and dwarf galaxies and as such are short and not as useful for constraining the Galactic potential. The majority of the listed streams were discovered using matched-filter star counts (Rockosi et al. 2002) on Sloan Digital Sky Survey data (SDSS, Eisenstein et al. 2011).

5.2.1 GD-1

Grillmair and Dionatos (2006b) detected a 63° stellar stream in SDSS data using star counts. This stream is referred to in the literature as GD-1. The stream is extremely narrow, from which the authors conclude that the progenitor was a globular cluster. However, the progenitor has not been identified suggesting that it has been completely disrupted. Because it is exceptionally long and thin, the GD-1 stream has been used by both Willett et al. (2009) and Koposov et al. (2010) to constrain the Galactic potential. Additionally, GD-1 is relatively close to the Sun for a tidal stream ($\sim$10 kpc), which allowed these authors to construct a full 6D phase-space map of the stream. Both sets of authors used the assumption that the stream delineates an orbit.

The data for GD-1 is currently the best data set for a tidal stream: Koposov et al. (2010) provides us with 6D phase-space coordinates for different fields along the stream. The authors fit an orbit to this stream using a 3-component potential. The best-fitting orbit has its pericentre at 14 kpc, apocentre at 26 kpc and reaches a maximum height above the Galactic plane of $\sim$11 kpc.

5.2.2 Orphan

The Orphan stream was discovered independently by both Grillmair (2006a) and Belokurov et al. (2007a) using SDSS photometry and spectroscopy. The nearest part of the 50°-long stream is $\sim$20 kpc from the Sun. The Orphan stream is so-called due to the lack of a progenitor. Belokurov et al. (2007a) suggested that Ursa Major II galaxy (UMa II) may be the progenitor. However, using distances and radial velocities Newberg et al. (2010) fitted an orbit to the stream which seemed to rule out UMa II as the progenitor. The more recently discovered star cluster Segue-1 (Belokurov et al. 2007b) seems a more-likely candidate.

Newberg et al. (2010) find a best-fitting orbit for the Orphan stream with pericentre at 16.4 kpc, apocentre at 90 kpc and reaching a maximum height above the Galactic plane of $\sim$45 kpc.

5.2.3 Anticenter

The Anticenter stream was detected by Grillmair (2006b) as a $\sim$65° long overdensity approximately $\sim$9 kpc away in the direction of the Galactic anticentre. Grillmair (2006b) concluded that it was not associated with the Monoceros Ring, despite lying in the same region of the sky, and this conclusion was strengthened by the kinematics measured by Carlin et al. (2010), who measured a 6D phase-space point on the stream. This single point may be used to construct an approximate orbit for the stream.

5.2.4 NGC 5466

A 45° stream was detected by Grillmair and Johnson (2006). It appears to coincide with the much smaller tidal tails of NGC 5466 found by Belokurov et al. (2006a), so it is believed to be associated with this extremely metal-poor globular cluster. In this case, we are in the fortunate position of confidently identifying the progenitor and we may use the orbit of the progenitor as a proxy for the path of the stream. An approximate orbit for the progenitor may be constructed from the 6D coordinates of NGC 5466 given by Harris (1996) and Dinescu et al. (1999).

5.2.5 Palomar 5

Palomar 5 (Pal 5) is a very low-mass, sparse halo cluster lying 18.6 kpc from the Sun, which was found to have short (~2.5°), strong leading and trailing tidal tails by Odenkirchen et al. (2001). It was the first example of tidal tails being resolved around a cluster and has received much attention in the literature as an example of the formation of tidal streams (Dehnen et al. 2004). Further observations found that the stream extended up to 22° (Grillmair and Dionatos 2006a). As with NGC 5466, we may use the 6D phase-space coordinates of the progenitor, given by Odenkirchen et al. (2001), to construct an approximate orbit for the progenitor, and hence for the stream members.

5.2.6 Sagittarius

The Sagittarius dwarf galaxy was discovered by Ibata et al. (1995) and is the third largest satellite of the Milky Way. Johnston et al. (1995) predicted that the Sagittarius dwarf would be heavily disrupted, and that debris might be observed in the solar neighbourhood. Majewski et al. (2003) observed extended leading and trailing tidal tails, which Belokurov et al. (2006b) found wrapped at least once around the Galaxy. Its length and number of constituent stars make the Sagittarius stream useful for constraining the Galactic potential. However, the Sagittarius stream is very broad and could potentially reflect the internal properties of the progenitor (Peñarrubia et al. 2010). Complex models, which account for dynamical friction (see later), are required to use the Sagittarius stream to constrain the Galactic potential.

Belokurov et al. (2006b) found that the Sagittarius stream had what they dubbed a bifurcation. These authors were limited to observing the stream in the northern Galactic hemisphere. Recently, Koposov et al. (2012) have extended the observations to the southern Galactic hemisphere and found that the bifurcation is also present there. It is believed that the bifurcation is actually due to a fainter stream that runs alongside the Sagittarius stream. This secondary stream is chemically distinct from the Sagittarius stream (Koposov et al. 2012), which seems to rule out the possibility that the secondary stream and the Sagittarius stream share a common progenitor. It is believed that the secondary stream originated from a different progenitor, presumably a companion of Sagittarius.

We take the sky coordinates of the Sagittarius dwarf galaxy from Majewski et al. (2003), the distance from Siegel et al. (2007), the line-of-sight velocity from Ibata et al. (1997) and the proper motions from Pryor et al. (2010), giving us a 6D phase-space point on the orbit of Sagittarius.

5.2.7 Acheron, Cocytos, Lethe and Styx

Four streams were discovered by Grillmair (2009) using a matched-filter technique and were named Acheron, Cocytos, Lethe and Styx in order of increasing distance from the Sun. The first three of these are very narrow and lie between 3 and 15 kpc from the Sun spanning between 37 and 84°. Styx is much more distant ($\sim$45 kpc), broader and spans at least 53°. None of the four streams has an identified progenitor, although the Styx stream is believed to be associated with the concurrently discovered cluster Bootes III. In the discovery paper, Grillmair (2009) fits orbits to the available data to predict a 6D phase-space point in each of the streams.

5.2.8 Aquarius

The Aquarius stream was detected as an overdensity in the line-of-sight velocity data from the Radial Velocity Experiment (RAVE) by Williams et al. (2011). The stream passes very close to the Sun (within 0.5 kpc), is particularly broad and has no identified progenitor. Williams et al. (2011) fits an orbit to this stream with pericentre at 1.8 kpc, apocentre at 9.0 kpc and reaching a maximum height above the Galactic plane of $\sim$5 kpc.

5.2.9 Cetus, Virgo and Triangulum

The Cetus stream was discovered by Newberg et al. (2009) in velocities from the Sloan Extension for Galactic Understanding and Exploration (SEGUE, Yanny et al. 2009). These observations were corroborated by Koposov et al. (2012), who observed the Cetus stream in the SDSS southern Galactic hemisphere data. The stream lies $\sim$34 kpc from the Sun and follows an approximately polar orbit.

Jurić et al. (2008) discovered a faint overdensity in the constellation of Virgo from SDSS stellar number counts. The Virgo overdensity was also observed as a velocity overdensity in measurements of RR Lyrae stars (Duffau et al. 2006). The overdensity has a large spatial extent but it is unclear whether it is a stream or not.

The Triangulum stream was very recently discovered by Bonaca et al. (2012) by searching SDSS data using a matched-filter technique. The stream extends over 12° and lies approximately 26 kpc from the Sun.

These three streams do not have sufficient data in the literature to reliably construct their 6D phase-space structure. A simple orbit fit may be possible but this is beyond the scope of this exercise.

5.3 Tidal Streams in Angle-Action Coordinates

Helmi and White (1999) and Tremaine (1999) explained the formation of tidal streams in angle-action space. In this formulation, streams are formed because stars do not share a common orbit. It is this formulation which we present here.

Given 6D phase-space information, the angle-action coordinates for each star along the tidal stream may be found. We assume that each star does not feel the gravitational influence of the stars in the stream but only the external Galactic potential. Angle-action variables provide a simple way to follow the dynamics of the stream as the actions are constants of the motion whilst the angles increase linearly in time. For a single star in the stream, the angle-action coordinates, $(\boldsymbol{\theta}, \boldsymbol{J})$, obey the equations

$$\boldsymbol{J} = \text{const.}, \quad \boldsymbol{\theta}(t) = \boldsymbol{\theta}(0) + \boldsymbol{\Omega} t, \tag{5.1}$$

where $\boldsymbol{\Omega} = \partial H/\partial \boldsymbol{J}$ are the frequencies of the Hamiltonian, H, and t is the time since the star was stripped from the progenitor. All the stars in the stream are assumed to derive from a progenitor with actions $\boldsymbol{J}_0$. The progenitor is assumed to be of low mass, so that we may neglect dynamical friction and the actions of the progenitor are constant throughout the motion. Also, we assume that, once a star has been stripped, the influence of the progenitor can be neglected, so the star's actions are constant from the time the star was stripped.

The stream is formed by the difference in angles between the progenitor and the stars in the stream, $\Delta\boldsymbol{\theta}$, increasing with time. For a single star, we have

$$\Delta\boldsymbol{\theta} = \boldsymbol{\theta} - \boldsymbol{\theta}_0 = \Delta\boldsymbol{\Omega} t + \Delta\boldsymbol{\theta}(0). \tag{5.2}$$

$\Delta\boldsymbol{\theta}(0)$ is the initial difference in angles between the progenitor and a given star. $\Delta\boldsymbol{\Omega}$ is the difference in frequencies. Both $\Delta\boldsymbol{\theta}(0)$ and $\Delta\boldsymbol{\Omega}$ depend upon the progenitor mass (see Sect. 5.5). When a stream has formed, $\Delta\boldsymbol{\theta}(0)$ has become small compared to the term $\propto t$ so we have

$$\Delta\boldsymbol{\theta} \approx \Delta\boldsymbol{\Omega} t. \tag{5.3}$$

As the frequencies and the angles depend on the potential, this equation provides a constraint for the potential. However checking whether this equation is obeyed for all stars in the stream in a given potential is complicated for two reasons: 1. The progenitor of the stream may be unknown, and 2. the time that the star left the progenitor, t, is not known.

Nevertheless, Eq. (5.3) provides a useful constraint on the potential. If the difference between the actions of the stream stars and those of the progenitor is small, the frequencies of a stream star are well approximated by the Taylor expansion

$$\boldsymbol{\Omega} \approx \boldsymbol{\Omega}_0 + \mathbf{D} \cdot \Delta \boldsymbol{J}, \tag{5.4}$$

where $\mathbf{D}$ is the Hessian matrix

$$D_{ij}(\boldsymbol{J}) = \frac{\partial^2 H}{\partial J_i \partial J_j}. \tag{5.5}$$

This matrix is symmetric, so at each point of action space it is characterised by three orthogonal eigenvectors, $\hat{\boldsymbol{e}}_i$, with associated real eigenvalues, λ_i. With the Taylor series for the frequencies given in Eq. (5.4), the difference in angles is related to the difference in actions by

$$\Delta \boldsymbol{\theta} \approx \Delta \boldsymbol{\Omega} t \approx \mathbf{D} \cdot \Delta \boldsymbol{J} t. \tag{5.6}$$

In this framework, we can understand the conditions required for a stream to form. Once a star has been stripped from the cluster, the action difference, $\Delta \boldsymbol{J}$, is frozen in and the angle difference increases with time. The Hessian determines along which directions the cluster spreads. For a long thin stream to form from an approximately isotropic cluster in action space, one eigenvalue of the Hessian must be much larger than the other two, $\lambda_1 \gg \lambda_2 \geqslant \lambda_3$. In this case, the stream will stretch along the eigenvector $\hat{\boldsymbol{e}}_1$ and

$$\frac{\Delta \boldsymbol{\theta}}{t} \approx \Delta \boldsymbol{\Omega} \approx \hat{\boldsymbol{e}}_1 (\lambda_1 \hat{\boldsymbol{e}}_1 \cdot \Delta \boldsymbol{J}). \tag{5.7}$$

Hence, the frequency difference should be aligned with the principal eigenvector of the Hessian for all stars in the stream, independent of their action. The structure of the stream distribution in action-space is not generically isotropic (Eyre and Binney 2011). In the absence of self-gravity, the action distribution of the cluster is constant and reflects the initial conditions used to seed it. As demonstrated in Eyre and Binney (2011) the action distribution of a cluster depends upon its orbital phase and can be elongated in the (J_R, L_z) plane. The distributions in (J_R, J_z) and (L_z, J_z) are near isotropic. When self-gravity is introduced, the anisotropy in the (J_R, L_z) plane is increased and anisotropy is introduced in the (J_R, J_z) and (L_z, J_z) planes. However, we will see later that the Hessian is highly anisotropic so small anisotropies in the action distribution are washed out. Additionally, we are not wholly reliant on the assumption of isotropy in later sections, but it helps to understand the stream geometry.

5.4 The Problem with Orbit-Fitting

We have seen that if the Hessian matrix is dominated by a single eigenvalue, λ_1, the stream will stretch along the corresponding eigenvector, $\hat{\boldsymbol{e}}_1$. In general, this vector will be misaligned with the progenitor frequency vector, $\boldsymbol{\Omega}_0$, with the angle, φ, between them given by

$$\varphi \equiv \arccos\left(\hat{\boldsymbol{\Omega}}_0 \cdot \hat{\boldsymbol{e}}_1\right). \tag{5.8}$$

The misalignment between these two vectors gives an indication of the error expected when the Galactic potential is constrained by assuming that the stream (aligned with $\hat{\boldsymbol{e}}_1$) delineates an orbit (aligned with $\hat{\boldsymbol{\Omega}}_0$). The potential in which the stream appears to delineate an orbit will, in general, be different to the true potential. Orbit-fitting algorithms also assume that the actions of all the constituent stars are the same which can also lead to errors. However, this effect is small as the stream spans a small range in actions, so we assume that the misalignment angle gives rise to all the error in orbit-fitting algorithms. Importantly, this misalignment is independent of the progenitor mass, and depends only on the progenitor orbit, and hence the underlying potential (see Sect. 5.5). Moving to lower-mass, and hence narrower streams, does not decrease the misalignment.

One key result of Eyre and Binney (2011) is that when this misalignment is $\varphi = 1.5°$ in the isochrone potential, the mass of the Galaxy is overestimated by approximately 20 % when using an orbit-fitting algorithm. Thus, even a small value of φ can lead to significant error in the potential. Eyre and Binney (2011) found the misalignment angle to be $\varphi \approx 1{-}3°$ at every point in action space in the isochrone potential, whereas in a Stäckel potential the angle could be as large as $\varphi \approx 20°$. These are special cases, so it is necessary to explore the magnitude of this angle for realistic Galaxy potentials to assess the need to go beyond orbit-fitting algorithms. Here we estimate φ for example realistic Galactic potentials.

We calculate φ by first finding the Hessian matrix at each point in action-space. This calculation is simply performed using the torus machine (McMillan and Binney 2008). The torus machine constructs orbital tori of given actions for a general potential. Position and velocity coordinates are determined as functions of the angles on the surface of the torus. Given a set of actions, the torus machine returns the corresponding frequencies. It is these properties which make it an appropriate tool for this task.

For each point in action-space, we use the torus machine to differentiate estimates of the frequency numerically. The error in the estimated actions of points that lie on a torus created by the torus machine is estimated as

$$\Delta J \approx \frac{\Delta H}{\sqrt{\Omega_R^2 + \Omega_z^2}} \tag{5.9}$$

where ΔH is the RMS variation in the energy across the torus and Ω_i are the frequencies. For each action-space point, we create a torus with the required actions,

J, and accuracy, ΔJ, as well as neighbouring tori that lie δJ away from the action-space point in each action-space direction. We use the calculated frequencies of these tori to construct numerically the Hessian matrix, $\partial\Omega_i/\partial J_j$, at the action-space point. We require $\Delta J \ll \delta J$ to ensure the numerical differentiation is accurate. The angle between the principal eigenvector of this matrix and the frequency vector at the action-space point is φ. We also calculate λ_1/λ_2, which gives a measure of the width of a stream formed at this action-space point. If this ratio is large, a long thin stream forms and Eq. (5.7) is satisfied. However, if the ratio is small, the stream will be broad.

For a given choice of ΔJ and δJ, the error in φ is estimated by calculating it for a known case. In the Kepler potential the angle is zero (Eyre and Binney 2011) and we use this fact to estimate the error in φ in a general potential.

We perform the above procedure on three potentials: the two-parameter logarithmic potential, the best-fitting potential from McMillan (2011) (hereafter referred to as PJM11), and a potential taken from Piffl et al. (2014) (hereafter referred to as Piffl). The logarithmic potential is defined as

$$\Phi(R,z) = \frac{V_c^2}{2}\ln\left(R^2 + \frac{z^2}{q^2}\right), \tag{5.10}$$

where V_c is the asymptotic circular speed and q is the flattening parameter. We choose $V_c = 220\,\mathrm{km\,s^{-1}}$ and $q = 0.9$, which gives a good representation of the potential of the Milky Way (Koposov et al. 2010). The PJM11 potential is a multi-component potential generated by a bulge, thick and thin discs and an NFW halo, which has been fitted to current experimental constraints. The Piffl potential has the same functional form as the PJM11 potential but with the addition of a gas disc of scale-height 40 pc. Additionally, the potential has a flattened halo ($Q = 0.8$ in the density profile) and the parameters have been chosen to match the constraints from PJM11 as well as match the dynamics of RAVE stars. Figures 5.1 and 5.2 show φ and the eigenvalue ratio in two action planes for these three potentials.[1]

For the logarithmic potential, φ is small (about 1.5°), but crucially non-zero at all points in the action-space planes explored. This is similar to the value found by Eyre and Binney (2011) for the isochrone potential. φ decreases with increasing J_z and increasing L_z as orbits move further out in the potential, but the trend is very subtle. The errors in φ are $\lesssim 0.005°$. The eigenvalue ratio, λ_1/λ_2, is greater than 20 for all action-space points shown. Therefore, we expect long thin streams which are misaligned with the orbit of their progenitor.

For the PJM11 potential, φ is $\gtrsim 1°$ for all the action-space points shown in Fig. 5.2, and can be as large as ~40° for orbits with low J_z. These low-J_z orbits are planar disc-like orbits, so clearly the discs in the PJM11 potential have a large effect on the misalignment angle. For these orbits, the ratio of eigenvalues is small so broad structures will form from debris stripped from disc-like orbits. φ decreases rapidly

[1] Throughout this chapter, the actions are stated in units of $\mathrm{kpc^2Myr^{-1}} = 977.8\,\mathrm{kpc\,km\,s^{-1}}$.

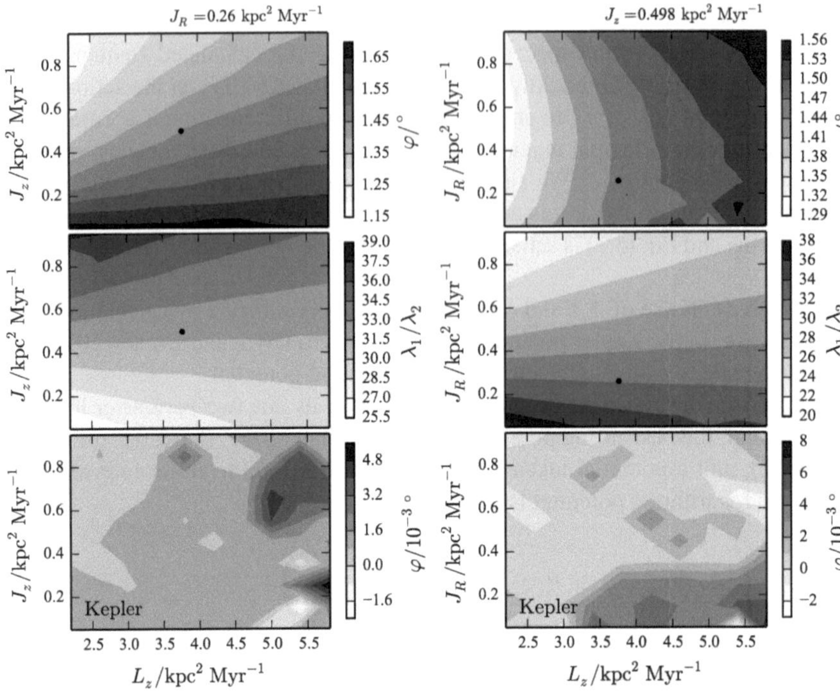

Fig. 5.1 Misalignment angle and ratio of two largest eigenvalues for the logarithmic potential with $V_c = 220\,\mathrm{km\,s^{-1}}$ and $q = 0.9$. Two planes are displayed: $J_R = 0.26$ and $J_z = 0.498\,\mathrm{kpcMyr^{-1}}$. The *bottom two panels* give the misalignment angle in the Kepler potential. This should be zero everywhere so gives a measure of the error in the misalignment angle calculated in the logarithmic potential. The *black dot* shows the approximate action coordinates of GD-1, which the simulations in Sect. 5.5 and Chap. 6 were chosen to emulate

with increasing J_z such that, for orbits that spend most of their time out in the halo, φ has a similar value to that found for the logarithmic and isochrone potentials. Similarly, the ratio of the eigenvalues increases with increasing J_z. In the halo, long thin streams will form. The planes of constant J_z show that orbits with large L_z will form narrower streams that are more aligned with their progenitor orbit. Interestingly, despite the magnitudes of φ being similar far out in the halo, the shapes of the φ surface and the λ_1/λ_2 surface are very different for the two potentials.

For the Piffl potential, the surfaces of φ and λ_1/λ_2 have a similar structure to those in the PJM11 potential. We find that even with a flattened halo ($Q = 0.8$ in the density profile) the misalignment angle is small, and in fact is smaller at high vertical action than in the PJM11 potential. It seems that the effect of the discs is more significant. The thin and thick discs in the Piffl potential have a shorter scale-height than in the PJM11 potential, but also there is a thin gas disc included.

In all three potentials examined here, φ is a few degrees far out in the halo. However, for the PJM11 and Piffl potentials, this can increase to several tens of

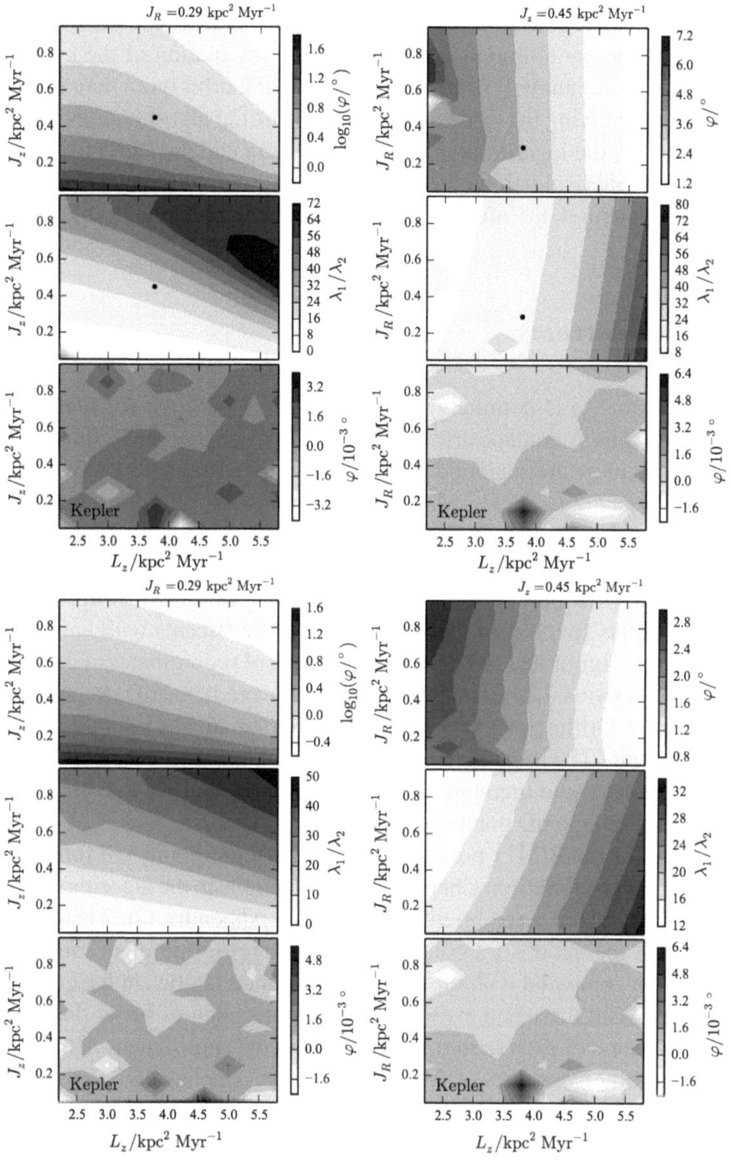

Fig. 5.2 Misalignment angle and ratio of two largest eigenvalues for the best potential from McMillan (2011, PJM11, *top six panels*) and a realistic Galactic potential with a flattened halo ($Q = 0.8$ in the density profile) from Piffl et al. (2014, Piffl, *bottom six panels*). Two planes are displayed: $J_R = 0.29$ and $J_z = 0.45\,\mathrm{kpcMyr}^{-1}$. In each section, the *bottom two panels* give the misalignment angle in the Kepler potential. This should be zero everywhere, so gives a measure of the error in the misalignment angle calculated in the potentials. The *black dot* in the *top section of panels* shows the approximate action coordinates of GD-1, which the simulations in Sect. 5.5 were chosen to emulate. Note we have plotted $\log_{10}\varphi$ in the *top left panels*

degrees for orbits that spend more time in the disc. Therefore, simulations which use the logarithmic potential may not give a good representation of the evolution of a tidal stream. The stream will potentially delineate the orbit more than it should in a realistic potential giving the impression that orbit-fitting algorithms are appropriate. We note that both the logarithmic potential used here and the PJM11 potential have approximately spherical halo potentials. However, even with a more flattened halo, as in the Piffl potential, the misalignment at high vertical action is still small.

5.4.1 Known Streams

We have seen that φ is definitely non-zero and can be large for realistic Galaxy potentials, and thus systematic errors can be made when an orbit-fitting algorithm is used. We now explore the magnitude of the misalignment angle for the known streams described in Sect. 5.2 to decide whether orbit-fitting algorithms are appropriate for analysing available stream data. For this task we must make estimates of their actions and then use the torus machinery as above.

With the information collected in Sect. 5.2, we can construct approximate 6D phase-space points for each of the streams. For those streams with known progenitors (Palomar 5, Sagittarius, NGC 5466), we use the 6D coordinates of the progenitor. The Acheron, Cocytos, Lethe and Styx streams have predicted 6D stream points from an orbit fitted by Grillmair (2009). The Anticenter stream has a single measured 6D stream coordinate. The remaining streams (GD-1, Orphan, Aquarius) have approximate orbit fits from the literature. In our chosen potential we can produce similar orbits and find a single 6D point on these orbits.

The actions in the PJM11 potential for each of these points are found using the Stäckel-fitting algorithm from Chap. 2 (small alterations to the algorithm appropriate for the problem at hand are briefly discussed in Appendix G). There is an error of ~5–10 % in the actions introduced by the Stäckel-fitting algorithm, but this is irrelevant when compared with the observational uncertainties in the coordinates of these streams. With these actions we find φ to a precision of better than 0.1° using the torus machine. For each stream, Table 5.1 gives the approximate actions, the ratio of the two largest eigenvalues of the Hessian and φ. The misalignment angle varies from $\varphi \approx 13^\circ$ to $\varphi \approx 0.15^\circ$ with the largest misalignment angle producing the smallest eigenvalue ratio, and hence the broadest streams. As a check, we performed the same calculations using the more accurate iterative torus procedure described in Sect. 2.7, which yielded very similar results.

The Anticenter, Aquarius and NGC 5466 streams all have $\varphi > 2.8^\circ$ and small eigenvalue ratios (~6). Therefore, just from their actions we anticipate that the streams formed will be broad. This is definitely true of the Aquarius stream, and

Table 5.1 Known tidal streams: the approximate actions of the progenitor (given in units of $\mathrm{kpc^2 Myr^{-1}}$), the ratio of the two largest eigenvalues, λ_1/λ_2, and the misalignment angle, φ, between the principal eigenvector and the frequency vector at the stream's action coordinates

	J_R	$\|L_z\|$	J_z	$\frac{\lambda_1}{\lambda_2}$	$\varphi/°$	k	Q	$\varphi^*/°$	k^*	Q^*	$\Delta\varphi/°$
Anticenter	0.06	3.4	0.15	6	13.0	0.72	0.15	6.18	1.01	0.42	7.0
Aquarius	0.34	0.61	0.28	7	8.0	2.06	0.10	7.14	2.20	0.06	2.1
GD-1	0.29	3.8	0.45	22	3.5	0.96	0.49	2.50	0.72	0.50	1.6
NGC 5466	3.4	0.30	2.8	6	2.8	0.92	1.05	1.42	1.10	1.15	2.9
Lethe	0.14	1.2	1.3	29	1.1	0.90	0.70	1.97	0.68	1.28	2.3
Cocytos	0.13	0.83	0.99	28	0.93	0.82	1.35	2.18	0.55	0.60	1.5
Palomar 5	0.24	1.2	1.7	30	0.89	1.14	1.61	1.13	0.74	0.75	1.6
Acheron	0.11	0.50	0.76	28	0.73	1.31	0.84	2.73	1.06	0.35	2.0
Orphan	4.0	5.9	0.88	34	0.64	1.03	1.10	0.65	0.72	0.89	0.3
Sagittarius	2.3	2.1	4.0	29	0.43	0.96	1.03	1.32	1.05	0.92	0.9
Styx	0.91	0.22	5.6	37	0.15	1.01	0.99	1.17	0.81	0.99	1.3

φ^* is the measured misalignment angle from the simulations presented in Sect. 5.6. k and Q give the potential parameters found using an orbit-fitting algorithm on a stream aligned with the principal eigenvector (Sect. 5.7), and k^* and Q^* give the parameters found using an N-body simulation. The true underlying potential has parameters $(k, Q) = (1, 1)$. The final column, $\Delta\varphi$, gives the angular difference between the principal eigenvector of the Hessian and the measured direction of the stream from the N-body simulation

the Anticenter stream is a complex which is believed to consist of three separate streams with the whole complex having a width of $\sim 5°$ (Grillmair 2006b). For all other streams, we find $\varphi \lesssim 1°$ and the eigenvalue ratio is large (>20) so narrow streams are expected.

5.5 Mass Dependence

When presenting the angle-action formalism of stream formation, we made little mention of the progenitor mass. We would like to know for what range of progenitor masses this approach and the above results are valid. Stream progenitor masses span a large range: GD-1 is observed to have a mass of $2 \times 10^4 M_\odot$ (Koposov et al. 2010), whilst the mass of the Sagittarius dwarf is believed to be 10^8–$10^9 M_\odot$ (Law et al. 2005; Fellhauer et al. 2006).

Here we discuss each of the assumptions made in the angle-action formalism in the context of mass-dependence:

1. The progenitor actions are assumed to be constant for all time. This is valid in the limit that dynamical friction is negligible. We can neglect dynamical friction if we are in the regime where

$$M_c \ll \frac{r_p V_c^2}{G}, \tag{5.11}$$

where V_c is the circular speed of the potential, and r_p is the pericentre radius. This mass is approximately $10^{11} M_\odot$ for a GD-1-like orbit, so we expect dynamical friction to be negligible for $M_c \lesssim 10^9 M_\odot$. Many streams lie much further out in the halo where this limit is expected to be much larger. Additionally, the effects of dynamical friction are expected to be comparable for the cluster and the stream, such that the relative structure of the stream is not affected, but the global cluster-stream complex is. We expect that dynamical friction is irrelevant for most streams, but its effects on the Sagittarius stream may be important (Jiang and Binney 2000).

2. We have assumed that a particle is instantaneously released from the cluster, and subsequently has constant actions. However, the self-gravity of the cluster will always be significant, regardless of mass. It is the self-gravity of the cluster that determines whether a particle leaves the cluster on each pericentric passage. A particle will always leave the cluster in the same way (approximately through the Lagrange points at pericentre), irrespective of the mass. In the absence of self-gravity, we do not have this restriction as particles leave the cluster more uniformly. Therefore, we expect that the inclusion of self-gravity will have an impact on the overall shape of the angle-action space structure of the stream, independent of the mass (see Sect. 5.6). We also expect that increased cluster self-gravity will produce broader streams as particles leave the cluster with a larger range of actions. Thus, the role of progenitor mass is to set the scale of the stream's structure without affecting its morphology in any other way. Despite the cluster self-gravity always being important, it should produce mass-independent effects on the overall shape of the stream (see below).
3. We assume that we can neglect the finite angle size of the cluster, $\Delta\boldsymbol{\theta}(0)$, after some time t as it is negligible compared to the contribution of secular evolution to $\Delta\boldsymbol{\theta}$. Assuming the secular evolution of the stream stretches the angle distribution along one direction (see below), $\Delta\boldsymbol{\theta}(0)$ will act to broaden the stream perpendicular to this principal direction. This is mass-dependent, and is related to the above self-gravity arguments, but when a stream has formed this term is always unimportant. As long as the initial spread in angles is symmetric about the stream path, this term will not affect the presented formalism.
4. We employ a Taylor expansion in $\Delta\boldsymbol{J}$ when finding the relationship between the frequencies and actions. This is important as it leads to the conclusion that $\Delta\boldsymbol{\Omega}$ for each particle will lie along the same vector $\hat{\boldsymbol{e}}_1$. This assumption is valid provided

$$\frac{\partial D_{ij}}{\partial J_k}\Delta J_k \ll D_{ij}. \tag{5.12}$$

If the Hamiltonian is a function of some low power of $\boldsymbol{J}$, this reduces to

$$\Delta J_i \ll J_i. \tag{5.13}$$

If we are in the progenitor-mass regime where we can neglect the above effects, what effect does the progenitor mass have on the resulting angle-action space distribution? $\Delta\boldsymbol{J}$, $\Delta\boldsymbol{\Omega}$ and $\Delta\boldsymbol{\theta}$ are all functions of the progenitor mass. We expect that larger progenitor masses produce larger spreads in the actions, frequencies and angles of the resulting stream, but we would like to know their exact mass dependence. Following Eyre and Binney (2011) we have that

$$\Delta J_i \approx \frac{1}{2\pi}\oint \Delta p_i \mathrm{d}x_i \approx \frac{1}{2\pi}\oint \sigma \mathrm{d}x_i, \tag{5.14}$$

where σ is the velocity dispersion of the progenitor. For axisymmetric systems, this is approximately

$$\Delta\boldsymbol{J} \approx \frac{1}{2\pi}(2\sigma\Delta R, 2\pi\sigma r_p, 4\sigma\Delta z) \tag{5.15}$$

where ΔR is the difference between the apocentric and pericentre radius, and Δz is the maximum height above the plane reached by the orbit. ΔL_z is calculated at pericentre as this is where the majority of particles are stripped. Under the assumption of Eq. (5.12), $\Delta\boldsymbol{\Omega}$ is linearly related to $\Delta\boldsymbol{J}$ via the (mass-independent) Hessian D_{ij}, and $\Delta\boldsymbol{\theta}$ is linearly related to $\Delta\boldsymbol{\Omega}$ via the time since stripping, t. Therefore, both $\Delta\boldsymbol{\Omega}$ and $\Delta\boldsymbol{\theta}$ will have the same dependence on mass as $\Delta\boldsymbol{J}$.

From the virial theorem, we relate the velocity dispersion of the cluster to its mass and radius via

$$\sigma^2 \approx \frac{GM_c}{r_t}. \tag{5.16}$$

and the tidal radius, r_t, is related to the mass of the cluster via

$$r_t = r_p\left(\frac{M_c}{M_g}\right)^{\frac{1}{3}}, \tag{5.17}$$

where M_g is the mass of the host galaxy contained within r_p. Therefore, the progenitor mass is proportional to the velocity dispersion cubed or $\sigma \propto M_c^{1/3}$. As $\Delta\boldsymbol{J}$, $\Delta\boldsymbol{\theta}$ and $\Delta\boldsymbol{\Omega}$ are proportional to σ in the regime we are considering, we expect all these quantities to also depend on $M_c^{1/3}$. Choi et al. (2007) showed from N-body simulations that the energy difference of stripped particles obeyed this same scaling with progenitor mass in a spherical halo. Similarly, Johnston (1998) demonstrated that the density profile along a stream was described by the same analytic form scaled by $M_c^{1/3}$, and Johnston et al. (2001) utilised this scaling relation to develop a semi-analytic formalism for predicting the morphology of a recently-formed stream.

These arguments convince us that the progenitor mass acts only to scale the angle-frequency distribution, and the shape is independent of the mass, provided we are in the aforementioned regime. Therefore, the misalignment angle is mass-independent.

Using these results, we relate the assumption of Eq. (5.12) to a constraint on the progenitor mass for a given orbit. We expect the neglected terms in the Taylor series to be non-negligible when

$$\Delta \boldsymbol{J} \approx \frac{1}{2\pi}(2\sigma\Delta R, 2\pi\sigma r_p, 4\sigma\Delta z) = \boldsymbol{J}. \tag{5.18}$$

Therefore, we expect the assumption to break down when

$$\sigma \gtrsim \min\left\{\frac{\pi J_R}{\Delta R}, \frac{L_z}{r_p}, \frac{\pi J_z}{2\Delta z}\right\}, \tag{5.19}$$

where all the quantities on the right-hand side depend only on the chosen orbit. We see that the cluster needs to be on a sufficiently eccentric orbit for the approximation to hold. However, we expect that the majority of tidal streams are formed from progenitors on eccentric orbits, so this constraint is not too restrictive.

All the above predictions may be tested by inspecting some N-body simulations. We construct a stream by placing a King cluster at apocentre on a stream-like orbit in the logarithmic potential defined by Eq. (5.10) and integrating with self-gravity until a stream has formed. King models (King 1966) are characterised by one shape parameter and two scale parameters: the ratio of central potential to squared-velocity parameter, $W_0 = \Psi_0/\sigma^2 = 2$, the cluster mass, M_c, and a tidal limiting radius, r_t, set by Eq. (5.17). Given W_0, a scaled tidal-limiting radius is found by integrating equation (4.112) from Binney and Tremaine (2008) with the boundary condition $\mathrm{d}W_0/\mathrm{d}r = 0$. This model is then scaled by the two parameters, M_c and r_t, setting the velocity dispersion parameter, σ, and the central density, ρ_0. We seed the clusters with $N = 10{,}000$ particles, and explore the range of masses $2\times10^4 \leqslant M_c \leqslant 2\times10^9 M_\odot$. The softening parameter, ϵ, is chosen to be $\epsilon = (M_c/\rho_0 N)^{1/3}$. The parameters for the simulations are given in Table 5.2.

The orbit was chosen to be similar to the orbit of the GD-1 stream (Koposov et al. 2010). The orbit has initial conditions $(R, z) = (26.0, 0.0)$ kpc and $(U, V, W) = (0.0, 141.8, 83.1)$ km s^{-1}, where positive U is towards the Galactic centre and positive V is in the direction of the Galactic rotation at the Sun. This orbit has $r_p \approx 14$ kpc. We evolve the simulation for $t = 4.27$ Gyr (just after the 11th pericentric passage) using the code GYRFALCON (Dehnen 2000, 2002), made available through the NEMO Stellar Dynamics Toolbox (Teuben 1995).

Table 5.2 Parameters of King models used in the simulations detailed in Sect. 5.5

N	W_0	r_p/ kpc	$\frac{M_c}{M_\odot}$	r_t/ kpc	σ/ km s^{-1}	ϵ/ pc
10,000	2.0	14	2×10^4	0.07	1.39	1.5
			2×10^5	0.14	3.01	3
			2×10^6	0.32	6.50	6
			2×10^7	0.69	14.0	14
			2×10^8	1.48	30.1	30
			2×10^9	3.20	65.0	66

For each resulting particle distribution, we cut out the remnant of the progenitor, and estimate the actions, angles and frequencies of the stream particles using the Stäckel-fitting algorithm from Chap. 2. We quantify the spread in each coordinate using the standard deviation. In Fig. 5.3, we plot the frequency difference as a function of progenitor mass. The correlation is very tight and, from the guiding line with slope 1/3, we see the data follow the expected trend. In Fig. 5.4, we plot the gradient of the frequency distribution. It is this quantity that gives the degree of stream-orbit misalignment. We see that, as expected, the gradient is constant with mass. There is a small deviation at the low-mass end, which is due to the numerical errors introduced by the Stäckel-fitting algorithm. We have near-perfect scaling of the results with mass, so there are no mass-dependent effects in the mass regime considered.

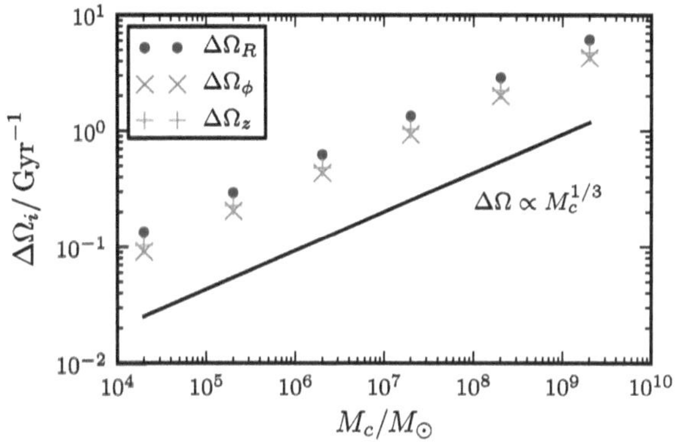

Fig. 5.3 The size of the frequency distribution against the progenitor mass. The size is estimated using the standard deviation in each frequency coordinate. As expected, $\Delta\Omega_i$ is proportional to $M_c^{1/3}$, shown by the *black line*

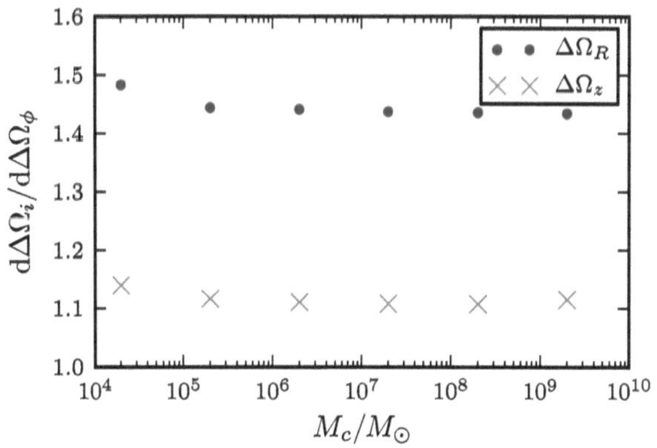

Fig. 5.4 The slope of the frequency distribution against the progenitor mass. We see the slopes are independent of mass. There is a small deviation at low mass due to errors introduced by the Stäckel-fitting algorithm used to estimate the frequencies. At the high-mass end, there is also a slight deviation that may be due to the higher-order action space structure

For this orbit, we use Eq. (5.19) to find the maximum progenitor velocity dispersion for the first order expansion of $\Delta\boldsymbol{\Omega}$ in terms of $\Delta\boldsymbol{J}$ to be valid. We find that radial and vertical actions give similar constraints of $\sigma_{\text{max}} \approx 70\,\text{km}\,\text{s}^{-1}$, which translates into a maximum mass of $M_{\text{max}} \approx 5 \times 10^9 M_{\odot}$. We have not quite reached this regime with the N-body simulations, but there is the suggestion of its impact at the high-mass-end of Fig. 5.4.

We have seen that the formalism is valid for $M_c \lesssim 10^9 - 10^{10} M_{\odot}$, when dynamical friction, the higher order action-space structure and perturbations from the cluster remnant become important. However, below this limit we find that the progenitor mass acts to scale the frequency, action and angle-distributions, such that the shapes of these distributions are essentially mass-independent. Therefore, we expect the angle-action formalism and the results of the previous sections to be valid for all observed streams, although dynamical friction may be relevant for the Sagittarius stream.

5.6 Anisotropies in the Action Distribution

In Sect. 5.4 we showed that φ is non-zero for the logarithmic potential. From the above simulations $\varphi = 0.18°$ but using the torus machine we find that $\varphi \approx 1.6°$. The source of this discrepancy is found by running the simulation without self-gravity. In that case, $\varphi \approx 1.8°$ in better agreement with the prediction. Therefore, the self-gravity of a cluster causes φ to decrease. The gradient of the frequency distribution is mass-independent, so this self-gravity effect is also mass-independent.

A similar experiment run in the PJM11 potential shows a similar $\sim$1.5° decrease in φ (see Fig. 5.5). However, in this case, the simulation with gravity included still shows a significant φ. It just seems a coincidence that, for the simulation in the logarithmic potential, the expected value of φ is almost cancelled by the inclusion of self-gravity.

This effect can be understood by considering the action-space structure of the cluster (Eyre and Binney 2011). In the formalism of Sect. 5.3, we showed that the stream would lie along the principal eigenvector of the Hessian, but only if the stream action-space distribution is isotropic. Eyre and Binney (2011) showed that the action-space distribution is not isotropic. Self-gravity introduces different anisotropies to those present when self-gravity is neglected. Different action-space distributions will give rise to different frequency-space distributions under the action of the Hessian. For Hessians with large eigenvalue ratios, a highly elongated frequency distribution will be produced, but its orientation will depend on the shape of the action-space distribution.

We can understand the difference between the self-gravity and no-self-gravity simulations by considering how the particles are stripped from the cluster when self-gravity is included. For a particle to be stripped, it must leave the cluster through the Lagrange points L1 and L2 at pericentre. For the orbit considered, the motion is dominated by the radial motion, and at pericentre the Lagrange points lie in a plane

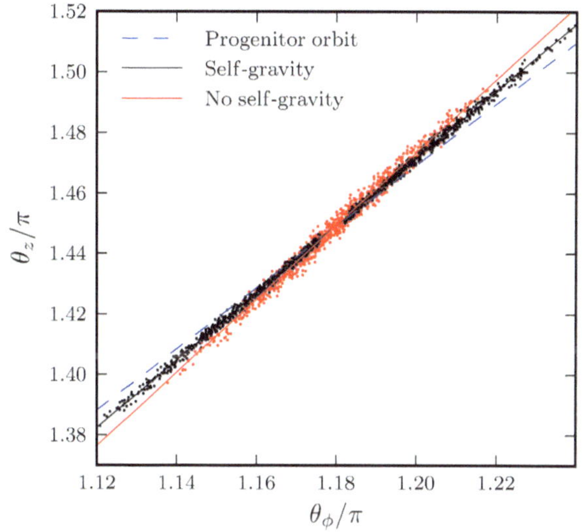

Fig. 5.5 Angle-angle plot for a similar simulation to those outlined in Sect. 5.5 but in the PJM11 potential run with and without self-gravity. This snapshot shows the clusters at the fifth apocentric passage. The *dashed blue line* gives the orbit of the progenitor. The *black* and *red lines* are *straight line* fits to the data for the self-gravity and no self-gravity simulations respectively. We note that there is a misalignment between the stream and the progenitor orbit which decreases when self-gravity is included. The same effect is observed in the logarithmic potential, but it is clearer to see in the PJM11 potential

which is nearly parallel to the plane $z = 0$. Therefore, a particle needs to have an increase in its radial velocity to have sufficient energy to be stripped. This increased radial motion will in turn increase/decrease the radial frequency Ω_R depending on whether the particle leaves through L1 or L2. Similarly, the angular frequency Ω_ϕ will increase/decrease as the particle moves to a smaller/larger radius without changing its transverse velocity. If we consider the motion in z to be completely decoupled from the radial motion, which in the orbit considered is a fair assumption, increased motion in R as the particle leaves the cluster will not alter the vertical action J_z and frequency Ω_z.

Now we can understand Fig. 5.5 as the result of this frequency-space evolution. Particles in the stream have increased/decreased angular frequency whilst their vertical frequency has remained constant. This causes the distribution in (θ_ϕ, θ_z) space to rotate clockwise thus decreasing the misalignment angle.

We investigate how the anisotropy of the action-space distribution affects the estimated misalignment angle for the presented streams. We could attempt to estimate the effects of the anisotropy analytically following a similar analysis to Eyre and Binney (2011). However, as we are only dealing with eleven streams, we choose to run some N-body simulations, which will fully account for these effects. For each known stream, we integrate the orbit in the PJM11 potential to find the pericentre radius, r_p, and a phase-space point at apocentre. We seed a 10,000 particle $2 \times 10^5 M_\odot$

King cluster with a tidal radius related to r_p via Eq. (5.17), and place it at the apocentre phase-space point. The simulation is then evolved in GYRFALCON, until a stream has formed. The misalignment angle is measured in angle-space as the angle between the angle distribution of the stream particles and the frequency vector of the progenitor. We note here that this result is independent of the mass of the progenitor, and the phase of the orbit of the progenitor.

We present the results of this procedure in Table 5.1, where we give the angular difference between the N-body stream structure and the principal eigenvector, $\Delta\varphi$. For all known streams, we find that, as with the simulation shown in Fig. 5.5, the angle-space distribution rotates by a few degrees. The Anticenter stream exhibits the largest angular change of $\sim 7°$. The misalignment between the streams and the progenitor orbit is still a few degrees, despite the anisotropies in the action-space distribution. Therefore, we expect that orbit-fitting algorithms will not be appropriate for real streams.

5.7 Errors in Potential Parameters

Whilst a good indicator of whether an orbit-fitting algorithm is appropriate or not, φ does not give a good measure of how much we will err if we use an orbit-fitting algorithm. We would like to know how the magnitude of the misalignment relates to the error in potential parameters found by simply fitting an orbit to the stream.

We use the suite of two-parameter potentials described in Appendix F. These are multi-component Galactic potentials that all have the same circular speed at the Sun, but which vary in two key respects: the flattening of the halo density, Q, and the ratio of the magnitude of the force due to the disc and the halo at the Sun, k, normalised such that the PJM11 potential is the potential with $(k, Q) = (1, 1)$.

An orbit fit is valid if all the particles have the same frequency, and the angles increase at this frequency along the stream. The first of these conditions is entirely dependent on the mass of the cluster, with high-mass progenitors producing large spreads in frequencies. However, the second condition is mass-independent, and it depends only upon the potential and the progenitor actions. Therefore, for low-mass progenitors the second condition is more significant, so we will focus on it here.

For each stream we use the Nelder-Mead algorithm (Nelder and Mead 1965) to adjust (k, Q), until the progenitor frequency vector is aligned with the angle distribution of the stream. We use the Stäckel-fitting algorithm to find the angles and frequencies. This simulates the operation of an orbit-fitting algorithm. The observed stream is misaligned with the orbit but by using an orbit-fitting algorithm we are requiring the 'best-fitting' potential to make this stream an orbit. This means we need to find a potential where the frequency vector of the stream members is aligned with the stream. This approach neglects the spread in frequencies of the stream members, which one might also want to minimise when orbit-fitting. Additionally, we have ignored the difference between the leading and trailing tails of the stream. The leading and trailing tails have distinct actions and hence are better fitted by two orbits as opposed to a single orbit. However, in the limit of low mass, the distinction between the actions of these tails becomes small, whereas the misalignment, which is mass-independent, remains significant.

We use two stream distributions for each known stream—one which is aligned with the principal eigenvector of the Hessian in the true potential, and one taken from an N-body simulation that includes the effects of the self-gravity. For the first of these we create a series of 100 $(\boldsymbol{\theta}, \boldsymbol{J})$ points with the same actions as the progenitor, and angles lying at regular intervals along the principal eigenvector of the Hessian. We then use the torus machine to find the corresponding $(\boldsymbol{x}, \boldsymbol{v})$ in the true potential $(k, Q) = (1, 1)$. For the second approach, we use a sample of 100 particles from each of the simulations given in the previous section.

The results of this experiment are shown in Table 5.1. If the orbit-fitting algorithm is appropriate for a given stream, we should recover $(k, Q) = (1, 1)$. In Fig. 5.6 we plot the errors, $(\Delta k, \Delta Q)$, in the parameters (k, Q) against the misalignment angle for all the streams using both the artificial stream distribution and the N-body distribution. We see that the error in the parameters scales approximately with the misalignment angle, so we expect large misalignment angles lead to large errors in the potential parameters using orbit-fitting algorithms. However, the scatter about

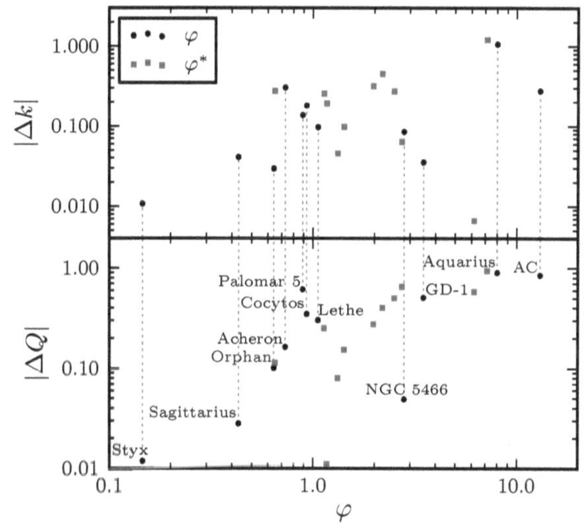

Fig. 5.6 Errors in the parameters (k, Q) of the two-parameter Galactic potentials obtained when using an orbit-fitting algorithm to analyse known streams plotted against the misalignment angle, φ, in the true potential with $(k, Q) = (1, 1)$. The potential parameters are adjusted until the frequency vector of the progenitor of the stream aligns with angle-space structure of the stream. This simulates the operation of an orbit-fitting algorithm. The *round black points* show the errors obtained when using an artificial stream perfectly aligned with the principal eigenvector of the Hessian in the true potential. The *square red points* show the errors when using an N-body simulation. The labels refer to the *round black points* only, and the *light grey lines* indicate the correspondence between the *round black points* in the *top* and *bottom panels*

this line is reasonably large so the relationship is not simple and other factors are at play.

We begin by discussing the results from the artificial stream distributions (the unstarred values). The Anticenter, Aquarius and GD-1 streams all have $\varphi > 3.5°$ and, as such, have large errors in the potential parameters, particularly the flattening. Notably, for the Aquarius stream the errors are of order one, due to the low actions of the stream. NGC 5466 also has large φ but the errors in the potential parameters are $<10\,\%$. However, this orbit is awkward to deal with on account of its low L_z, yet high J_R and J_z. It is in this regime where the largest errors in the actions are expected (Sanders 2012), and correspondingly the largest errors in the frequencies and Hessian. Therefore, the error in φ is expected to be large for NGC 5466. Only the Orphan, Sagittarius and Styx streams have small enough misalignment angles that their potential parameters are accurate to $<10\,\%$. We therefore expect orbit-fitting algorithms to be appropriate for these streams. The other streams have intermediate potential parameter errors which range from 10 to 60 %, and the use of orbit-fitting algorithms may be appropriate depending on the quality of the data.

From analysing the N-body simulations, we find a similar set of results. The Orphan, Styx, Palomar 5 and Sagittarius streams all have small potential parameter errors of $<30\,\%$, so orbit-fitting algorithms should be appropriate for these streams. Again NGC 5466 has very small errors of $<15\,\%$. For those streams with $\varphi^* \gtrsim 2°$, the errors in the parameters are $\gtrsim 30\,\%$.

The results of Fig. 5.6 have not accounted for the masses of the progenitors. We note that some streams have very massive progenitors, particularly the Sagittarius stream. For these streams the spread in frequencies will be more important than the misalignment in angle-space. In the case of Sagittarius, Fig. 5.6 suggests that the misalignment is small enough for an orbit fit to be valid. However, an orbit fit on the entire Sagittarius stream will be flawed due to the spread in the frequencies. Still, for low-mass progenitors on similar orbits to the Sagittarius stream (e.g. perhaps the neighbouring bifurcation) an orbit fit to the stream is appropriate.

Finally, we show the error in the flattening as a function of the vertical action of the progenitor in Fig. 5.7. We see that lower vertical action corresponds to larger error in the halo flattening. For a progenitor with low vertical action the resulting stream has shallower gradient in (θ_ϕ, θ_z) space due to the stripping through the Lagrange points barely affecting the vertical action. When we fit an orbit to this distribution, we attempt to make the vertical frequency smaller such that is aligned with the shallow angle distribution. This corresponds to increasing the vertical action and is achieved through increasing the mass in the plane. This can be achieved through increased halo flattening.

We note that it is not clear how the phase of the progenitor affects the recovery of the potential parameters. Our procedure for constructing streams similar to the known streams did not take into consideration the observed position of the stream and only ensured the progenitor orbit was realistic. A small test with a second snapshot of the simulated GD-1 stream produces similar biases in the recovered potential parameters to those found from the first snapshot. Therefore, it seems that the results are not too sensitive to the phase of the progenitor. There may be specific choices of progenitor

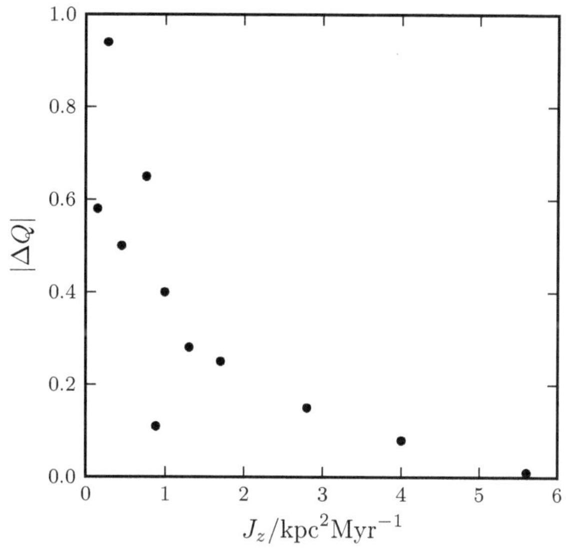

Fig. 5.7 Error in halo flattening as a function of vertical action of progenitor

phase for which the biases are smaller, but a method that relies on special conditions is not suitable for general use.

In conclusion, we have found that for a realistic Galactic potential, order one errors in the parameters of the potential can arise from naively using an orbit-fitting algorithm on known streams.

5.8 Conclusions

In the next few years, more tidal stream data will be collected by surveys of the Galactic halo, so there is considerable scope for using tidal streams to constrain the Galactic potential at these large scales, and it is imperative that appropriate algorithms are developed and tested for this end. Here we have provided an in-depth discussion of the applicability of orbit-fitting algorithms, which rely on the assumption that a stream delineates an orbit. We have shown that this assumption is necessarily false and can lead to systematic biases.

We presented the angle-action formalism of stream formation, in which streams form due to their member stars being on different orbits. We demonstrated that in the angle-action framework streams do not delineate orbits, and the degree of misalignment depends only on the progenitor orbit and the Galactic potential, not the mass of the progenitor.

The degree of misalignment was quantified for the logarithmic potential, which is used in many simulations, and two multi-component realistic Galactic potentials. We found that the misalignment angle is small but non-zero for the logarithmic potential.

For the realistic Galactic potentials, we found similar results for orbits that lie far out in the halo, but the misalignment increases significantly as we approach the disc, where the potential flattens. We concluded that tests of orbit-fitting methods that use the logarithmic potential may give unrealistically good results due to its very small misalignment angles.

We have presented a summary of known streams that may be useful for constraining the Galactic potential. For each of these streams, we have estimated the actions of the progenitor using data from the literature. At each of these action-space points, we quantified the expected misalignment between the stream and the underlying progenitor orbit for a realistic Galactic potential.

Whilst a useful indicator as to whether an orbit-fitting algorithm is appropriate or not, the misalignment angle does not quantify the error involved in estimating the potential parameters from orbit-fitting. We introduced a family of two-parameter realistic Galactic potentials described by the halo-flattening and the halo-to-disc force ratio at the Sun. For each of the known streams, we explored this space of potentials until we found the potential which fits an orbit to the stream. As expected, the error in the potential parameters correlates approximately with the magnitude of the misalignment angle. We showed that this can introduce order one errors in the potential parameters.

We demonstrated that all these results are essentially independent of the mass of the progenitor up to the mass scale where dynamical friction becomes relevant. Mass acts to scale the angle-action distributions, whilst leaving the shape unaffected. We therefore expect that even for large progenitor masses, the results are valid. We also showed from N-body simulations that anisotropies in the action distribution introduced by the self-gravity of the cluster cause the misalignment of stream particles to change by a few degrees. However, the misalignment for the known streams is non-negligible when the effects of self-gravity are included.

The angle-action formalism is a clear framework in which to view and discuss stream formation. It has enabled us to quantify the errors involved in orbit-fitting methods for interesting potentials and led to the conclusion that orbit-fitting algorithms are not appropriate when analysing many streams in the Milky Way. Hence streams need to be modelled without resort to orbit fitting, and in the next chapter we present such an alternative algorithm.

5.8.1 Future Work

We have seen that the structure of the Hessian matrix $\mathbf{D}$ is critical to the structure of streams and that significant flattening leads to stream-orbit misalignment. However, in general, it is not clear how the structure of the potential governs the structure of $\mathbf{D}$. Eyre and Binney (2011) inspected the isochrone potential, which has a small misalignment angle. For the isochrone, the misalignment angle is governed by the scale-radius b, and reduces to zero when $b = 0$ i.e. the Kepler case. However, we must rely, in general, on numerical results like those presented in this chapter.

One class of potentials for which we can gain some insight is the class of scale-free potentials, $\Phi(r) \propto r^{\alpha}$. These potentials admit self-similar orbits. Under the scaling $\boldsymbol{x} \rightarrow a\boldsymbol{x}$ and $\boldsymbol{v} \rightarrow a^{\alpha/2}\boldsymbol{v}$, the energy scales as $E \rightarrow a^{\alpha}E$. Similarly, the actions must scale like $\boldsymbol{J} \rightarrow a^{1+\alpha/2}\boldsymbol{J}$, and so we must have Hamiltonians of the form

$$H = [h(\boldsymbol{J})]^{\beta}, \tag{5.20}$$

where $\beta = \alpha/(1+\alpha/2)$ and $h(\boldsymbol{J})$ is a homogeneous function of order one. If $h(\boldsymbol{J})$ is a linear function of the actions, $\mathbf{D}$ has a principal eigenvector aligned with the frequency vector. Williams et al. (2014) show that a linear function is a good approximation for $h(\boldsymbol{J})$ but small corrections are required. The form of these corrections is not clear. Williams et al. include a term of the form $\epsilon\sqrt{J_R L}$. Given the potential, it is unclear what the structure of the Hamiltonian and hence the Hessian is, and whether the Hessian has its principal eigenvector parallel to the frequency vector. This is a direction for further work.

References

Belokurov V, Evans NW, Irwin MJ, Hewett PC, Wilkinson MI (2006a) ApJL 637:L29
Belokurov V et al (2006b) ApJL 642:L137
Belokurov V et al (2007a) ApJ 658:337
Belokurov V et al (2007b) ApJ 654:897
Binney J (2008) MNRAS 386:L47
Binney J, Tremaine S (2008) Galactic dynamics, 2nd edn. Princeton University Press, Princeton
Bonaca A, Geha M, Kallivayalil N (2012) ApJL 760:L6
Carlin JL, Casetti-Dinescu DI, Grillmair CJ, Majewski SR, Girard TM (2010) ApJ 725:2290
Choi J-H, Weinberg MD, Katz N (2007) MNRAS 381:987
Dehnen W (2000) ApJL 536:L39
Dehnen W (2002) J Comput Phys 179:27
Dehnen W, Odenkirchen M, Grebel EK, Rix H-W (2004) AJ 127:2753
Dinescu DI, Girard TM, van Altena WF (1999) AJ 117:1792
Duffau S, Zinn R, Vivas AK, Carraro G, Méndez RA, Winnick R, Gallart C (2006) ApJL 636:L97
Eisenstein DJ et al (2011) AJ 142:72
Eyre A, Binney J (2009a) MNRAS 399:L160
Eyre A, Binney J (2009b) MNRAS 400:548
Eyre A, Binney J (2011) MNRAS 413:1852
Fellhauer M et al (2006) ApJ 651:167
Grillmair CJ (2006a) ApJL 645:L37
Grillmair CJ (2006b) ApJL 651:L29
Grillmair CJ (2009) ApJ 693:1118
Grillmair CJ, Dionatos O (2006a) ApJL 641:L37
Grillmair CJ, Dionatos O (2006b) ApJL 643:L17
Grillmair CJ, Johnson R (2006) ApJL 639:L17
Harris WE (1996) AJ 112:1487
Helmi A, White SDM (1999) MNRAS 307:495
Ibata RA, Gilmore G, Irwin MJ (1995) MNRAS 277:781
Ibata RA, Wyse RFG, Gilmore G, Irwin MJ, Suntzeff NB (1997) AJ 113:634
Jiang I-G, Binney J (2000) MNRAS 314:468

Jin S, Lynden-Bell D (2007) MNRAS 378:L64
Johnston KV (1998) ApJ 495:297
Johnston KV, Hernquist L, Bolte M (1996) ApJ 465:278
Johnston KV, Sackett PD, Bullock JS (2001) ApJ 557:137
Johnston KV, Spergel DN, Hernquist L (1995) ApJ 451:598
Johnston KV, Zhao H, Spergel DN, Hernquist L (1999) ApJL 512:L109
Jurić M et al (2008) ApJ 673:864
King IR (1966) AJ 71:64
Koposov SE et al (2012) ApJ 750:80
Koposov SE, Rix H-W, Hogg DW (2010) ApJ 712:260
Law DR, Johnston KV, Majewski SR (2005) ApJ 619:807
Majewski SR, Skrutskie MF, Weinberg MD, Ostheimer JC (2003) ApJ 599:1082
McGlynn TA (1990) ApJ 348:515
McMillan PJ (2011) MNRAS 414:2446
McMillan PJ, Binney JJ (2008) MNRAS 390:429
Nelder JA, Mead R (1965) Comput J 7:308
Newberg HJ, Willett BA, Yanny B, Xu Y (2010) ApJ 711:32
Newberg HJ, Yanny B, Willett BA (2009) ApJL 700:L61
Odenkirchen M et al (2001) ApJL 548:L165
Peñarrubia J, Belokurov V, Evans NW, Martínez-Delgado D, Gilmore G, Irwin M, Niederste-Ostholt M, Zucker DB (2010) MNRAS 408:L26
Piffl T. et al.: 2014, ArXiv e-prints
Pryor C, Piatek S, Olszewski EW (2010) AJ 139:839
Rockosi CM et al (2002) AJ 124:349
Sanders J (2012) MNRAS 426:128
Sanders JL, Binney J (2013) MNRAS 433:1813
Siegel MH et al (2007) ApJL 667:L57
Teuben P (1995) In: Shaw RA, Payne HE, Hayes JJE (eds) Astronomical data analysis software and systems IV, vol 77. Astronomical society of the pacific conference series, p 398
Tremaine S (1999) MNRAS 307:877
Willett BA, Newberg HJ, Zhang H, Yanny B, Beers TC (2009) ApJ 697:207
Williams AA, Evans NW, Bowden A: 2014, ArXiv e-prints
Williams MEK et al (2011) ApJ 728:102
Yanny B et al (2009) AJ 137:4377

Chapter 6
Stream Modelling in Angle-Frequency Space

The formation of tidal streams reflects the underlying Galactic potential, such that the resulting structures may be used to constrain the potential. One way of using streams to constrain the Galactic potential is to assume that a stream delineates an orbit. This assumption was discussed in Chap. 5, and the results indicate that an improvement over orbit-fitting is required, which accounts for the stream-orbit misalignment.

Several authors have constructed modelling approaches for constraining the Galactic potential using streams, without assuming the stream delineates an orbit. Johnston et al. (1999) accounted for the misalignment between the stream and progenitor orbit by calculating a progenitor-mass-dependent energy offset at pericentre. The observed stream stars are assigned an energy in this range, and the stream is integrated backwards for a Galactic lifetime. The quality of the trial potential is assessed by the number of 'captured' particles at time $t = 0$. This algorithm has been updated and applied to simulations by Price-Whelan and Johnston (2013) and set in a truly probabilistic framework by Price-Whelan et al. (2014). Varghese et al. (2011) developed a similar method that took into account the misalignment between the stream and progenitor orbit by correcting a proposed orbit track in real-space with a progenitor-mass dependent term. Similarly, Küpper et al. (2012) produced model streams by continuously releasing particles from the Lagrange points of an assumed cluster model. Bonaca et al. (2014) have applied this method to stream data taken from the Via Lactea II simulation to demonstrate the recovery of potential parameters from a non-analytic potential model.

Peñarrubia et al. (2012) present an entropy-based algorithm for constraining the Galactic potential with tidal streams. Assuming the distribution function is separable in energy and position, the best-fitting potential is the one that minimises the entropy of the energy distribution, or, equivalently, minimises the range of energies of the stream members. Sanderson et al. (2014) employ a similar algorithm, which seeks the potential that minimises the Kullback-Leibler divergence of the action distribution of a disrupted cluster. Both these methods use the expected clumping of a stream in integral-of-motion space, but neglect to use the phase information of the stream members. A stream is highly non-uniform in its phase coverage, and this fact is not used by these authors.

J. Sanders, *Dynamics of the Milky Way*, Springer Theses,
DOI 10.1007/978-3-319-18772-3_6

To fully model stream formation with limited assumptions, we must turn to N-body models. Law et al. (2005), Fellhauer et al. (2006) and Law and Majewski (2010) have all employed N-body models of the tidal disruption of the Sagittarius dwarf to produce constraints on the Galactic potential. However, full searches over the entire parameter space, both potential parameters and initial cluster conditions, are very expensive as we must create a new simulation each step, and it is difficult to assess how well a given simulation fits the data.

Angle-action coordinates are very useful quantities in galactic dynamics, and, in Chaps. 2, 3 and 4, we detailed several methods for their practical calculation in realistic Galactic potential. As we saw in Chap. 5, stream formation is very simply expressed in the angle-action formalism. The structure of a tidal stream in angle-action space will only be physically correct if we have used the correct potential, regardless of whether the stream delineates an orbit or not. This formalism provides us with a route to develop a new algorithm to constrain the Galactic potential: we search for the potential that produces the correct correlations in angle-action space. In Sect. 6.1, we recap the angle-action framework of stream formation, and more closely inspect an N-body simulation to motivate and understand this formalism. In Sect. 6.2, we discuss how we can improve on orbit-fitting by exploiting the known angle-action correlations for streams and arrive at a suitable algorithm to constrain the potential. In Sect. 6.3, we test the algorithm by attempting to recover the potential from the simulation.

Due to their large distances from the Sun, observations of tidal streams have large errors. It is therefore important that any stream-fitting algorithm can cope with appropriate observational errors. Some authors have tested their methods on simulated observational data with appropriate errors (Binney 2008; Eyre and Binney 2009). In Sect. 6.4, we conclude with a full discussion of the effect of observational errors on our proposed algorithm, and compare to the errors from current and future surveys.

At the end of the chapter (Sect. 6.6), we improve on our proposed algorithm by setting it in a probabilistic framework. We use insights from the angle-action structure of a stream when viewed in the correct potential to write down a probabilistic model expressed in the observable space. We demonstrate how this framework can be used to constrain the parameters of a simple two-parameter potential, and how this improved method can more capably deal with errors as well as the possibility of contaminants in our stream sample. Concurrent to this research, Bovy (2014) presented a machinery very similar to that shown here for constructing models of tidal streams. However, Bovy did not demonstrate the ability to infer potential parameters from a stream using the model.

The work in this chapter is based on that published in Sanders and Binney (2013) and Sanders (2014).

6.1 Streams in Angle-Action Space

In Chap. 5, we presented the angle-action formalism of stream formation. In this formalism, streams form due to stars being on different orbits. Importantly, this can lead to a misalignment, φ, between the stream and the underlying progenitor orbit given by

$$\varphi \equiv \arccos\left(\hat{\boldsymbol{\Omega}}_0 \cdot \hat{\boldsymbol{e}}_1\right), \tag{6.1}$$

where $\boldsymbol{\Omega}_0$ is the frequency vector of the progenitor orbit. $\hat{\boldsymbol{e}}_1$ is the principal eigenvector of the Hessian, $\mathbf{D}$, given by

$$D_{ij}(\boldsymbol{J}) = \frac{\partial^2 H}{\partial J_i \partial J_j}. \tag{6.2}$$

This misalignment only depends upon the progenitor orbit, and hence the potential, and not the progenitor mass. In Chap. 5, we demonstrated that a non-zero misalignment angle leads to biases in the estimation of potential parameters when fitting orbits to streams. We, therefore, require an alternative to orbit-fitting, and the angle-action formalism provides us with a clear route.

For long narrow streams to form, the Hessian must be dominated by a single eigenvalue, λ_1, with corresponding eigenvector, $\hat{\boldsymbol{e}}_1$. Under this approximation, both the angle and frequency differences of all the stars in the stream lie along the same straight line:

$$\frac{\Delta\boldsymbol{\theta}}{t} \approx \Delta\boldsymbol{\Omega} \approx \hat{\boldsymbol{e}}_1(\lambda_1\hat{\boldsymbol{e}}_1 \cdot \Delta\boldsymbol{J}). \tag{6.3}$$

In the correct potential, this correlation between the angle and frequency-space structure of the stream should be apparent. Inspecting the stream in angle-action space for a trial potential should tell us whether this potential is the true potential.

Throughout this chapter we use the Stäckel-fitting method of Chap. 2 to find the angles, actions and frequencies. In Appendix G we discuss adjustments to this algorithm appropriate to the problem at hand.

6.1.1 A Simulation

In Chap. 5, we introduced a range of N-body simulations to show that the angle-action formalism is mass-independent for all interesting progenitor mass scales. Before discussing the details of the proposed algorithm, we briefly reintroduce one of these N-body stream simulations to motivate the discussion. We take the lowest mass stream from Chap. 5, which was produced as follows: we construct a stream by evolving an N-body simulation of a King cluster on a stream-like orbit in the logarithmic potential using the code GYRFALCON (Dehnen 2000, 2002). The simple two-parameter logarithmic potential is given by

$$\Phi(R, z) = \frac{V_c^2}{2} \ln\left(R^2 + \frac{z^2}{q^2}\right), \tag{6.4}$$

where V_c is the asymptotic circular speed and q is the flattening parameter. We choose $V_c = 220\,\mathrm{km\,s^{-1}}$ and $q = 0.9$ which gives a good representation of the potential of the Milky Way (Koposov et al. 2010). We set $W_0 = 2.0$ and $M_c = 2 \times 10^4 M_\odot$ and seed the cluster with $N = 10{,}000$ particles. Following Dehnen et al. (2004), we relate r_t to the pericentric radius of the orbit, r_p, via

$$r_t^3 \approx \frac{GM_c}{V_c^2} r_p^2. \tag{6.5}$$

We place the cluster at the apocentre of the orbit shown in Fig. 6.1, which has initial conditions $(R, z) = (26.0, 0.0)$ kpc and $(U, V, W) = (0.0, 141.8, 83.1)\,\mathrm{km\,s^{-1}}$.

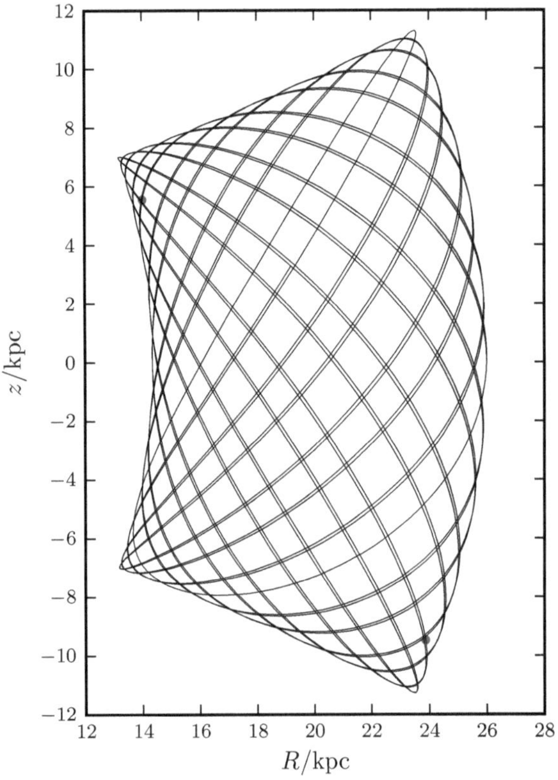

Fig. 6.1 Progenitor orbit used for the simulation in Sect. 6.3. The positions of the cluster at $t = 4.02$ Gyr and $t = 4.27$ Gyr are marked by *red dots*

Positive U is towards the Galactic centre and positive V is in the direction of the Galactic rotation at the Sun. This orbit has $r_p \approx 14$ kpc, and was chosen due to its similarity to the GD-1 orbit found by Koposov et al. (2010). We evolve the simulation for $t = 5$ Gyr (approximately 12 radial oscillations of the progenitor). The chosen cluster model parameters along with quantities derived from these parameters are given in the first row of Table 5.2.

Figure 6.2 shows the time evolution of the angle differences between the progenitor and several particles in the simulation. Each particle oscillates in the cluster until it escapes at pericentre, after which it moves as a free particle in the Galactic potential. The slope of the particle's motion in angle-space is given by the frequency. We see that the first particles to leave have a higher frequency than those released at later times. This is because the first particles to leave require a higher energy to escape the cluster. One unexpected feature of Fig. 6.2 is the noise in the angle calculation when the particles are moving freely. There are small bumps in the angle difference at each orbit pericentre, which alternate in sign and increase in magnitude with time. This is due to errors introduced by the Stäckel-fitting algorithm for estimating the angles and frequencies, which we discuss in Appendix G.

In the right panel of Fig. 6.3 we show the stream in real-space at $t = 4.27$ Gyr (just after the 11th pericentric passage). Figure 6.4 shows the corresponding angle-space and frequency-space structures calculated using the correct potential. We see that, as predicted by Eq. (6.3), the stream stars all lie along a straight line in both angle and frequency-space. In Fig. 6.4 we see that the spread in frequencies in the cluster is much larger than the spread in the estimated frequencies along the progenitor orbit (given by

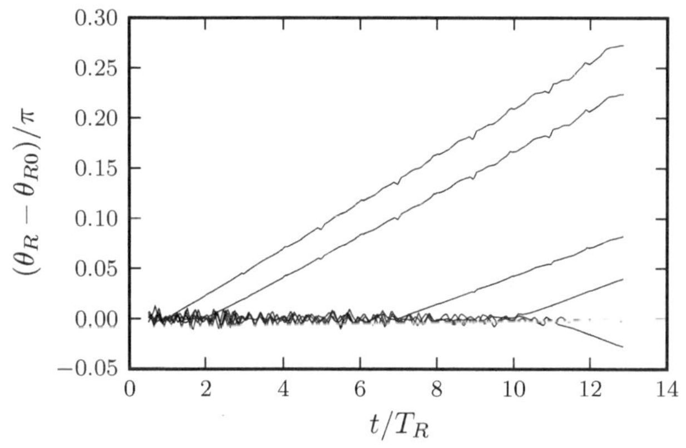

Fig. 6.2 Difference between the radial angle of the progenitor and six particles selected from the simulation. At early times, the particles oscillate inside the cluster until they are released at pericentre (given by the units on the x-axis). Particles are stripped symmetrically in angle space. After release, the particles orbit freely in the external Galactic potential. The small blips in the plot are due to numerical errors introduced by the Stäckel-fitting algorithm. The particles that are released first have larger frequency differences (i.e. steeper slopes in the plot) than those released at later times

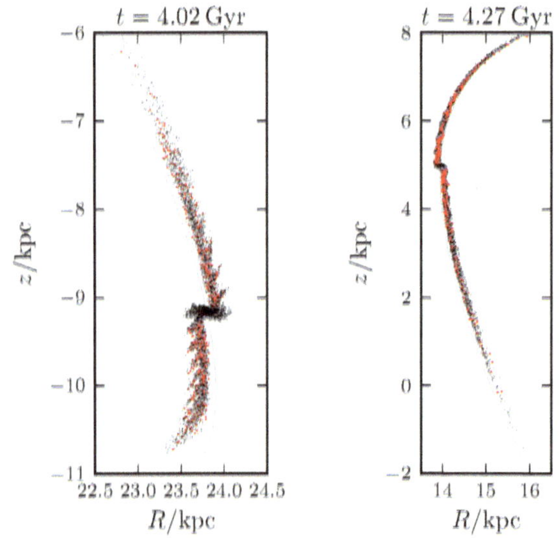

Fig. 6.3 Cluster at $t = 4.02$ Gyr (just after 10th apocentric passage) and $t = 4.27$ Gyr (just after subsequent pericentric passage). The *solid line* is the orbit of the progenitor and the *red dots* are the sample of 500 stars used to constrain the potential in Sect. 6.3

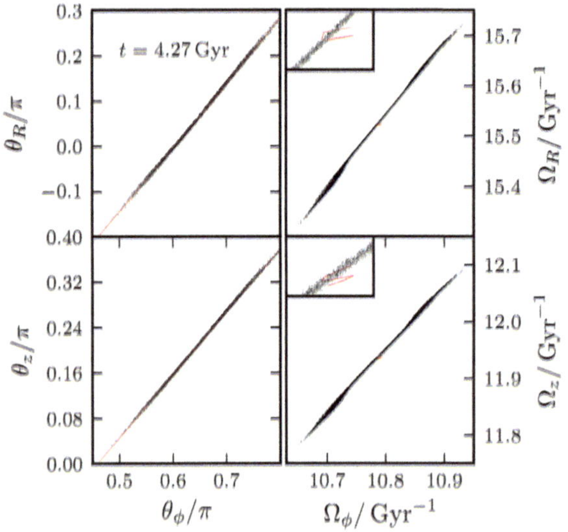

Fig. 6.4 Cluster at $t = 4.27$ Gyr (just after the 11th pericentre passage) in angle and frequency space. The *red line* is the projection of the progenitor orbit shown in Fig. 6.3 into angle space and frequency space. The angles and frequencies were calculated using the true potential. The inset plots show a zoom-in of the frequency space, so the orbit projections are more easily visible. In frequency space, errors in the determination of frequencies cause the *red points* to form a line rather than all coincide as they should

the red line). This suggests that the Stäckel-fitting algorithm is sufficiently accurate. However, the broadening at the extremes of frequency distributions indicates that there is a systematic error present that will skew the gradient estimation. A fuller discussion of the errors is given later.

6.2 Algorithm

The N-body stream observations now provide us with a way to utilise the angle-action formalism to constrain the potential. Equation (6.3) states that the frequency and angle differences of all the stars in the stream must lie along $\hat{\boldsymbol{e}}_1$, and this is observed in the above N-body simulation (Fig. 6.4). For all the stars, we can calculate the angles and frequencies and then obtain independent estimates $\hat{\boldsymbol{e}}_1^\theta$ and $\hat{\boldsymbol{e}}_1^\Omega$ of $\hat{\boldsymbol{e}}_1$ by performing linear fits to the angles and frequencies. The potential closest to the Galaxy's potential is then the potential that maximises $\hat{\boldsymbol{e}}_1^\theta \cdot \hat{\boldsymbol{e}}_1^\Omega$. The present kinematics are given by $\hat{\boldsymbol{e}}_1^\Omega$—it gives us the direction that stars will move in if the true potential suddenly changed to the trial potential. However, $\hat{\boldsymbol{e}}_1^\theta$ gives a measure of the positions the stars have reached by moving in the true potential i.e. the history of the stars. In the true potential, the current motions of stars (along $\hat{\boldsymbol{e}}_1^\Omega$) will match the positions they have in fact reached (along $\hat{\boldsymbol{e}}_1^\theta$).

Equation (6.3) was derived on the assumption that the stream structure is isotropic in action-space. In Chap. 5, we saw, in general, there are anisotropies in the action-space distribution, which cause a deviation of the frequency structure from the principal eigenvector of the Hessian. For long thin streams, the Hessian must have a large principal-to-second-eigenvalue ratio. Therefore, we expect that, whatever the shape of the action-space distribution, the resulting frequency-distribution will be highly elongated, although not necessarily along the principal eigenvector of the Hessian. The approach suggested here is comparing the angle and frequency distributions, which should have the same principal axis in the correct potential. Therefore, the approach should be insensitive to anisotropies in action-space if we are observing long thin streams.

This approach has several clear advantages:

1. We haven't assumed that the stars in the stream delineate an orbit.
2. The position of the progenitor is irrelevant. It reduces to a constant in the linear fits.
3. We can use any subsection of the stream.
4. The time since each star was stripped does not matter. Changing the assumed stripping time only moves the progenitor angles up and down the straight line in angle space and so does not affect the linear fit.
5. Only a single calculation is required for each trial potential, whereas when orbits are fitted in real space, orbits must be generated in each potential until the best fit is found.
6. The quality of the potential is determined solely by the degree of alignment, so it is an unambiguous single measure of the fit. When fitting orbits, the quality of the

Table 6.1 Algorithm to find the best-fitting potential given stream data

Given a set of 6D phase-space points $\boldsymbol{x}_i$, we

1. pick a trial potential Φ_{trial},
2. calculate the angles $\boldsymbol{\theta}_i(\boldsymbol{x}_i, \Phi_{\text{trial}})$ and frequencies $\boldsymbol{\Omega}_i(\boldsymbol{x}_i, \Phi_{\text{trial}})$ using the local Stäckel-fitting algorithm,
3. fit four straight lines to the graphs of
   1. Ω_ϕ against Ω_R,
   2. Ω_ϕ against Ω_z,
   3. θ_ϕ against θ_R,
   4. and θ_ϕ against θ_z

   to find the gradients denoted by a_1, a_2, b_1, b_2,
4. calculate the quantity
$$\cos\psi \equiv \frac{a_1 b_1 + a_2 b_2 + 1}{\sqrt{(a_1^2 + a_2^2 + 1)(b_1^2 + b_2^2 + 1)}},$$
5. pick a new trial potential and repeat until the angle, ψ, between the vectors is minimised.

potential is evaluated by measuring the proximity of the orbit found to the data in the 6D phase space, so an arbitrary metric must be introduced to relate position and velocity differences.

However, our approach suffers from the disadvantages that

1. the angles and frequencies can only be found with full six-dimensional phase-space information,
2. stream data are often poor and it is not clear how the errors affect the accuracy of the method,
3. and unlike orbit fitting, we must process the data before assessing the quality of the potential, as opposed to directly checking the quality of the fits in observable space.

In what follows, we will assume that we have full phase-space information for all the stars in the stream, so we are not concerned with the first of these points. In Sect. 6.4, we explore how errors in the 6D data for the members of the stream affects the accuracy of the method. The last point is addressed in the final part of this chapter, when we construct a fully probabilistic model in observable space. We summarise the above discussion into the simple algorithm presented in Table 6.1.

6.3 Test

We test the algorithm of Table 6.1 by seeing how accurately the method recovers the chosen parameters of the potential from the N-body simulation. We examine the stream at two times—once at $t = 4.02\,\text{Gyr}$, just after the tenth apocentric passage of the progenitor, and once at $t = 4.27$ Gyr, just after the subsequent pericentric passage. From these snapshots, we form samples of stream stars by first removing the remnant of the progenitor with a spatial cut and then randomly sampling 500 of the remaining ~5000 stars. The real-space structures of the disrupted cluster and these samples are shown in Fig. 6.3.

We now use the two samples to diagnose the potential. For each trial pair of (V_c, q), we find the misalignment, ψ, between the directions of the angle difference vector and the frequency difference vector. We pick trial pairs of parameters by exploring a regular grid around the true point ($V_c = 220\,\mathrm{km\,s^{-1}}$, $q = 0.9$). Figure 6.5 is a plot of φ in the plane of the parameters.

In the apocentric case (left panel), the global minimum occurs at $V_c = 220.3\,\mathrm{km\,s^{-1}}$, $q = 0.899$. The landscape around the minimum is quite complex. There are a series of local minima for $q > 1.0$. Therefore, care will be needed when searching for the minimum automatically. In the pericentric case (right panel), we do not recover the true potential parameters as successfully: the global minimum lies at $V_c = 220.4\,\mathrm{km\,s^{-1}}$, $q = 0.915$ and the ψ landscape is significantly more complex. Again, for $q > 1.0$ there are a series of local minima. Worryingly, there is another deep local minimum at $V_c \approx 212\,\mathrm{km\,s^{-1}}$, $q \approx 0.74$. Clearly, there are sources of error in the algorithm that are systematically shifting the best-fitting value of q in the pericentric case. Before applying this algorithm to more realistic data sets, we must understand the cause of this systematic shift. More generally, we need a method to assess the magnitude of the systematic error in the algorithm when the underlying potential parameters are not known. In the following section, we discuss the causes of the error and present a method to estimate its magnitude.

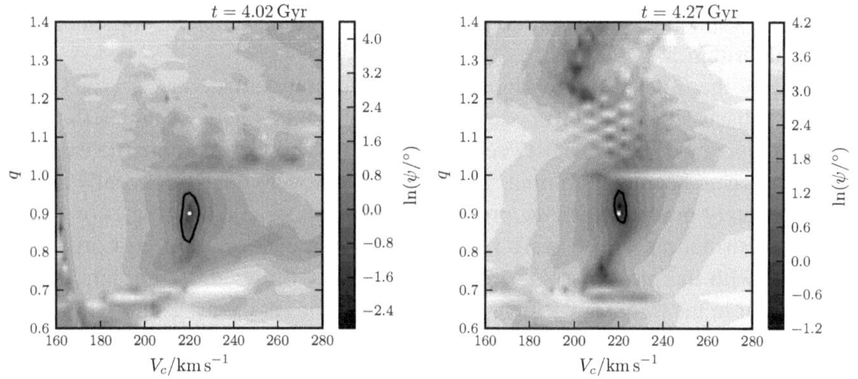

Fig. 6.5 Misalignment, φ, between the angle difference and frequency difference vectors. The results at $t = 4.02$ Gyr (just after the 10th apocentric passage) are shown in the *left panel* and at $t = 4.27$ Gyr (just after the subsequent pericentric passage) in the *right panel*. The true potential parameters are $V_c = 220\,\mathrm{km\,s^{-1}}$, $q = 0.9$ marked by a *white dot*. The *thick black lines* mark an error contour showing the uncertainty in the position of the found minimum

6.3.1 Error Estimation

Even in the correct potential with perfect data, the particles will not lie on exact straight lines in angle and frequency space on account of systematic errors introduced by the approximations used including the Stäckel-fitting algorithm. In the above test, we have seen that this can lead to a slight misdiagnosis of the underlying potential parameters. When we include observational errors, the error ellipse obtained by resampling the random errors may not encompass the true parameters due to these systematic errors. We need to quantify the accuracy of the method even when perfect data are used. A naive error may be taken from the error in the linear regression. However, this error only takes account of the random errors and ignores any systematics. In Chap. 5, we discussed the effects of the progenitor mass on the angle-action framework, and hence the algorithm presented in this chapter. It was shown that the approach is mass-independent up to about $10^9 M_\odot$. Therefore, we expect the self-gravity, the finite cluster size and the linear approximation (see Eq. (6.3)) to not affect the algorithm.

There are still two sources of systematic error in the method. The first is that the action-space structure is not perfectly isotropic. In Chap. 5, we saw that particles are not isotropically stripped, but rather escape through the Lagrange points. This led to the entire stream structure being rotated in the angle and frequency space. Eyre and Binney (2011) showed that the action-space of a tidally-disrupted cluster has a 'bow-tie' structure. When this action-space structure is transformed into frequency space via the Hessian (see Eq. (6.3)), the broadness of the wings of the action-space distribution is still apparent at the extremes of the frequency structure. This is due to the smaller eigenvalues of the Hessian. This effect is small as we are considering a Hessian with a ratio of eigenvalues, $\lambda_1/\lambda_2 \approx 30$. However, this effect should also be observed in the angle-space structure. As we are comparing the angle and frequency structures, this shouldn't have a significant impact on the implementation of the algorithm.

Second, the Stäckel-fitting algorithm introduces systematic errors (see Chap. 2 for more discussion). The errors depend upon the orbital phase, such that the errors in the frequency differences between the progenitor and the stream particles exhibit a characteristic beating with time. The particles that were released recently are still approximately in phase with the progenitor, whilst those at the extremes of the frequency distribution, which were released many pericentres ago, have drifted out of phase with the progenitor. Therefore, the Stäckel-fitting algorithm introduces a larger error in the frequency difference for the particles at the extremes, which leads to a broadening of the distribution at the extremes. In Appendix G.2, we give a full discussion.

In conclusion, the extremes of the frequency distribution are broader, and so less reliable, than the central portion of the stream, due to systematic errors introduced by the Stäckel-fitting algorithm and anisotropies in the action-space distribution. Therefore, we estimate the magnitude of the error in the parameters from the spread in gradients that can be obtained by considering the angle and frequency structure of different parts of the stream in the best-fitting potential. Specifically, we take the error in the gradient to be the difference between this gradient and that calculated using the central particles nearest to the progenitor. We may then find a 'threshold angle', ψ_T, as the angle between the two particle distributions. A threshold angle may be found for both the angle- and frequency-space distributions. However, in the simulation, the frequency distribution is the main cause of error so we only calculate a threshold angle for this distribution. This threshold angle gives an estimate of the minimum angle that we can reliably calculate. The range of valid parameters are those for which $\psi < \psi_T$, which define an error contour in the potential parameter space. Inside the error contour, the angle between the frequency and angle distributions is smaller than the width of the frequency distribution.

This also helps to solve an additional problem with the algorithm. As we are blindly fitting straight lines to distributions which may not have a linear structure, there may be some data sets for which the best-fitting straight lines fortuitously give a minimum in $\cos\psi$. However, if we calculate an error contour associated with this minimum, we expect the error to be very large. Therefore, we should be able to rule out this fortuitous minimum without explicit inspection.

Applying this error estimation method to the data in Fig. 6.4, we plot an error contour in the parameter plane, given by the thick black curve in Fig. 6.5. Assuming the errors in each parameter estimate are independent, we estimate the parameters in the apocentric case as $V_c = (220 \pm 4)\,\mathrm{km\,s^{-1}}$, $q = (0.90 \pm 0.07)$ in good agreement with the truth. Similarly, for the pericentric case, the parameter estimates are $V_c = (220 \pm 3)\,\mathrm{km\,s^{-1}}$, $q = (0.92 \pm 0.04)$.

We have shown that we can use our algorithm to constrain the parameters of a simple potential using an error-free sample of 500 stars from a stream simulation. When the stream is observed at either apocentre or pericentre, the error contour is elongated along the q-direction. The relative error in the circular speed of the potential is smaller than the relative error in the shape of the potential. The recovery of the parameters at apocentre is slightly more successful than at pericentre. However, the errors at apocentre are slightly larger.

6.4 Errors in Stream Data

We have assumed in the above test that the input data are perfect. Obviously this check that the algorithm works in the most optimistic situation is necessary, but it does not give an indication of how the algorithm will perform on a real data set. To give a more realistic test, we now add errors to the data and rerun the algorithm. We simulate an observation of the particles at $t = 4.27\,\mathrm{Gyr}$ in the simulation from the position of the Sun $(R, z) = (8.0, 0.0)\,\mathrm{kpc}$, which has velocity $(U, V, W) = (11.1, 232.4, 7.25)\,\mathrm{km\,s^{-1}}$ (Schönrich et al. 2010). There is considerable uncertainty in the circular speed at the Sun (see Bovy et al. 2012 for a summary), but for a fixed solar position, the solar V is well constrained by the motion of Sgr A* (Reid and Brunthaler 2004). Therefore, in what follows, we fix the velocity of the Sun, irrespective of the choice of potential as this more accurately simulates a realistic application of the above algorithm.

The pericentric snapshot is chosen as the GD-1 stream is currently around pericentre. At pericentre, the stellar density in the stream is increased, so the overdensity is more likely to be observed. However, this effect is counteracted by the short time a stream spends around pericentre. We project the positions and velocities of the particles relative to the Sun into observable space on the sky—sky position (l, b), distance s, proper motions (μ_l, μ_b) and line-of-sight velocity $v_{||}$. The particles are then scattered in observable space by appropriate Gaussian errors $(\sigma_{l,b}, \sigma_s, \sigma_\mu, \sigma_{||})$ to form an 'observed' data set.

The most accurate data are obtained from streams that are closest to the Sun. For instance, GD-1 lies between 8 and 12 kpc away from the Sun. Therefore, when we perform the above data scattering we place the Sun as close to the centre of the stream as possible. This involves rotating the stream around the Galaxy until the centre of the stream lies at the same azimuthal angle as the Sun. This better simulates observations that could feasibly be performed.

We make 'observations' with the errors listed in Table 6.2, along with errors in current and future data. The data from the O2 observation are shown in Fig. 6.6. We now use each of these data sets to constrain the potential. We transform the data set back into the Galactocentric coordinate system, and determine the angle and frequency distributions and ψ for each trial potential. We use the Nelder Mead algorithm (Nelder and Mead 1965) to find the maximum of $\cos\psi$ as a function of q and V_c. We restrict the range of parameter space explored[1] to $0.7 < q < 1.1$ and $170\,\mathrm{km\,s^{-1}} < V_c < 270\,\mathrm{km\,s^{-1}}$. The initial point passed to the algorithm is the true minimum position. We estimate the error in the position of the minimum for each set of observational errors by repeating this process 100 times and finding the minimum of each data set. These points sample the distribution of the parameters given the observational errors. Using these points, we can reconstruct the distribution and estimate the mean and error of the distribution. We have seen in the previous section that the systematic error estimates around pericentre are $\Delta V_c \sim 3\,\mathrm{km\,s^{-1}}$ and $\Delta q \sim 0.04$. We sum these in quadrature with the estimated impact of observational error on the resulting parameter estimates.

We first explore the effects of including an observational error to each observable independently (O1). The current position errors $\sigma_{l,b} = 100$ mas do not produce a significant change to the position of the minimum found using perfect data. Therefore, for all other observations we use a more optimistic position error of 10 mas. A distance error of $\sim$5 % (O1b) does not alter the estimate of the circular speed significantly but slightly decreases the estimate of q, such that it is not consistent with the truth within the quoted error. A proper-motion error of 0.21 mas yr^{-1} (which corresponds to a transverse velocity error of $4.7\,\mathrm{km\,s^{-1}}(\mathrm{kpc\,mas\,yr^{-1}})^{-1} \times 0.21\,\mathrm{mas\,yr^{-1}} \times 10\,\mathrm{kpc} \approx 10\,\mathrm{km\,s^{-1}}$) applied independently produces a result that is consistent with the truth

[1]Note that, as discussed in Evans (1993), the logarithmic potential produces negative densities for $q < 1/\sqrt{2}$ or $q > 1.08$. Therefore, for a realistic application, we would want to restrict the parameters to ensure we only explore physical models.

Table 6.2 Observational errors used to make observations of the stream near pericentre at $t = 4.27$ Gyr, along with the estimates of the potential parameters (V_c, q) using the stream-fitting algorithm outlined in this chapter

	$\sigma_{l,b}$/ mas	σ_s/s	σ_μ/ mas yr^{-1}	$\sigma_{\|\|}$/ km s^{-1}	V_c/ km s^{-1}	q	$\frac{\Delta V_c}{\sigma_{V_c}}$	$\frac{\Delta q}{\sigma_q}$
O1a	100	0.0	0.0	0	220 ± 3	0.92 ± 0.04	0.1	0.4
O1b	0	0.04	0.0	0	220 ± 3	0.85 ± 0.04	0.1	1.1
O1c	0	0.0	0.21	0	221 ± 3	0.88 ± 0.04	0.2	0.4
O1d	0	0.0	0.0	10	222 ± 3	0.83 ± 0.04	0.7	1.6
O2	10	0.04	0.21	10	224 ± 4	0.87 ± 0.04	1.2	0.7
O3	10	0.04	0.42	10	228 ± 4	0.94 ± 0.05	1.9	0.9
O4	10	0.08	0.42	10	231 ± 5	0.92 ± 0.05	2.3	0.4
O5	10	0.04	0.21	1	221 ± 3	0.88 ± 0.04	0.4	0.5
O6	10	0.16	0.1	1	261 ± 8	1.05 ± 0.08	5.3	1.9
OA1	10	0.04	0.21	10	224 ± 4	0.9 ± 0.04	1.1	0.1
OA2	10	0.08	0.42	10	224 ± 4	0.9 ± 0.05	0.9	0.0
OA3	10	0.2	1.2	10	223 ± 10	0.91 ± 0.09	0.3	0.1
O4M	10	0.08	0.42	10	225 ± 4	0.91 ± 0.05	1.5	0.2
OA3M	10	0.2	1.2	10	223 ± 6	0.9 ± 0.07	0.5	0.0
SEGUE	50–100	0.15–0.2	3	10				
RAVE	$\lesssim 100$	0.133	$\lesssim 8$	<2				
Pan-STARRS	~ 10	15 (s/kpc)	1.2	–				

(continued)

Table 6.2 (continued)

	$\sigma_{l,b}$/ mas	σ_s/s	σ_μ/ mas yr^{-1}	$\sigma_{\|\|}$/ km s^{-1}	V_c/ km s^{-1}	q	$\frac{\Delta V_c}{\sigma_{V_c}}$	$\frac{\Delta q}{\sigma_q}$
LEGUE	$\lesssim$100	–	–	1–7				
LSST	15	0.8 (s/kpc)	0.3	–				
Gaia	0.14–0.44	<0.5 (s/kpc)	0.12–0.38	~1–5				

$\sigma_{l,b}$ is the observational error in the sky position, σ_s/s the fractional distance error, σ_μ the proper motion error, and $\sigma_{||}$ the line-of-sight velocity error. The O# observations use all 500 stars in the stream sample to determine the potential parameters whilst the OA# observations bin the observational data in 10 bins on the sky and average before performing the algorithm. The O4M and OA3M observations use a simulation which uses a progenitor cluster which is ten times more massive, but uses the same observational errors as O4 and OA3 respectively. The final two columns give the deviation in standard deviations between the found parameters and the true parameters. At the bottom of the table, we show current and future observational errors for comparison. SEGUE data taken from Pier et al. (2003), Jurić et al. (2008), Munn et al. (2004) and Yanny et al. (2009). RAVE data taken from Siebert et al. (2011) and Burnett et al. (2011) (RAVE uses position data from 2MASS, Skrutskie et al. 2006). Pan-STARRS data taken from Kaiser et al. (2002) and Magnier et al. (2008). LSST data taken from Ivezić et al. (2008) for an $r = 22$ star. LEGUE (LAMOST Experiment for Galactic Understanding and Exploration) data taken from Deng et al. (2012) (LAMOST uses SDSS data as an input catalogue). *Gaia* data taken from Perryman et al. (2001) for $G = 20$–21 mag stars. The line-of-sight velocity errors from *Gaia* are $\approx$5 km s^{-1} for cool stars. The Gaia-ESO survey (Gilmore et al. 2012) aims to record spectra of $\geq 10^5$ *Gaia* stars, which will yield line-of-sight velocity errors of $\approx$1 km s^{-1} for cool stars. The fractional error in the distance for Pan-STARRS, LSST and *Gaia* are determined from the error in the parallax under a linear approximation

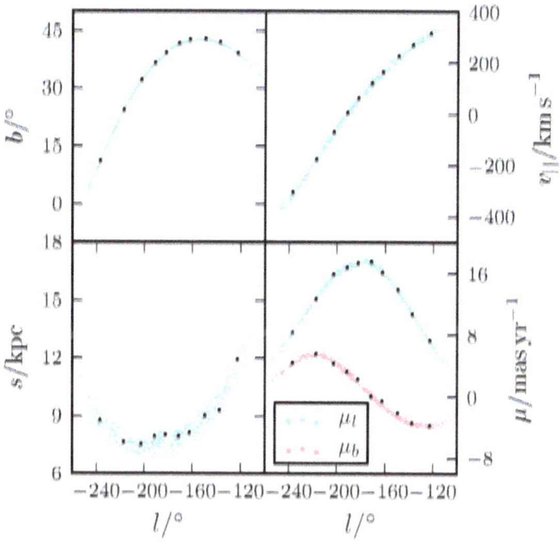

Fig. 6.6 Stream sample at $t = 4.27$ Gyr (just after 11th pericentric passage). The small *blue* and *red points* show the sample observed from the Sun with observational errors O2 given in Table 6.2, and the larger *black points* show the reduced data set produced by averaging the data in observable space on the sky with errors OA3

(O1c). An equivalent line-of-sight velocity error (O1d) shifts the minimum in the circular speed up slightly, and moves the minimum in q down significantly. The effects of errors in these three coordinates are different and probably reflect the observation geometry. We are observing at pericentre at the same azimuthal angle as the centre of the stream. Therefore, the line-of-sight velocity around the centre of the stream (the densest part) will be very small so small errors can have a significant effect. The effect of observational errors on the Galactocentric Cartesian coordinates is also reasonably complex, particularly for the distance error which affects both the observed position and velocity of a star.

When these three errors are combined (O2), the circular speed is shifted upwards and q is recovered within the errors. Increasing the error in the proper motion (O3) shifts the circular speed upwards further and also systematically shifts the minimum in q upwards. An increase in the distance error (O4) again shifts the circular speed estimate upwards but the recovery of q is good. O5 shows a combination of errors that recovers the underlying potential parameters. These errors are perhaps too optimistic and unlikely to be achieved in the near future. The line of sight velocity error may be achieved by RAVE, whilst the proper motion error may be achieved by *Gaia*. However, it is unlikely that such a small distance error will be achieved soon from either parallax measurements, or spectroscopic or photometric distances. However, if one were able to identify standard candles, such as RR Lyrae stars, in a stream, this distance error is achievable now (Drake et al. 2013). With a low-mass stream, such as GD-1, we expect very few RR Lyrae stars, so using this method may only be useful for more massive streams. Also, we require 500 observations of this quality from a tidal stream to replicate this test, which makes such a data set even more improbable. O6 shows an example of large errors. Both the circular speed and the q estimate are shifted to very large values. However, these results do not give a good indication of the parameter estimates with these errors as we are approaching the edge of the allowed range of parameters. We can still see that large errors degrade the landscape significantly, so that no minimum is found within the appropriate region.

The effect of experimental errors on the position of the minimum is clearly very complex and to use this method with confidence one would need to investigate the systematics on an appropriate simulation before application to data. We see that, when large errors are added, the minimum can shift away from the true parameters significantly with very small associated formal errors. The error in the location of these minima does not encompass the true parameter values. In conclusion, it seems that current and near-future data could lead to limited use of this method: to use the method with confidence we require accuracy in the observables that may not be attained for many tidal streams for some while.

6.5 Data Averaging in Observable Space

Despite the results of the previous section, the method could still prove relevant if slightly adjusted. The motion of the stars in a stream in real space is coherent, and the above algorithm does not utilise this fact. This additional information may be exploited by first binning in real space. The stream is essentially a one dimensional structure regardless of whether it delineates an orbit or not. Therefore, binning along the stream is valid even when the stream does not delineate the orbit. We form bins along the stream and estimate the observables at the centre of the bin by the average of the observables of the stars in the bin. One of the advantages of the above routine is that the linear regression in angle space and frequency space beats down the random errors in the estimated angles and frequencies. Averaging the data in observable space on the sky provides an alternative method for reducing the random errors, which, for large observational errors, is probably the preferable technique. Indeed, when analysing the GD-1 stream Koposov et al. (2010) mitigate the problem of large errors by binning and averaging the data along the stream in observable space to obtain a handful of much more accurate data points—the proper-motion errors for instance can be reduced by a factor of five.

When large observational errors are added, particles can be scattered to very high actions. Chapter 2 showed that the systematic errors in the angle-action variables increases with the actions. Therefore, larger observational errors will also introduce larger systematic errors. In the extreme case, observational errors can make orbits unbound.[2] These orbits have infinite actions and must be ignored when performing the linear regression in angle and frequency space. Averaging data in observable space should remove these issues as we will have a data set which spans a smaller range in actions.

[2] A logarithmic potential does not tend to an asymptotic limit at large R or z, so technically every orbit in a logarithmic potential is bound. However, in practice, orbits that reach very large R and z take an extremely long time to integrate. Therefore, we consider all orbits which stray more than 200 kpc from the Galactic centre to be 'unbound'.

Unfortunately, averaging in observable space results in a smaller number of data points with which to perform the linear regression. In the above examples, we have used 500 data points which provides a very tight linear regression. However, even just a handful of data points should provide a good estimate of the gradient.

Here we explore the results of observational data averaging for three observations listed in Table 6.2. We bin the 'observed' data in each observational coordinate using ten equally populated bins in Galactic longitude, l. The bins are sufficiently small for the stream to be approximately linear in each subspace for each bin so averaging within a bin works well. This procedure reduces the error in the input coordinates by a factor of $\sim 1/\sqrt{50} \approx 0.14$. After transforming back to the Galactocentric Cartesian coordinates, we perform the algorithm on the averaged data to find the best combination of potential parameters. We then repeat the scattering and binning process 100 times to estimate the error. In Fig. 6.6 we plot the binned data for the observation OA3.

We can see from Table 6.2 that averaging in observable space produces superior results. First, the systematic shifts in the minima are removed, and we recover an average for the parameters which is close to the truth, particularly for q. The average position of the minimum does not depend as sensitively on the magnitude of the observational errors (compare OA1 and OA2). If we compare O4 and OA2, the observation errors are the same, but the recovery of the parameter estimates with the two methods are different: the circular speed is better recovered by averaging in observable space. This is achieved without an increase in the quoted error of the parameter estimates. Second, we can weather much larger observational errors. The errors in observation OA3 seem achievable with future data and we correctly recover the potential parameters using the averaging method. For all three observations, the estimated errors encompass the true potential parameters.

In conclusion, we can achieve more accurate results by first averaging in observable space and then performing the algorithm on the reduced data set. This represents an improvement over the previously presented method as we remove unphysical measurements (those with infinite actions) and hence we only spend time calculating the angles and frequencies for reliable phase-space points. Similarly, we need only calculate the angles and frequencies—a time-consuming process—for a handful of points. Scattering and binning is, in comparison, very fast. These principles apply to any process trying to calculate actions, or any other non-trivial variable, from noisy data: binning the data in the space of observables removes outliers and results in fewer, more reliable points to run through the complicated calculation.

Finally, we briefly investigate how the recovery of the potential parameters depends upon the initial mass of the cluster. In Chap. 5, we showed that the stream-orbit misalignment is independent of the cluster mass. The effect of increasing the cluster mass is to scale the angle and frequency distributions without altering the morphology. We take the $M_c = 2 \times 10^5 M_\odot$ simulation from Chap. 5, which was evolved on the same orbit for the same time as the above simulation. Due to the higher mass, the resulting stream is longer, spanning $\sim 190°$ in Galactic longitude, so we expect that the estimates of the parameters will be superior as the stream probes a larger range in both angles and frequencies. In Table 6.2, we show the estimates of the parameters from this higher-mass simulation using the observational errors from O4 and OA3 with the observable-space binning (these are labelled in the table as O4M and OA3M). For O4M, the increased stream length has weakened the systematic shift of the circular velocity found in O4, producing a mean value which is closer to the truth. Despite the increased width of the stream, the error in the parameters is small. For OA3M, the recovery of parameters is similar to the OA3 observation, but the error in these estimates has been reduced without excluding the truth from the error ellipse. These two experiments show that longer streams, which have a larger spread in angle space, produce superior parameter estimates.

6.6 A Probabilistic Approach

The method presented above suffers several drawbacks: we were doing the inference in model space, not the data space; it was difficult to assess the errors in the obtained potential parameters; it was awkward to handle errors in the data, and the method did not behave well for large errors. This state of affairs leads us to analyse the data by constructing probabilistic models that correctly handle large errors, missing data, and contaminants. Such an approach is much more robust than the previous effort, and lends itself perfectly to being combined with other independent measurements of the Galactic potential. Here we present a probabilistic model for tidal streams that may be used to infer the properties of the Galactic potential. The model is expressed in the space of observables, but relies heavily on the expected structure of streams in angle-action space. In this section, we motivate our choice of model by considering an idealised case of a Gaussian structure in angle-action space evolving in time. In Sect. 6.7, we use these insights to write down a practical model for the stream, and discuss how it may be used to infer properties of the Galactic potential. In Sect. 6.8, we infer the parameters of a simple two-parameter potential from mock stream observations using our model. In Sect. 6.9, we discuss proposed improvements to the approach taken in this paper.

Given a set of observations, D, of N stars believed to be members of a stream, what can we infer about the Galactic potential? For star i, we have a 6D set of Galactic coordinates $\boldsymbol{L}_i = (l, b, s, v_{||}, \boldsymbol{\mu})$ with associated errors described by the covariance matrix $\mathbf{S}_i$. Note that we can fit any missing data into this formalism by taking the associated error with the data point to be infinite.

Given the data, we want to know the posterior distribution of the potential, Φ, given by $p(\Phi|D)$. From Bayes' theorem, we have

$$p(\Phi|D) = \frac{p(D|\Phi)p(\Phi)}{p(D)}, \tag{6.6}$$

where $p(\Phi)$ is the prior on the potential, and the evidence $p(D)$ is not important for the present exercise. We wish to evaluate the likelihood $p(D|\Phi)$.

The probability of the data given the potential is related to the properties of the stream progenitor, $\mathscr{C}$. $\mathscr{C}$ contains information about the current phase-space coordinates of the progenitor (i.e. the progenitor's actions, $\boldsymbol{J}_0$, frequencies, $\boldsymbol{\Omega}_0$, and angles, $\boldsymbol{\theta}_0$), as well as the size (and internal properties) of the progenitor. Therefore, we write

$$
\begin{aligned}
p(D|\Phi) &= \int \mathrm{d}\mathscr{C}\, p(\mathscr{C})p(D|\Phi,\mathscr{C}),\\
p(D|\Phi,\mathscr{C}) &= \prod_i^N p(\boldsymbol{L}_i|\Phi,\mathscr{C},\mathbf{S}_i),\\
p(\boldsymbol{L}_i|\Phi,\mathscr{C},\mathbf{S}_i) &= \int \mathrm{d}\boldsymbol{L}_i'\, p(\boldsymbol{L}_i|\boldsymbol{L}_i',\mathbf{S}_i)\det\Big(\frac{\partial(\boldsymbol{x}_i,\boldsymbol{v}_i)}{\partial\boldsymbol{L}_i'}\Big)p(\boldsymbol{x}_i,\boldsymbol{v}_i|\Phi,\mathscr{C}),
\end{aligned}
\tag{6.7}
$$

where

$$
p(\boldsymbol{L}_i|\boldsymbol{L}_i',\mathbf{S}_i) = \frac{1}{\sqrt{(2\pi)^k\det(\mathbf{S}_i)}}\exp\Big(-\frac{1}{2}(\boldsymbol{L}_i-\boldsymbol{L}_i')^T\mathbf{S}^{-1}(\boldsymbol{L}_i-\boldsymbol{L}_i')\Big), \tag{6.8}
$$

the Jacobian factor is given by

$$
\det\Big(\frac{\partial(\boldsymbol{x}_i,\boldsymbol{v}_i)}{\partial\boldsymbol{L}_i'}\Big) = s'^4\cos b', \tag{6.9}
$$

and the $(\boldsymbol{x},\boldsymbol{v})$ coordinates are related to the Galactic coordinates in the usual way.

We want to work with actions, angles and frequencies. As in the previous sections, we use the Stäckel-fitting method from Chap. 2. Therefore, we write

$$
p(\boldsymbol{x}_i,\boldsymbol{v}_i|\Phi,\mathscr{C}) = \det\Big(\frac{\partial(\boldsymbol{\Omega}_i,\boldsymbol{\theta}_i)}{\partial(\boldsymbol{x}_i,\boldsymbol{v}_i)}\Big)p(\boldsymbol{\theta}_i,\boldsymbol{\Omega}_i|\Phi,\mathscr{C}), \tag{6.10}
$$

where the angles and frequencies are related to $(\boldsymbol{x},\boldsymbol{v})$ via the potential Φ using the Stäckel-fitting approximation and the Jacobian is given by

$$
\begin{aligned}
\det\Big(\frac{\partial(\boldsymbol{\Omega}_i,\boldsymbol{\theta}_i)}{\partial(\boldsymbol{x}_i,\boldsymbol{v}_i)}\Big) &= \det\Big(\frac{\partial(\boldsymbol{\Omega}_i,\boldsymbol{\theta}_i)}{\partial(\boldsymbol{J}_i,\boldsymbol{\theta}_i)}\Big)\det\Big(\frac{\partial(\boldsymbol{J}_i,\boldsymbol{\theta}_i)}{\partial(\boldsymbol{x}_i,\boldsymbol{v}_i)}\Big)\\
&= \det\Big(\frac{\partial\boldsymbol{\Omega}_i}{\partial\boldsymbol{J}_i}\Big) = \det(\mathbf{D}_i).
\end{aligned}
\tag{6.11}
$$

We have used the fact that $(\boldsymbol{J},\boldsymbol{\theta})$ are canonical coordinates, such that the phase-space volume is conserved under the transformation, and introduced the Hessian matrix $\mathbf{D}$ defined as

$$
\mathbf{D} \equiv \frac{\partial^2 H}{\partial\boldsymbol{J}^2}. \tag{6.12}
$$

This matrix can be calculated analytically in a Stäckel potential, so we extend the Stäckel-fitting algorithm of Sanders (2012) to estimate $\mathbf{D}$. We give details of this in Appendix G. We proceed by splitting $p(\boldsymbol{\theta}_i, \boldsymbol{\Omega}_i|\Phi, \mathscr{C})$ into two components

$$p(\boldsymbol{\theta}_i, \boldsymbol{\Omega}_i|\Phi, \mathscr{C}) = p(\boldsymbol{\theta}_i|\boldsymbol{\Omega}_i, \Phi, \mathscr{C})p(\boldsymbol{\Omega}_i|\Phi, \mathscr{C}). \tag{6.13}$$

To proceed further, we must consider what we know about stream formation in angle-action space (Tremaine 1999). Assuming the spread in actions in the stream is small, for each star in the stream we have

$$\begin{aligned}\Delta\boldsymbol{\Omega}_i &= \boldsymbol{\Omega}_i - \boldsymbol{\Omega}_0 \approx \mathbf{D}_0 \cdot (\boldsymbol{J}_i - \boldsymbol{J}_0) = \mathbf{D}_0 \cdot \Delta\boldsymbol{J}_i \\ \Delta\boldsymbol{\theta}_i &= \boldsymbol{\theta}_i - \boldsymbol{\theta}_0 = t_i\Delta\boldsymbol{\Omega}_i + \Delta\boldsymbol{\theta}_i(0),\end{aligned} \tag{6.14}$$

where t_i is the time since the particle was stripped from the progenitor and $\Delta\boldsymbol{\theta}_i(0)$ is the separation between the ith particle and the progenitor when the particle is released. $\mathbf{D}_0$ is the Hessian from Eq. (6.12) evaluated at the progenitor actions, $\boldsymbol{J}_0$.

To motivate our choice of model, we begin by assuming that $\boldsymbol{J}_i$ follows an isotropic normal distribution such that

$$\begin{aligned}p(\boldsymbol{J}_i|\Phi, \mathscr{C}) &\approx p(\boldsymbol{J}_i|\mathscr{C}) \\ &= \frac{\sqrt{\det(\mathbf{A})}}{(2\pi)^{\frac{3}{2}}} \exp\Big(-\frac{1}{2}\Delta\boldsymbol{J}_i^T \cdot \mathbf{A} \cdot \Delta\boldsymbol{J}_i\Big) \\ &= \Big(\frac{a}{2\pi}\Big)^{\frac{3}{2}} \exp\Big(-\frac{a}{2}|\Delta\boldsymbol{J}_i|^2\Big),\end{aligned} \tag{6.15}$$

where a gives the spread of the action distribution. This is related to the progenitor mass, M, by $a \propto M^{-2/3}$ (see Chap. 5). Such a simple model for the action distribution is unrealistic (Eyre and Binney 2011), but our understanding of this simplistic model will aid in the construction of a more realistic model. Similarly, we assume that $\Delta\boldsymbol{\theta}_i(0)$ is distributed as an isotropic Gaussian such that

$$p(\Delta\boldsymbol{\theta}_i(0)) = \Big(\frac{b}{2\pi}\Big)^{\frac{3}{2}} \exp\Big(-\frac{b}{2}|\Delta\boldsymbol{\theta}_i(0)|^2\Big). \tag{6.16}$$

This Gaussian model for a stream in actions and initial angles was studied by Helmi and White (1999). From Eq. (6.14), the frequency is linearly related to the actions via the Hessian, $\mathbf{D}_0$, so we can write down the distribution for the frequencies as

$$\begin{aligned}p(\boldsymbol{\Omega}_i|\Phi, \mathscr{C}) &= \det\Big(\frac{\partial\boldsymbol{J}_i}{\partial\boldsymbol{\Omega}_i}\Big)p(\boldsymbol{J}_i|\Phi, \mathscr{C}) \\ &\approx \det(\mathbf{D}_0^{-1})\Big(\frac{a}{2\pi}\Big)^{\frac{3}{2}} \exp\Big(-\frac{a}{2}\Delta\boldsymbol{\Omega}_i^T\mathbf{D}_0^{-1}\mathbf{D}_0^{-1}\Delta\boldsymbol{\Omega}_i\Big),\end{aligned} \tag{6.17}$$

where, as the spread in actions is small, we have approximated the Jacobian by its value at the progenitor actions.

This distribution is a multivariate normal distribution with principal axes along the principal eigenvectors of $\mathbf{D}_0$ and with width given by the corresponding eigenvalues. $\mathbf{D}_0$ is a symmetric matrix so has real eigenvalues and orthogonal eigenvectors. Note here that, for long thin streams to form, $\mathbf{D}_0$ has one eigenvalue much greater than the other two. In Chap. 5, we demonstrated that in a realistic Galactic potential this condition was satisfied for a large volume of action space. Therefore, we write

$$\begin{aligned}\mathbf{D}_0^{-1} &= \sum_j^3 \frac{1}{\lambda_j}\hat{\boldsymbol{e}}_j \cdot \hat{\boldsymbol{e}}_j^T \\ &\approx \frac{1}{\lambda_2}\hat{\boldsymbol{e}}_2 \cdot \hat{\boldsymbol{e}}_2^T + \frac{1}{\lambda_3}\hat{\boldsymbol{e}}_3 \cdot \hat{\boldsymbol{e}}_3^T\end{aligned} \tag{6.18}$$

where λ_j and $\hat{\boldsymbol{e}}_j$ are the eigenvalues and eigenvectors of $\mathbf{D}_0^{-1}$, and we have $\lambda_1 \gg \lambda_2 > \lambda_3$. Therefore, we find that

$$p(\boldsymbol{\Omega}_i|\Phi, \mathscr{C}) \approx \lambda_1\lambda_2\lambda_3\Big(\frac{a}{2\pi}\Big)^{\frac{3}{2}} \exp\Big(-\frac{a}{2\lambda_2^2}(\Delta\boldsymbol{\Omega}_i \cdot \hat{\boldsymbol{e}}_2)^2 - \frac{a}{2\lambda_3^2}(\Delta\boldsymbol{\Omega}_i \cdot \hat{\boldsymbol{e}}_3)^2\Big). \tag{6.19}$$

The distribution of frequencies is 2D Gaussian perpendicular to a straight line in frequency space defined by $\hat{\boldsymbol{e}}_1$. In the simple model presented here, the distribution along $\hat{\boldsymbol{e}}_1$ is also Gaussian (with a very large dispersion). However, we will later adopt a superior distribution along $\hat{\boldsymbol{e}}_1$ which better reflects the stream distribution.

Next we address the angle distribution. The angles depend upon the additional variables, t_i and $\Delta\boldsymbol{\theta}_i(0)$. Therefore, we write

$$p(\boldsymbol{\theta}_i|\boldsymbol{\Omega}_i, \Phi, \mathscr{C}) = \int \mathrm{d}t_i\, \mathrm{d}^3\Delta\boldsymbol{\theta}_i(0)\; p(\boldsymbol{\theta}_i|\boldsymbol{\Omega}_i, \Delta\boldsymbol{\theta}_i(0), t_i, \mathscr{C})p(t_i)p(\Delta\boldsymbol{\theta}_i(0)). \tag{6.20}$$

Given a time since stripping, a frequency separation, and an initial angle separation, the present angle separation is completely determined by Eq. (6.14) so

$$p(\boldsymbol{\theta}_i|\boldsymbol{\Omega}_i, \Delta\boldsymbol{\theta}_i(0), t_i, \mathscr{C}) = \delta^3(\Delta\boldsymbol{\theta}_i - t_i\Delta\boldsymbol{\Omega}_i - \Delta\boldsymbol{\theta}_i(0)). \tag{6.21}$$

Substituting this and Eq. (6.16) into Eq. (6.20) and performing the integral over $\Delta\boldsymbol{\theta}_i(0)$ using the δ-function, we have that

$$p(\boldsymbol{\theta}_i|\boldsymbol{\Omega}_i, \Phi, \mathscr{C}) = \int \mathrm{d}t_i\; p(t_i)\Big(\frac{b}{2\pi}\Big)^{\frac{3}{2}} \exp\Big(-\frac{b}{2}|\Delta\boldsymbol{\theta}_i - t_i\Delta\boldsymbol{\Omega}_i|^2\Big). \tag{6.22}$$

Now we rearrange the argument of the exponential as

$$|\Delta\boldsymbol{\theta}_i - t_i\Delta\boldsymbol{\Omega}_i|^2 = |\Delta\boldsymbol{\Omega}_i|^2\Big(t_i - \frac{\Delta\boldsymbol{\theta}_i \cdot \Delta\boldsymbol{\Omega}_i}{|\Delta\boldsymbol{\Omega}_i|^2}\Big)^2 - \frac{(\Delta\boldsymbol{\theta}_i \cdot \Delta\boldsymbol{\Omega}_i)^2}{|\Delta\boldsymbol{\Omega}_i|^2} + |\Delta\boldsymbol{\theta}_i|^2, \tag{6.23}$$

and note that $\Delta\boldsymbol{\Omega}_i \approx \lambda_1 \hat{\boldsymbol{e}}_1(\hat{\boldsymbol{e}}_1 \cdot \Delta\boldsymbol{J}_i)$ so

$$\frac{(\Delta\boldsymbol{\theta}_i \cdot \Delta\boldsymbol{\Omega}_i)^2}{|\Delta\boldsymbol{\Omega}_i|^2} \approx (\Delta\boldsymbol{\theta}_i \cdot \hat{\boldsymbol{e}}_1)^2, \tag{6.24}$$

and

$$-\frac{(\Delta\boldsymbol{\theta}_i \cdot \Delta\boldsymbol{\Omega}_i)^2}{|\Delta\boldsymbol{\Omega}_i|^2} + |\Delta\boldsymbol{\theta}_i|^2 \approx (\Delta\boldsymbol{\theta}_i \cdot \hat{\boldsymbol{e}}_2)^2 + (\Delta\boldsymbol{\theta}_i \cdot \hat{\boldsymbol{e}}_3)^2. \tag{6.25}$$

Therefore, Eq. (6.22) becomes

$$\begin{aligned} p(\boldsymbol{\theta}_i|\boldsymbol{\Omega}_i, \Phi, \mathscr{C}) \approx & \left(\frac{b}{2\pi}\right)^{\frac{3}{2}} \exp\Big(-\frac{b}{2}\sum_{k=2,3}(\Delta\boldsymbol{\theta}_i \cdot \hat{\boldsymbol{e}}_k)^2\Big) \\ & \times \int \mathrm{d}t_i\, p(t_i) \exp\Big(-\frac{b|\Delta\boldsymbol{\Omega}_i|^2}{2}\Big(t_i - \frac{\Delta\boldsymbol{\theta}_i \cdot \hat{\boldsymbol{e}}_1}{\Delta\boldsymbol{\Omega}_i \cdot \hat{\boldsymbol{e}}_1}\Big)^2\Big). \end{aligned} \tag{6.26}$$

The first part is a 2D Gaussian perpendicular to the eigenvector $\hat{\boldsymbol{e}}_1$ (as with the frequencies), whilst the second part depends on when the particles were stripped from the progenitor and only affects the angle distribution along the vector $\hat{\boldsymbol{e}}_1$ i.e. $(\Delta\boldsymbol{\theta}_i \cdot \hat{\boldsymbol{e}}_1)$. If we assume that $p(t_i)$ is uniform, the integral can be performed analytically to give

$$\begin{aligned} p(\boldsymbol{\theta}_i|\boldsymbol{\Omega}_i, \Phi, \mathscr{C}) \simeq & \frac{b}{4\pi|\Delta\boldsymbol{\Omega}_i|}\Big[\mathrm{erf}\Big(\frac{\Delta\boldsymbol{\theta}_i \cdot \Delta\boldsymbol{\Omega}_i}{|\Delta\boldsymbol{\Omega}_i|}\sqrt{\frac{b}{2}}\Big) \\ & + \mathrm{erf}\Big(|\Delta\boldsymbol{\Omega}_i| t_{\max}\sqrt{\frac{b}{2}} - \frac{\Delta\boldsymbol{\theta}_i \cdot \Delta\boldsymbol{\Omega}_i}{|\Delta\boldsymbol{\Omega}_i|}\sqrt{\frac{b}{2}}\Big)\Big] \\ & \times \exp\Big(-\frac{b}{2}[(\Delta\boldsymbol{\theta}_i \cdot \hat{\boldsymbol{e}}_2)^2 - (\Delta\boldsymbol{\theta}_i \cdot \hat{\boldsymbol{e}}_3)^2]\Big). \end{aligned} \tag{6.27}$$

If we assume $\Delta\boldsymbol{\theta} \gg \Delta\boldsymbol{\theta}(0)$ (i.e. the particle was stripped long enough ago that the time part of Eq. (6.14) dominates the initial angle separation from the progenitor (this is an assumption we made in earlier in the chapter), this expression is approximately

$$p(\boldsymbol{\theta}_i|\boldsymbol{\Omega}_i, \Phi, \mathscr{C}) \approx \frac{b}{2\pi|\Delta\boldsymbol{\Omega}_i| t_{\max}} \exp\Big(-\frac{b}{2}\sum_{k=2,3}(\Delta\boldsymbol{\theta}_i \cdot \hat{\boldsymbol{e}}_k)^2\Big) \text{ if } 0 < \frac{\Delta\boldsymbol{\theta}_i \cdot \hat{\boldsymbol{e}}_1}{\Delta\boldsymbol{\Omega}_i \cdot \hat{\boldsymbol{e}}_1} < t_{\max}, \tag{6.28}$$

where the condition ensures that the stripping time for each stream member is positive, and less than some maximum stripping time, $t_{\max}$. This expression demonstrates explicitly that the distribution perpendicular to $\hat{\boldsymbol{e}}_1$ is independent of the distribution along $\hat{\boldsymbol{e}}_1$. In conclusion, in this model both the angle and frequency distributions are highly elongated along the vector $\hat{\boldsymbol{e}}_1$. This validates the procedure followed earlier in the chapter.

The assumption of a uniform stripping-time distribution, $p(t_i)$, does not well model the highly-concentrated stripping events around pericentric passage observed in N-body simulations of clusters on eccentric orbits. However, as shown in Eq. (6.26) the exact form adopted for $p(t_i)$ only affects the structure of the angle distribution along $\hat{\boldsymbol{n}}$. For diagnosis of the Galactic potential and mass distribution, we are interested in the shape of the stream, so the real diagnostic power comes from the clumping in frequency space (i.e. the width of the frequency distribution, described above by a) and the alignment of the frequency and angle distributions (described by the eigenvector $\hat{\boldsymbol{e}}_1$). Our simple assumption of uniform stripping times should not affect the recovery of the potential parameters significantly as the stripping time describes the phase-space distance along the stream path for each stream particle and not the shape of the stream path, which is the true probe of the potential. We will see in Sect. 6.8 that $p(t_i)$ for a stream generated from an N-body simulation is not uniform. However, the potential parameters are recovered successfully using this stream data when the assumption of a uniform stripping-time distribution is made.

6.7 Model

In the formalism of the previous section, we made several assumptions that, while useful for illustrative purposes, we would like to relax. The assumption of isotropic $\Delta\boldsymbol{J}$ distribution is not valid as evidenced in Eyre and Binney (2011). For a general action distribution, we would still expect a highly anisotropic frequency distribution but the principal eigenvector of this distribution will not be that of the Hessian matrix $\mathbf{D}$, but some other vector $\hat{\boldsymbol{n}}$, with vectors $\hat{\boldsymbol{d}}_1$ and $\hat{\boldsymbol{d}}_2$ perpendicular to this. The intricacies of the action distribution will be reflected in the frequency distribution along the vector $\hat{\boldsymbol{n}}$. Additionally, the angle distribution will also be highly elongated along this direction $\hat{\boldsymbol{n}}$. Analogous to a combination of Eqs. (6.19) and (6.26), we write

$$\begin{aligned} p(\boldsymbol{\theta}_i, \boldsymbol{\Omega}_i|\Phi, \mathscr{C}) = &\frac{K_\theta(\boldsymbol{\theta}_i|\boldsymbol{\Omega}_i)}{2\pi u^2}\exp\Big[-\sum_{j=1,2}\frac{1}{2u^2}(\boldsymbol{\theta}_i\cdot\hat{\boldsymbol{d}}_j-\gamma_j)^2\Big] \\ &\times\frac{K_\Omega(\boldsymbol{\Omega}_i)}{2\pi w^2}\exp\Big[-\sum_{j=1,2}\frac{1}{2w^2}(\boldsymbol{\Omega}_i\cdot\hat{\boldsymbol{d}}_j-\omega_j)^2\Big], \end{aligned} \tag{6.29}$$

where the functions K_i define the stream distribution along the vector $\hat{\boldsymbol{n}}$ in the angle or frequency space. The quantities u and w are the widths perpendicular to this vector in angle and frequency space respectively. Note we have assumed the stream is isotropically distributed perpendicular to the vector $\hat{\boldsymbol{n}}$. ω_j and γ_j are related to the present frequency and angle coordinates of the progenitor.

We define the angles ϕ and ψ such that

$$\hat{\boldsymbol{n}} = (\sin\phi\cos\psi, \cos\phi\cos\psi, \sin\psi), \tag{6.30}$$

and we choose

$$\hat{\boldsymbol{d}}_1 = (\cos\phi, -\sin\phi, 0),$$
$$\hat{\boldsymbol{d}}_2 = (\sin\phi\sin\psi, \cos\phi\sin\psi, -\cos\psi). \tag{6.31}$$

Note this choice of vectors perpendicular to $\hat{\boldsymbol{n}}$ is arbitrary. We have set the distribution perpendicular to $\hat{\boldsymbol{n}}$ to be isotropic, so our choice of vectors is unimportant. We now specify the functions K_i defining the stream distribution along the vector $\hat{\boldsymbol{n}}$. The frequency distribution along $\hat{\boldsymbol{n}}$ consists of two separated peaks corresponding to the leading and trailing tails of the stream (see next section). For simplicity, we assume that each of these peaks is Gaussian. The angle distribution depends upon both the frequency distribution and the distribution of stripping times. As in Eq. (6.28), we make the simple first-order assumption of a uniform stripping time distribution such that the angle distribution along the stream given a frequency separation is also uniform between zero and some maximum stripping time, $t_{\max}$ (see next section). Therefore, we write

$$K_\Omega(\boldsymbol{\Omega}_i) = \frac{1}{\sqrt{2\pi w_0^2}}\sum_{k=\pm1}\exp\Big[-\frac{1}{2w_0^2}(\boldsymbol{\Omega}_i\cdot\hat{\boldsymbol{n}} - \omega_0 + k\Omega_s)^2\Big],$$
$$K_\theta(\boldsymbol{\theta}_i|\boldsymbol{\Omega}_i) = \begin{cases}\dfrac{1}{|\boldsymbol{\Omega}_i\cdot\hat{\boldsymbol{n}} - \omega_0|t_{\max}}, & \text{if } 0 < \dfrac{(\boldsymbol{\theta}_i\cdot\hat{\boldsymbol{n}} - \gamma_0)}{(\boldsymbol{\Omega}_i\cdot\hat{\boldsymbol{n}} - \omega_0)} < t_{\max},\\ 0, & \text{otherwise.}\end{cases} \tag{6.32}$$

$2\Omega_s$ gives the separation between the Gaussian peaks along $\hat{\boldsymbol{n}}$ in frequency space. When Eq. (6.29) is combined with Eqs. (6.7) and (6.10), we have completely specified our model. Given a set of 6D stream data with associated errors, we can assess the likelihood of a given potential by evaluating the integral of Eq. (6.7). It is defined by 13 progenitor parameters given by $\mathscr{C} = \{\phi, \psi, \gamma_j, \omega_j, u, w, w_0, t_{\max}, \Omega_s\}$ and $\mathscr{N}$ potential parameters.

6.7.1 MCMC

We sample from the posterior using Markov chain Monte Carlo (MCMC). We use an affine-invariant sampler implemented in the *emcee* package from Foreman-Mackey et al. (2013). For each of the following tests, we use a group of 144 walkers, and vary the nuisance parameters $\mathscr{C}$ as well as the potential parameters. For all scale parameters (i.e. $u, w, w_0, t_{\max}$) we use a logarithmic flat prior, whilst for the other parameters we use uniform flat priors.

To perform the integral over the errors in the calculation of the likelihood we use the Vegas Monte Carlo integration algorithm (Lepage 1978) implemented in the Gnu Science Library (Galassi et al. 2009). Our stream model is typically very narrow, whilst the error distribution for each observable coordinate can be very broad.

Therefore, there is a very small region of the 4D integrand which has any support. Using an adaptive integration scheme, such as Vegas, means we can rapidly focus on this small region.

6.8 Tests of Probabilistic Approach

We test the above procedure using particles taken from the $t = 4.2$ Gyr snapshot of the stream simulation detailed in Sect. 6.1.1. We work again with the two-parameter ($\mathcal{N} = 2$) logarithmic potential of Eq. (6.4). The two parameters of this potential are V_c and q, and we set $V_c = 220\,\mathrm{km\,s^{-1}}$ and $q = 0.9$ for the simulation.

We take the resulting distribution of particles, remove the progenitor remnant with a spatial cut, and rotate the coordinate frame such that the Sun is placed at the same azimuthal angle as the progenitor. We show the resultant stream simulation in Fig. 6.7. From the resulting particles, we randomly select 30 particles that lie in the range $-200° < l < -140°$. The chosen sample of particles is shown in Fig. 6.8. For the tests shown below, we include the observational errors by scattering the observed coordinates by the appropriate Gaussian errors.

In the top panels of Fig. 6.9, we plot the angles and frequencies in the correct potential, along with a cross-section through the model specified in Eq. (6.29). The parameters of the model were chosen as those that maximised the likelihood in the correct potential. We see that, in the correct potential, the angle and frequency structure of the stream takes on a simple linear distribution. In the bottom panels of Fig. 6.9, we plot the projections of the frequency histograms along the vectors $\hat{\boldsymbol{n}}$, $\hat{\boldsymbol{d}}_1$ and $\hat{\boldsymbol{d}}_2$ for all particles in the stream from the simulation. We see that the distribution

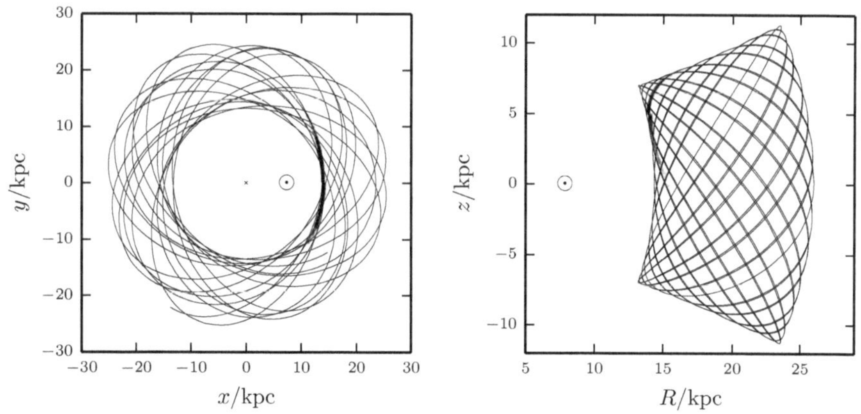

Fig. 6.7 Stream simulation in real space—the stream particles are shown in *black*, and the progenitor orbit is given by the *black line*. The Sun is marked by ⊙, and the *black* cross in the *left panel* is the Galactic Centre

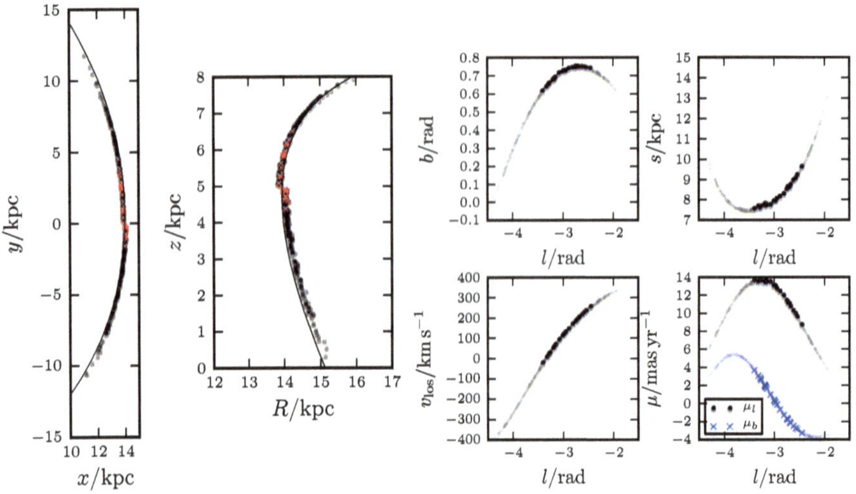

Fig. 6.8 Stream simulation used in Sect. 6.8. The *left two panels* show the simulation in real space—the stream particles are shown in *black* with the 30 selected particles shown in *red*, and the progenitor orbit is given by the *black line*. The *right panels* show the stream particles in Galactic coordinates with the 30 selected particles indicated by the heavy markers

perpendicular to $\hat{\boldsymbol{n}}$ is approximately Gaussian. The distribution along $\hat{\boldsymbol{n}}$ consists of two peaks due to the leading and trailing tail. Each peak is skewed such that there is a longer tail towards larger values of $|\Delta\boldsymbol{\Omega}\cdot\hat{\boldsymbol{n}}|$. The structure of these peaks was discussed by Johnston (1998).

Our model assumes a uniform stripping-time distribution. In Fig. 6.10, we plot the time since release for the particles in the stream, which we have estimated as $t_i = |\Delta\boldsymbol{\theta}_i|/|\Delta\boldsymbol{\Omega}_i|$ found in the true potential. We see that the distribution is peaked around pericentric passage with slightly more particles being stripped at later times as the cluster mass decreases. If the stream has undergone several stripping events, the distribution of stripping times is approximately uniform, if we average on a time-scale comparable to the radial period. As mentioned previously, we expect the assumption of a uniform stripping-time distribution to be appropriate for measuring the Galactic potential, but more detailed modelling is required if we wish to reproduce the density distribution along the stream.

By only sampling a portion of the stream track on the sky, we have limited the range of available angles of the stream particles, but we expect that the range of frequencies sampled is fair. We will miss some high frequency separations corresponding to particles stripped earliest. In this situation, the $t_{\max}$ parameter tells us about the time since the first of the observed particles were stripped (~2 Gyr).

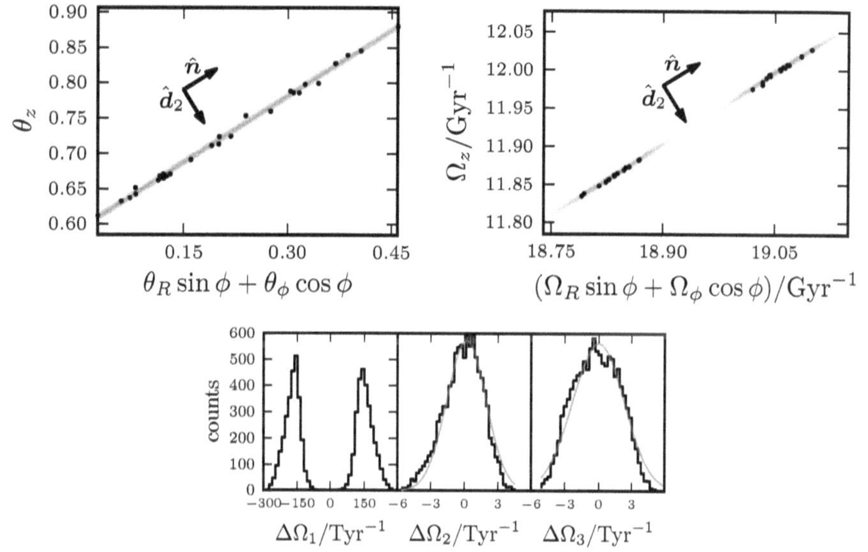

Fig. 6.9 Mock data and model in angle-frequency space. In the *top two panels*, we show two planes in angle and frequency space. The *black points* give the angles and frequencies of the 30 selected particles calculated using the correct potential. The *green* shows the logarithm of the value of the model pdf given in Eq. (6.29) on a plane through angle-frequency space that contains the vectors $\hat{\boldsymbol{n}}$ and $\hat{\boldsymbol{d}}_2$. We see that, in both angle and frequency space, the stream has a linear structure aligned with the vector $\hat{\boldsymbol{n}}$. In the *bottom panels*, we show histograms of the frequencies for *all* stream particles projected along the vectors $\hat{\boldsymbol{n}}$, $\hat{\boldsymbol{d}}_1$ and $\hat{\boldsymbol{d}}_2$ such that $\Delta\Omega_1 = (\Delta\boldsymbol{\Omega} \cdot \hat{\boldsymbol{n}})$, $\Delta\Omega_2 = (\Delta\boldsymbol{\Omega} \cdot \hat{\boldsymbol{d}}_1)$, and $\Delta\Omega_3 = (\Delta\boldsymbol{\Omega} \cdot \hat{\boldsymbol{d}}_2)$. The *middle* and *right panels* show Gaussian fits to the histograms in *red*

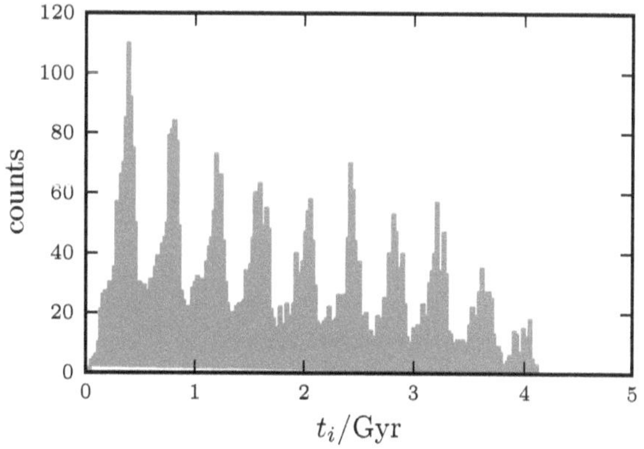

Fig. 6.10 Time since release for the particles in the stream estimated as $t_i = |\Delta\boldsymbol{\theta}_i|/|\Delta\boldsymbol{\Omega}_i|$ in the true potential. The distribution is peaked around pericentric passage, and, in general, more particles are stripped at later times

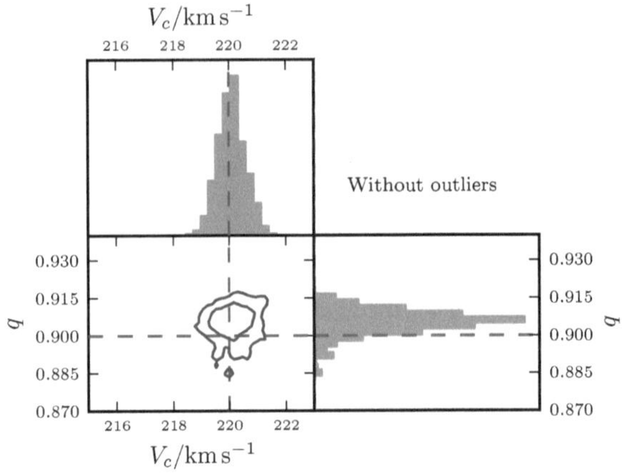

Fig. 6.11 Posterior distributions of the potential parameters for error-free data. The contours contain 68 and 95 % of the samples from an MCMC chain. The *red dashed lines* show the parameters used to produce the simulation

6.8.1 No Contaminants

We test the above method by considering a data set where each of our stream particles has full 6D data, $(l, b, s, v_{||}, \boldsymbol{\mu})$. We first demonstrate that the method works for error-free data. In Fig. 6.11 we show the posterior distributions of the potential parameters. The correct potential is recovered with $\sim$0.5 km s^{-1} and 0.005 errors in V_c and q respectively.

Next, we consider an *optimistic* data set where we assume each of the stars is an RR Lyrae observed by Spitzer (Price-Whelan and Johnston 2013). We assume that the covariance matrix is diagonal and identical for all data such that $S_{jk} = \delta_{jk}\sigma_j^2$. We adopt 2 % distance errors (σ_s/s), 5 km s^{-1} line-of-sight velocity errors ($\sigma_{||}$), and 0.1 mas yr^{-1} proper motion errors (σ_μ). Such a data set is unrealistic for such a low mass stream. However, it is suitable for demonstrating the method. We fix the parameters $w_0 = 0.08\,\mathrm{Gyr}^{-1}$, $u = 0.02$ rad and $w = 0.006\,\mathrm{Gyr}^{-1}$, and let the others vary as before. This prior essentially sets the mass of the cluster. The posterior distribution of the recovered potential parameters is shown in Fig. 6.12. We recover the correct potential parameters as $V_c = (219.4 \pm 1.4)\,\mathrm{km\,s^{-1}}$ and $q = (0.909 \pm 0.009)$, where we have approximated the posterior as an uncorrelated Gaussian.

6.8.2 Inclusion of Outliers

Stream data are inevitably contaminated with stars that are not associated with the stream. Many authors attempt to remove these outliers by performing cuts in the observable space. However, it is much better to model the outliers. To include m outliers in our test, we randomly select m stars from our simulation sample. We use

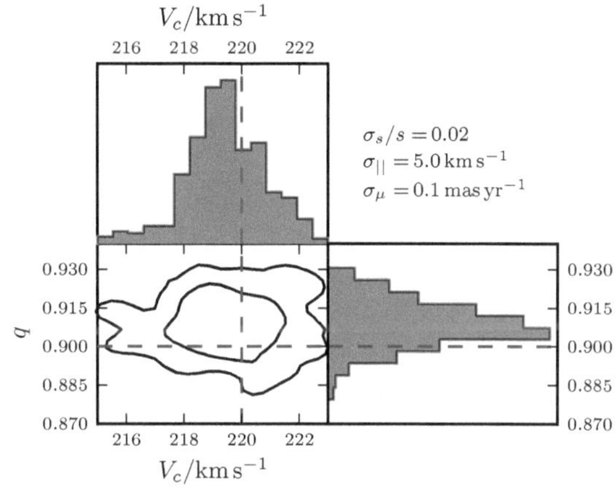

Fig. 6.12 Posterior distributions of the potential parameters for data with the errors shown in the *top right corner*. The *black* contours enclose 68 and 95 % of the samples. The *red dashed lines* show the parameters used to produce the simulation

the l and b values of these m stars for our outlier stars. For each star, we sample a distance, s, from a uniform distribution between 6 and 12 kpc. We convert the tuple (l, b, s) to a Cartesian position, $\boldsymbol{x}$. At this position $\boldsymbol{x}$, we draw a set of velocities $\boldsymbol{v}$ from an isothermal distribution function given by

$$p_{\text{iso}}(\boldsymbol{x}_i, \boldsymbol{v}_i|\Phi) = p_h(E|\Phi) \propto \exp(-E/\sigma_h^2), \tag{6.33}$$

with dispersion $\sigma_h = 100\,\text{km s}^{-1}$.

In our model, we assume that the likelihood of a star at angle-frequency coordinates $(\boldsymbol{\Omega}_i, \boldsymbol{\theta}_i)$ is given by

$$p(\boldsymbol{\Omega}_i, \boldsymbol{\theta}_i|\Phi) = (1-\epsilon)p_S(\boldsymbol{\Omega}_i, \boldsymbol{\theta}_i|\Phi) + \epsilon p_h(\boldsymbol{\Omega}_i, \boldsymbol{\theta}_i|\Phi), \tag{6.34}$$

where p_S is the likelihood given it is a member of the stream as outlined above, and p_h is the likelihood given that it is not a member of the stream (either a member of a smooth background population or another structure). ϵ is the probability of being an outlier, which is given a logarithmic uniform prior. We choose to specify the outlier model in angle-frequency space as it is simpler to normalize in these coordinates, and we are less sensitive to systematics arising from the Stäckel approximation. We set p_h to be uniform in both the angle and frequency space, such that

$$p_h(\boldsymbol{\Omega}_i, \boldsymbol{\theta}_i|\Phi) = \begin{cases} \frac{\epsilon}{(2\pi)^3\Omega_{\text{max}}^3}, & \text{if } \Omega_i < \Omega_{\text{max}}, \\ 0, & \text{otherwise.} \end{cases} \tag{6.35}$$

The natural prior on Ω_{max} is a logarithmic prior with a large cut-off. However, we choose to set $\Omega_{\text{max}} = 30\,\text{Gyr}^{-1}$. This model is clearly simplistic, and we are not

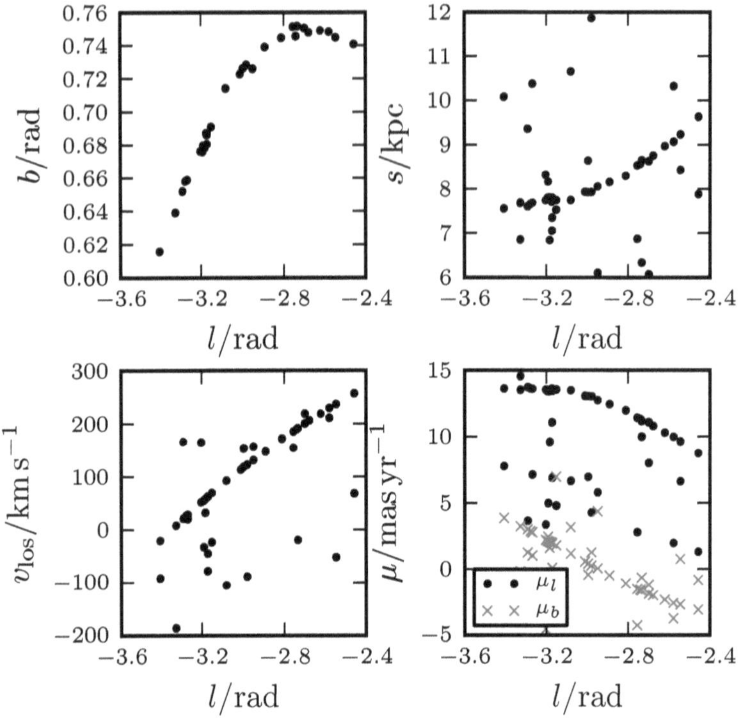

Fig. 6.13 Stream and outlier particles in Galactic coordinates

correctly accounting for selection effects in the angles due to observing stars in some region of the Galaxy.

To our 30-particle stream data set, we add 20 outliers from our halo model, such that $\epsilon_{\rm true} = 0.4$, and consider the simple case where we have an error-free data set. The input data set is shown in Fig. 6.13 and the resulting posterior distributions for the potential parameters are given in the left panel of Fig. 6.14. In the right panel of Fig. 6.14, we show the posterior distribution for the outlier fraction. It peaks nicely around the input outlier fraction, but has fairly large scatter around this value. This is probably due to the simplistic nature of the background model, but also, as we can see in Fig. 6.13, there is significant overlap in observable space between the stream and outlier distributions. Further information such as metallicities would potentially provide a simpler way to disentangle the stream from the background.

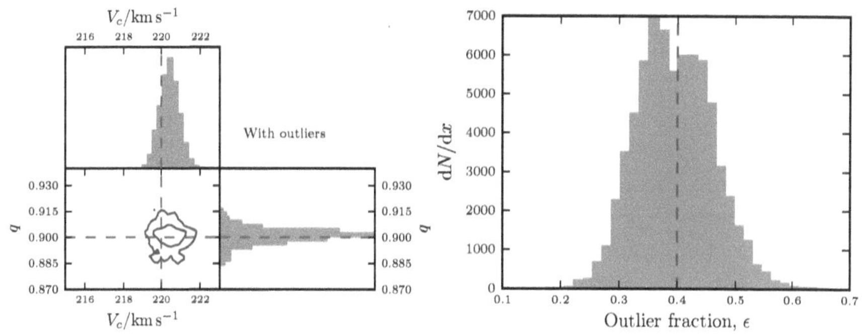

Fig. 6.14 Posterior distributions of the potential parameters (*left*) and outlier fraction (*right*) for error-free data with outliers included. The contours contain 68 and 95 % of the samples from an MCMC chain. The *red dashed lines* show the parameters used to produce the simulation

6.9 Discussion

The formalism presented in this chapter provides the first steps towards constraining the potential of the Galaxy from a generative stream model expressed in angle-frequency space. Through the course of developing the necessary machinery, it has become apparent that many improvements can be made. These are as follows:

(i) The model for a cold stream is very narrow, whilst the error distribution for the observational quantities is expected to be large. The convolution of the error and model distributions only has support over a very narrow range in each observable coordinate. Here we have performed the integration using the sophisticated Monte-Carlo integration method (Vegas) to concentrate rapidly on this small region. However, a more efficient scheme would be full forward modelling. To do this we would perform the Monte Carlo integration by sampling 'true' coordinates from the stream distribution and sum the appropriate Gaussian error distributions for each of the observed coordinates. For this we need an efficient scheme for going from a frequency-angle sample to observable space. The torus machine (McMillan and Binney 2008) was developed to convert from angle-actions to $(\boldsymbol{x}, \boldsymbol{v})$ efficiently. Such machinery seems ideal for our purpose. A torus is constructed by finding the coefficients of the generating function from a toy torus. We can construct a series of nearby tori that cover the small range in action space occupied by the stream particles. For a given set of angles, we can find a corresponding $(\boldsymbol{x}, \boldsymbol{v})$ on each torus. For any action point that lies between tori, we can construct an appropriate torus on-the-fly by interpolating the generating function coefficients of the nearby tori. A small complication here is that we wish to work with frequencies instead of actions so a scheme for finding $\boldsymbol{J}(\boldsymbol{\Omega})$ is required. We expect that, for the region of interest, this function is simple and may be obtained by interpolation between tori. With this function, we may also calculate the Hessian as a function of $\boldsymbol{\Omega}$. Bovy (2014) exploits the narrow range of frequencies in the stream to construct a simple linear map between $(\boldsymbol{\Omega}, \boldsymbol{\theta})$ and $(\boldsymbol{x}, \boldsymbol{v})$. The Gaussian structure in $(\boldsymbol{\Omega}, \boldsymbol{\theta})$-space can then be simply translated into a Gaussian structure in $(\boldsymbol{x}, \boldsymbol{v})$ making marginalization over missing data simpler. Helmi and White (1999) present a similar approach to analysing a stream model consisting of a Gaussian structure in action space and initial angle space.

(ii) The distributions in frequency and angle, K_i, along the principal stream direction, $\hat{\boldsymbol{n}}$, were taken in this chapter to be simple Gaussians and a uniform distribution. This procedure is adequate for our purposes, but more realistic distributions are required to reproduce the peaky distribution from Fig. 6.10, and the expected feathering in the stream (e.g. Fig. 6.3). For instance, we could adjust the distribution over stripping times to be more concentrated around pericentric passage of the progenitor.

(iii) Our choice of model parametrisation is perhaps more flexible than it should be. All the widths (w, u, w_0) are related to $M^{1/3}$ (see Chap. 5), so there could be a way to link all three parameters into a single mass parameter. Also, we have allowed the stream to be oriented along some random direction, $\hat{\boldsymbol{n}}$. We gain information about $\hat{\boldsymbol{n}}$ through the use of the angle and frequency structure as discussed earlier in this chapter. However, we could instead choose to make $\hat{\boldsymbol{n}}$ a function of the progenitor actions in the chosen potential. For a cluster with an isotropic action distribution, $\hat{\boldsymbol{n}}$ is aligned with $\hat{\boldsymbol{e}}_1$. However, as shown in Eyre and Binney (2011), the action distribution of an orbiting cluster is not isotropic, such that $\hat{\boldsymbol{n}}$ may not be aligned with $\hat{\boldsymbol{e}}_1$. These authors show that we can estimate the shape of the action distribution, $\Delta \boldsymbol{J}$, using Eq. (5.15), which depends on the velocity dispersion and orbit of the progenitor. The direction of $\hat{\boldsymbol{n}}$ is then along the vector $\mathbf{D} \cdot \Delta \boldsymbol{J}$. This would constrain the models in a more physically motivated fashion, and would provide more information when the quality of the data is reduced.

(iv) We have analysed a simulated stream in a near-spherical logarithmic potential. As discussed in Chap. 5, this does not exhibit substantial offset between the stream and orbit tracks. Such an offset is more apparent in flattened potentials. Therefore, more tests are required with more realistic potentials with disc components to validate the modelling approach presented here.

6.10 Conclusions

Tidal streams are very attractive structures for probing the Galactic potential on large scales, but it is essential that appropriate algorithms are developed for their study. In Chap. 5, we showed that orbit-fitting is inappropriate for many streams in the Milky Way, and can lead to order one errors in estimation of parameters of the Galactic potential. Motivated by the need for an improvement over orbit-fitting, we have presented an algorithm for using tidal-stream data to constrain the potential of the Galaxy without assuming that the stream delineates an orbit. Instead, it identifies the true potential as the one that yields corresponding patterns in angle and frequency space. The algorithm was tested by evolving an N-body simulation of a King cluster in a two-parameter logarithmic potential until a stream is formed. The degree of correlation between the angle and frequency structure was maximised with respect to trial potentials. The correct parameters were recovered within the estimated systematic errors of the method.

As tidal streams are very distant structures, we expect large errors in the observables. Therefore, it is imperative that any stream-fitting method is shown to function for large observational errors. We have shown that the observational errors in the distances and proper motions for individual stars in tidal streams are currently too large to use the above technique with any confidence. However, if the data are first binned and averaged in observable space on the sky, we can recover the correct potential parameters even for large observational errors. The current observational errors may be small enough for close streams such as GD-1 and it seems promising that in the era of *Gaia* the data for more streams will be sufficiently accurate to use this algorithm.

We have shown that longer streams produce superior potential parameter estimates, so there is hope for using higher-mass streams to produce better constraints on the potential. We have seen that observing a stream at different orbital phases of the progenitor results in different constraints on the potential. Therefore there is much to be gained by using several streams simultaneously to constrain the potential. Hopefully, this would also remove local minima and make a global minimum more apparent. Similarly the approach taken here uses a constant prior for the potential parameters. In reality, a more informative prior could be used, that would rule out regions of the parameter space which are populated by local minima.

Inspired by our naive algorithm, we went on to present a probabilistic model for a tidal stream and used this to constrain the potential from a simulation. The generative model produces streams in observable coordinates by using the simple structure of a stream in frequency and angle space. The presented Bayesian formalism improved on the previous approach by: 1. performing the inference in the data space, not the model space; 2. naturally accounting for the errors in stream data; 3. incorporating the possibility of stream data being contaminated with stars from a smooth halo population or another tidal structure; and 4. allowing the possibility of partial 6D data. The results from this formalism may be simply combined with other constraints on the potential. For example, we might produce many independent constraints on the potential from different streams that may then be combined by multiplying their likelihoods. We have successfully recovered the potential parameters used to run an N-body simulation of a GD-1-like stream from error-free data, data with small errors included, and data with outliers included.

As currently formulated, the computational cost of implementing our approach increases significantly with the magnitude of the observational errors. We have described modifications that promise to mitigate this effect, and thus to make the approach a powerful technique for constraining the Galaxy's gravitational potential.

6.10.1 Future Work

The approach presented in this chapter seems promising but there is much future work to be done. We detail the intended future work here in order of importance.

1. The high computational cost of calculating the likelihoods with large errors was due to our inability to find $(\boldsymbol{x}, \boldsymbol{v})$ coordinates given $(\boldsymbol{\Omega}, \boldsymbol{\theta})$. The torus machinery (McMillan and Binney 2008) naturally provides this transformation, and, as detailed in the discussion section, stream models can be rapidly generated using this machinery.
2. We have only investigated how well a very simple two-parameter potential may be constrained. As we have seen in the previous chapter, a more realistic Galactic potential with highly flattened disc components produces more interesting stream geometries. The machinery presented in the second half of this chapter needs testing on these more realistic Galactic potentials. Also, the question of exactly what constraints on the potential a given stream produces is still an open question.
3. After these two steps are complete, we will then be in a position to analyse real data, for example the GD-1 data from Koposov et al. (2010). Whether we will see the power of these improved techniques in the short term whilst the quality of data is poor is unclear, but with improved data from focussed surveys (e.g. Sesar et al. 2013), the hope is that these new methods will be crucial to extracting important diagnostics of the Galactic potential.
4. Finally, we can simply incorporate incomplete 6D data into our presented scheme. It would be interesting to investigate the quality of the recovery of the potential parameters when we have, for instance, no line-of-sight velocities, or no proper motions.

References

Binney J (2008) MNRAS 386:L47
Bonaca A, Geha M, Kuepper AHW, Diemand J, Johnston KV, Hogg DW (2014) ArXiv e-prints
Bovy J (2014) ArXiv e-prints
Bovy J et al (2012) ApJ 759:131
Burnett B et al (2011) A & A 532:A113
Dehnen W (2000) ApJL 536:L39
Dehnen W (2002) J Comput Phys 179:27
Dehnen W, Odenkirchen M, Grebel EK, Rix H-W (2004) AJ 127:2753
Deng L-C et al (2012) Res Astron Astrophys 12:735
Drake AJ et al (2013) ApJ 763:32
Evans NW (1993) MNRAS 260:191
Eyre A, Binney J (2009) MNRAS 399:L160
Eyre A, Binney J (2011) MNRAS 413:1852
Fellhauer M et al (2006) ApJ 651:167
Foreman-Mackey D, Hogg DW, Lang D, Goodman J (2013) emcee: The MCMC Hammer. PASP 125:306–312

Galassi M, Davies J, Theiler J, Gough B, Jungman G (2009) GNU scientific library—reference manual for GSL version 1.12, 3rd edn. Network Theory Ltd, pp 1–573
Gilmore G et al (2012) Messenger 147:25
Helmi A, White SDM (1999) MNRAS 307:495
Ivezić Ž et al (2008) SPIE 248:537
Johnston KV (1998) ApJ 495:297
Johnston KV, Zhao H, Spergel DN, Hernquist L (1999) ApJL 512:L109
Jurić M et al (2008) ApJ 673:864
Kaiser N et al (2002) SPIE 4836:154
Koposov SE, Rix H-W, Hogg DW (2010) ApJ 712:260
Küpper AHW, Lane RR, Heggie DC (2012) MNRAS 420:2700
Law DR, Johnston KV, Majewski SR (2005) ApJ 619:807
Law DR, Majewski SR (2010) ApJ 714:229
Lepage GP (1978) J Comput Phys 27:192
Magnier EA, Liu M, Monet DG, Chambers KC (2008) IAU Symp 248:553
McMillan PJ, Binney JJ (2008) MNRAS 390:429
Munn JA et al (2004) AJ 127:3034
Nelder JA, Mead R (1965) Comput J 7:308
Peñarrubia J, Koposov SE, Walker MG (2012) ApJ 760:2
Perryman MAC et al (2001) A & A 369:339
Pier JR, Munn JA, Hindsley RB, Hennessy GS, Kent SM, Lupton RH, Ivezić Ž (2003) AJ 125:1559
Price-Whelan AM, Hogg DW, Johnston KV, Hendel D (2014) ArXiv e-prints
Price-Whelan AM, Johnston KV (2013) ApJL 778:L12
Reid MJ, Brunthaler A (2004) ApJ 616:872
Sanders J (2012) MNRAS 426:128
Sanders JL (2014) MNRAS 443:423
Sanders JL, Binney J (2013) MNRAS 433:1826
Sanderson R, Helmi A, Hogg DW (2014) In: American astronomical society meeting abstracts, vol 223. American Astronomical Society Meeting Abstracts, p 346.10
Schönrich R, Binney J, Dehnen W (2010) MNRAS 403:1829
Sesar B et al (2013) ApJ 776:26
Siebert A et al (2011) AJ 141:187
Skrutskie MF et al (2006) AJ 131:1163
Tremaine S (1999) MNRAS 307:877
Varghese A, Ibata R, Lewis GF (2011) MNRAS 417:198
Yanny B et al (2009) AJ 137:4377

Chapter 7
Determining the Velocity Dispersion of the Thick Disc

7.1 Introduction

We are now going to shift focus from the modelling of streams, which act as mere tracers of the potential, to one of the Galaxy's main components, the Galactic stellar disc.

Ever since Oort (1932) measured the local velocity dispersions of stars, it has been known that dark matter forms a key contribution to the local vertical force. There have been many attempts to measure the local dark matter density using tracer populations (Kuijken and Gilmore 1989; Moni Bidin et al. 2012b; Garbari et al. 2012; Bovy et al. 2012; Zhang et al. 2013; Piffl et al. 2014). The route to constraining the local dark matter density is through constructing consistent dynamical models of the Galactic disc. Only in the correct potential will the velocity structure and the density structure of a dynamical model match those of a tracer population. These distribution functions must obey the collisionless Boltzmann equation, and by the strong Jeans' theorem this implies we can use the actions, $\boldsymbol{J}$, as arguments for the distribution function.

Distribution functions of the Galaxy are incredibly useful for constraining the Galactic potential. Additionally, they provide a model from which one can generate mock catalogues. Currently, the most popular mock catalogue generator for the Milky Way is the Besançon model (Robin et al. 2003), which is implemented in an efficient way in the *Galaxia* code (Sharma et al. 2011). However, these models are essentially kinematic, so do not provide the crucial physical links between the spatial and velocity structure. In this Z, we present the dynamically consistent distribution functions of Binney (Binney 2010, 2012a; Binney et al. 2014), and go on to test a recent method applied to data by applying the method to mock catalogues generated from the distribution function.

Moni Bidin et al. (2012b) used the Jeans equation to constrain the stellar mass density at heights $1.5\,\mathrm{kpc} \leqslant z \leqslant 4\,\mathrm{kpc}$ from the Galactic plane using a sample of 412 red giants and concluded that there is a lack of dark matter in the solar neighbourhood. Bovy and Tremaine (2012) have pointed out that one of the assumptions

J. Sanders, *Dynamics of the Milky Way*, Springer Theses,
DOI 10.1007/978-3-319-18772-3_7

made by Moni Bidin et al. (2012b) in their Jeans-equation analysis (specifically the assumption that the mean azimuthal velocity is independent of Galactocentric radius at all heights) is false, and a reanalysis of the data without this assumption leads to a non-zero local dark matter density. Despite the fact that the data now appear to conform with standard expectation, it is wise to check that all parts of the analysis are sound before the matter is put to rest. To use the Jeans equation with confidence, one needs to be sure that an underlying population has been isolated and that the mean velocities and velocity dispersions of that population can be reliably calculated. Here we investigate the method used by Moni Bidin et al. (2012a; hereafter MB) to calculate the mean velocities and velocity dispersions of the thick disc. It is these quantities which are then used by Moni Bidin et al. (2012b) (along with literature values for the density profile of the thick disc) and Bovy and Tremaine (2012) in their Jeans-equation analysis, so it is crucial that they are calculated correctly and that their associated errors are realistic.

MB use the probability plot method, which we detail in Sect. 7.2. We proceed by producing a sample drawn from the realistic Galaxy distribution functions of Binney (2012b) chosen to be similar to the sample of MB. These distribution functions are briefly discussed in Sect. 7.3. In Sect. 7.4, we attempt to recover the known mean velocity and velocity dispersions of the thick disc using the method presented by MB. We also implement the error analysis used by MB, and show that it does not include any estimates of the Poisson noise. In Sect. 7.5.2, the analysis is repeated for a large number of samples to investigate the effects of the Poisson noise and to uncover any systematics introduced by the method. Finally in Sect. 7.6, the same data set is analysed by a similar method inspired by Girard et al. (2006) and the results are compared.

The work of this chapter was published in Sanders (2012).

7.2 Probability Plot Method

The data analysis used by MB uses the probability plot to determine the mean and standard deviation of a sample. Here we briefly present the method and give a simple example to demonstrate its use.

Suppose we have N ordered data points. If these N data points are drawn from a normal distribution of mean μ and standard deviation σ, then, for large N, the ith data point, x_i, should satisfy

$$F(x_i;\mu,\sigma) \simeq \frac{i}{N+1}. \tag{7.1}$$

$F(x;\mu,\sigma)$ is the cumulative distribution function for a normal distribution, which is given by

$$F(x;\mu,\sigma) = \frac{1}{2}\Big[1+\mathrm{erf}\Big(\frac{x-\mu}{\sqrt{2}\sigma}\Big)\Big], \tag{7.2}$$

where erf is the error function. Hence the ith data point should lie at c_i standard deviations from the mean, where

$$c_i \equiv \sqrt{2}\,\mathrm{erf}^{-1}\Big(\frac{2i}{N+1} - 1\Big) \approx \Big(\frac{x_i - \mu}{\sigma}\Big). \tag{7.3}$$

Therefore, a plot of x_i against c_i should have gradient σ and intercept μ, which may be found by linear regression. Such a plot is termed a probability plot.

Clearly, this approach only works exactly if N is large and the data have been drawn from a single normal distribution. However, we can use it to estimate the mean and standard deviation of a sample drawn from any underlying distribution that is approximately Gaussian. Here we show how the method operates for a simple case. We draw 50 data points from a normal distribution of mean $\mu = 0$ and standard deviation $\sigma = 1$, and assess how well the above method can recover these quantities. Figure 7.1 shows the result for one randomly drawn sample. The measured standard deviation for this sample is $s = 1.061$. From the linear regression, the mean is estimated as $m = 0.061 \pm 0.141$ and the standard deviation $s = 1.130 \pm 0.152$, where the errors are given by the deviations of the points away from a straight line. With N data points, the expected error in the mean is $\sqrt{\sigma^2/N}$ and the expected error in the variance is approximately $\sqrt{2\sigma^4/N}$. Therefore, for this sample we expect an error in the mean of 0.141 and an error in the standard deviation of 0.1. This is well represented by the errors in the linear regression. Thus, the method does work well when the data are drawn from an underlying Gaussian distribution, and the errors from the linear regression are comparable to the expected Poisson noise.

When the data have been drawn from the sum of two Gaussian distributions, we can estimate the means and standard deviations by fitting straight line segments to different parts of the plot. This is the procedure followed by Bochanski et al. (2007) to calculate the velocity dispersions of the thin and thick discs using a tracer population of 7398 M dwarfs.

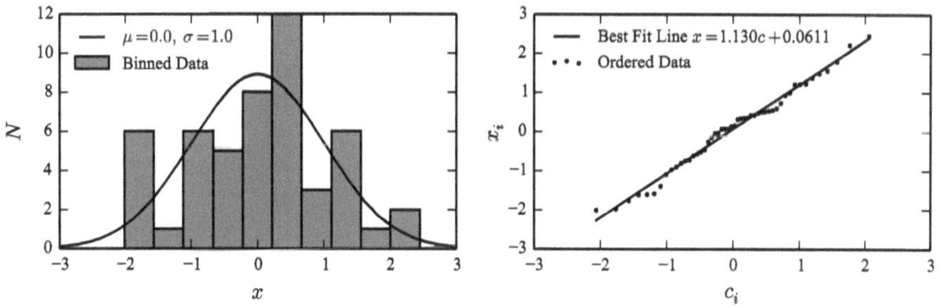

Fig. 7.1 Demonstration of finding the mean and standard deviation of a sample. The left panel shows a histogram of data drawn from a normal distribution of mean zero and unit standard deviation shown by the line. The *right panel* shows the ordered data plotted against the expected deviation in units of the standard deviation, c_i, given by Eq. 7.3. A *straight line* fit to this plot yields an estimate for the standard deviation as the gradient and an estimate for the mean as the intercept. For this example, the standard deviation is estimated as 1.130 and the mean as 0.0611

7.3 Dynamical Galaxy Models

To test the method used by MB, we require a sample drawn from a realistic Galaxy distribution function (DF) for which we know the underlying velocity moments. We use the models of Binney (2012b). These models were developed and discussed by Binney (2010) and Binney and McMillan (2011). The distribution function is a function of the actions in an axisymmetric potential: the radial action J_R, the vertical action J_z and the z-component of the angular momentum, L_z (in this chapter and the next, we set positive L_z as the direction of Galactic rotation). The advantage of this approach is that the distribution function clearly satisfies the Jeans' theorem as the actions are isolating integrals. The transformation from $(\boldsymbol{x}, \boldsymbol{v})$ to $\boldsymbol{J}$ is performed using the algorithm from Binney (2012a), which was described in Chap. 2. These models consist of a thick and thin disc composed of quasi-isothermal distribution functions. A single quasi-isothermal distribution function is given by

$$f_{\sigma_R,\sigma_z}(J_R, J_z, L_z) = f_{\sigma_R}(J_R, L_z) f_{\sigma_z}(J_z, L_z), \tag{7.4}$$

where

$$f_{\sigma_R}(J_R, L_z) \equiv \left.\frac{\Omega}{2\pi^2 R_d^2 \sigma_R^2 \kappa}\right|_{R_c} [1 + \tanh(L_z/L_0)] \mathrm{e}^{-\kappa J_R/\sigma_R^2} \mathrm{e}^{-R_c/R_d} \tag{7.5}$$

and

$$f_{\sigma_z}(J_z, L_z) \equiv \frac{\nu}{2\pi\sigma_z^2} \mathrm{e}^{-\nu J_z/\sigma_z^2}. \tag{7.6}$$

Here $R_c(L_z)$ is the radius of a circular orbit with z-component of angular momentum, L_z. $\kappa(L_z)$, $\nu(L_z)$ and $\Omega(L_z)$ are the radial, vertical and circular epicycle frequencies respectively. The factor of $[1 + \tanh(L_z/L_0)]$ eliminates retrograde stars as $L_0 = 10\,\mathrm{km\,s^{-1}\,kpc}$. Note that the DF is correctly normalised such that $\int \mathrm{d}^3\boldsymbol{x}\,\mathrm{d}^3\boldsymbol{v}\, f = (2\pi)^3 \int \mathrm{d}^3\boldsymbol{J}\, f = 1$. These distribution functions are called 'quasi-isothermal' as, in the epicyclic approximation, we can introduce the radial energy $E_R = \kappa J_R$ and the vertical energy $E_z = \nu J_z$ such that the DF reduces to an isothermal DF (i.e. solely a function of the radial and vertical energies). In this case, σ_i are the height-independent velocity dispersions. $\sigma_R(L_z)$ and $\sigma_z(L_z)$ are exponentially decaying functions of R_c that control the radial and vertical velocity dispersions:

$$\begin{aligned} \sigma_R(L_z) &= \sigma_{R0} \mathrm{e}^{(R_0 - R_c)/R_\sigma}, \\ \sigma_z(L_z) &= \sigma_{z0} \mathrm{e}^{(R_0 - R_c)/R_\sigma}, \end{aligned} \tag{7.7}$$

where σ_{R0} and σ_{z0} are approximately equal to the radial and vertical velocity dispersions at the solar radius, R_0, and R_d is the scale length of the disc. Each quasi-isothermal is controlled by four parameters: R_d, R_σ, σ_{R0} and σ_{z0}.

Binney (2012a) showed that a good fit to the GCS data and the Gilmore-Reid density curve was afforded by modelling the thin and thick discs as quasi-isothermal components. We choose to model the thick disc as a single old quasi-isothermal, and the thin disc as a superposition of differently-aged quasi-isothermals. Aumer and Binney (2009) found from Hipparcos data that the velocity dispersion of the thin disc increases with age, τ, as τ^{β_i}. Therefore, we set

$$\sigma_i(L_z,\tau)=\sigma_i(L_z)\Big(\frac{\tau+\tau_1}{\tau_1+\tau_T}\Big)^{\beta_i}, \tag{7.8}$$

where $\tau_1 = 0.11\,\text{Gyr}$ sets the velocity dispersion for stars at birth, and τ_T is the maximum age of the thin disc, chosen to be 10 Gyr. Here we set $\beta_R = 0.33$ and $\beta_z = 0.33$. The age of the Galaxy, τ_m, is assumed to be 12 Gyr such that the thick disc is defined as the population of stars born in the first 2 Gyr of the lifetime of the Galaxy. The thin and thick discs have differing velocity dispersion parameters such that the velocity dispersion against age for the disc is discontinuous at 10 Gyr ago. We assume that this time corresponds to the last strong heating event in the disc.

Additionally, Aumer and Binney (2009) showed that the star-formation rate in the thin disc near the Sun varied approximately exponentially with age with a time-scale of $\tau_f = 8\,\text{Gyr}$. Therefore, the full distribution for the thin disc is given by

$$\begin{aligned} f_{\text{thin}}(\boldsymbol{J}) &= \int_0^{\tau_T} \mathrm{d}\tau\; \Sigma_{\text{thin}}(\tau) f_{\text{thin},\sigma_R(\tau),\sigma_z(\tau)}(\boldsymbol{J}) \\ &= \int_0^{\tau_T} \mathrm{d}\tau\, \frac{\mathrm{e}^{\tau/\tau_f}}{\tau_f(\mathrm{e}^{\tau_T/\tau_f}-1)} f_{\text{thin},\sigma_R(\tau),\sigma_z(\tau)}(\boldsymbol{J}), \end{aligned} \tag{7.9}$$

whilst, for the thick disc, we have

$$f_{\text{thick}}(\boldsymbol{J}) = \int_0^{\tau_T} \mathrm{d}\tau\; \Sigma_{\text{thick}}(\tau) f_{\text{thick},\sigma_R,\sigma_z}(\boldsymbol{J}) = \int_{\tau_T}^{\tau_m} \mathrm{d}\tau\, \frac{1}{\tau_m-\tau_T} f_{\text{thick},\sigma_R,\sigma_z}(\boldsymbol{J}). \tag{7.10}$$

The full DF is given by

$$f(\boldsymbol{J}) = (1-\mathcal{F}) f_{\text{thin}}(\boldsymbol{J}) + \mathcal{F} f_{\text{thick}}(\boldsymbol{J}), \tag{7.11}$$

where $\mathcal{F}$ gives the global fraction of thick disc stars. We note that the thin/thick disc divide in age is discrete, and represents a more continuous violent heating event at some point in the early life of the Galaxy.

The fraction of stars in the disc belonging to the thick disc is controlled by the parameter $\mathcal{F} = 0.387$, and implies a Solar neighbourhood thick disc fraction of 29 %. We set the parameters to the values given in Table 7.1. The density profile and velocity dispersion for this distribution function as a function of Galactic height are shown in Fig. 7.2.

Table 7.1 Parameters for the Binney distribution function used in this section

Thin	σ_{R0}	42.3
	σ_{z0}	20.3
	R_d	2.17
	R_σ	50
Thick	σ_{R0}	26.3
	σ_{z0}	34.0
	R_d	3.66
	R_σ	3.43
	$\mathcal{F}$	0.387

Velocity dispersions have units km s^{-1} and scale lengths have units kpc

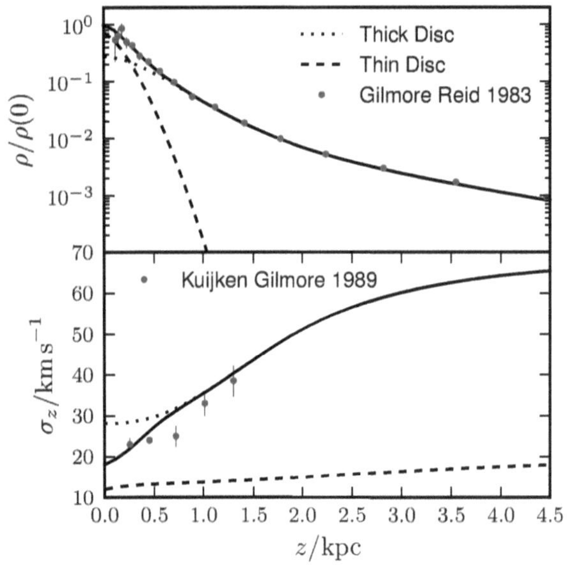

Fig. 7.2 Density profile and velocity dispersion for the Binney distribution function used in this section as a function of Galactic height. The *dashed line* gives the contribution from the thin disc and the *dotted line* from the thick disc. The data points are taken from Gilmore and Reid (1983) and Kuijken and Gilmore (1989)

We limit our investigation to just the W component of the velocity (the component along the z direction). We are not interested in the full distribution function, so we marginalise over the other two velocity components (U and V) to find the number of stars per unit W velocity per unit volume as

$$n_W(R, z, W) = \int_{-\infty}^{\infty} \mathrm{d}U \int_{-\infty}^{\infty} \mathrm{d}V f(J_R, J_z, L_z) \tag{7.12}$$

where, in practice, the limits of the integrals are finite as the distribution function falls off rapidly at large velocities. The transformation from polar positions and velocities to actions is carried out by the algorithm presented by Binney (2012a). We use an adjusted version of Potential II from Dehnen and Binney (1998) that consists of a thin and thick disc, a gas disc and two spheroids representing the bulge and the halo.

Table 7.2 Parameters of Galactic potential used in this chapter

Thin	R_d/kpc	2.4
	z_d/kpc	0.36
	$\Sigma_d/M_\odot\,pc^{-2}$	1106
Thick	R_d/kpc	2.4
	z_d/kpc	1.
	$\Sigma_d/M_\odot\,pc^{-2}$	73
Gas	R_d/kpc	4.8
	z_d/kpc	0.04
	$\Sigma_d/M_\odot\,pc^{-2}$	114
	R_m/kpc	4
Bulge	$\rho_0/M_\odot\,pc^{-3}$	0.76
	r_0/kpc	1
	γ	1.8
	β	1.8
	q	0.6
	r_t/kpc	1.9
Halo	$\rho_0/M_\odot\,pc^{-3}$	1.26
	r_0/kpc	1.09
	γ	−2
	β	2.21
	q	0.8
	r_t/kpc	∞

We have increased the scale-height of the thin disc to 360 pc and increased the mass of the thin disc such that the circular velocity at the solar radius is 220 km s^{-1}. The functional form for the discs is given by

$$\rho_d(R,z) = \frac{\Sigma_d}{2z_d}\exp\left(-\frac{R_m}{R} - \frac{R}{R_d} - \frac{|z|}{z_d}\right), \tag{7.13}$$

where R_d is the scale-length, z_d the scale-height, Σ_d is the central surface density and R_m controls the size of the hole at the centre of the disc which is only non-zero for the gas disc. The spheroids obey the functional form

$$\rho_s(m) = \rho_0\left(\frac{m}{r_0}\right)^{-\gamma}\left(1+\frac{m}{r_0}\right)^{\gamma-\beta}\exp\left(-\frac{m^2}{r_t^2}\right), \tag{7.14}$$

where $m = (R^2 + q^{-2}z^2)^{1/2}$. ρ_0 is the central density, r_0 a scale-length, q a flattening, γ and β control the inner and outer slopes, and r_t is a truncation radius. The adopted parameters are given in Table 7.2.

From the distribution function of Eq. (7.12), we are able to draw a sample of stars. The sample is selected by following a rejection algorithm. If we wish to draw stars that all lie at the solar radius, R_0, and that lie in the ranges $z_{\min} \leqslant z \leqslant z_{\max}$ and $|W| \leqslant W_{\max}$ then we first note that the maximum of the distribution function in this range occurs at $z = z_{\min}$ and $W = 0$. This gives us a normalisation. We then proceed by drawing trial values of $z = z_t$ and $W = W_t$ from uniform distributions over the required ranges and accepting this trial as a data point with probability $n_W(R_0, z_t, W_t)/n_W(R_0, z_{\min}, 0)$.

The obvious advantage of drawing sample data from a known distribution function is that we know exactly the underlying properties of the distribution. In this section, we focus on calculating the mean W velocity and the W velocity dispersion of the thick disc, σ_z. As the distribution function is a symmetric function of W, we expect $\langle W \rangle = 0$. σ_z is given by

$$\sigma_z^2(R, z) = \frac{\int_{-\infty}^{\infty} \mathrm{d}U \int_{-\infty}^{\infty} \mathrm{d}V \int_{-\infty}^{\infty} \mathrm{d}W\, W^2 f_{\text{thick}}(J_R, J_z, L_z)}{\int_{-\infty}^{\infty} \mathrm{d}U \int_{-\infty}^{\infty} \mathrm{d}V \int_{-\infty}^{\infty} \mathrm{d}W\, f_{\text{thick}}(J_R, J_z, L_z)}. \tag{7.15}$$

7.4 MB Method

MB use a sample of 412 red giants that lie in the range $1.3\,\text{kpc} \leqslant z \leqslant 5\,\text{kpc}$ and $|W| \leqslant 150\,\text{km s}^{-1}$. The cut in W is performed to remove contamination from the halo. MB estimate an error of $0.7\,\text{km s}^{-1}$ in the line-of-sight velocity measurements and an error of approximately 20 % in the distances.[1] We use the procedure outlined above to draw 412 stars from the Binney distribution function which all lie at the solar radius and inside the range probed by MB. As the W velocity is nearly entirely line-of-sight velocity error we include a random Gaussian error of $0.7\,\text{km s}^{-1}$ to the W velocities and we assume that the full distance error of 20 % corresponds to a 20 % error in the z values. In order to correctly account for stars that may have entered our sample due to the error in their distances, we increase the sample range to $1\,\text{kpc} \leqslant z \leqslant 6\,\text{kpc}$ and then cut out any stars which lie outside $1.3\,\text{kpc} \leqslant z \leqslant 5\,\text{kpc}$ after the error has been included. Histograms of the resulting sample are shown in Fig. 7.3.

We now follow the same procedure as MB to extract the mean W velocity and the W velocity dispersion, σ_z. We first bin the data in z with bin centres spaced by 0.1 kpc in the range $1.5\,\text{kpc} \leqslant z \leqslant 4.5\,\text{kpc}$. The bin sizes are allowed to vary such that we have 100 data per bin for $z \leqslant 2.1\,\text{kpc}$, 80 data per bin for $2.2\,\text{kpc} \leqslant z \leqslant 2.4\,\text{kpc}$ and 50 data per bin for $z \geqslant 2.5\,\text{kpc}$. For each binned subset of the sample, we follow the probability plot method outlined in Sect. 7.2 by ordering the sample in W and performing a linear regression between the sample velocities and the expected deviations to find the mean velocity of the bin and the velocity dispersion.

[1] MB state that there is an additional 10–20 % systematic error in the distances to thin disc stars due to the thin disc stars not following the assumed age and metallicity distributions. We ignore this additional error here. This error will increase thin disc contamination at low z, but should not affect the determination of the velocity dispersion or conclusions presented here significantly.

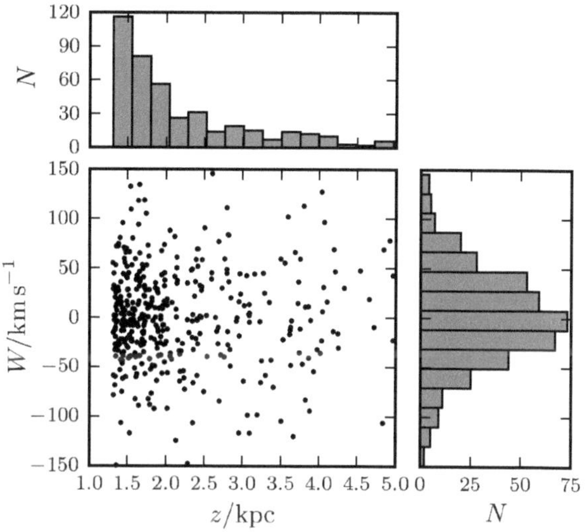

Fig. 7.3 Sample of 412 stars drawn from the Binney distribution function that emulates the MB sample

For $z \leqslant 2.5$ kpc, we expect a non-negligible thin disc contamination. MB try to isolate the thick disc contribution by only fitting the wings of the distribution, where, as the thick disc has a higher velocity dispersion than the thin disc, the data are assumed to be contributed by thick disc stars. Therefore, we first sort all the data in the bin and assign each an expected deviation but only fit a straight line to the data with $|W| > 30\,\mathrm{km\,s^{-1}}$. Each wing is fitted separately. MB do not make it completely clear how they combine the fits of each wing but here we adopt the procedure of fitting each wing independently and then calculating the mean and standard deviation by an average of the intercepts and gradients respectively.

MB also ignore any points that seem to be outliers in the probability plot when performing the linear regression. We simulate this effect by ignoring the most negative and most positive data point when fitting a straight line to the probability plot.

7.5 Errors

7.5.1 Observational Errors

MB estimate the errors in their calculated W moments by essentially only considering the observational error in the W velocity as follows. In each bin, MB add random Gaussian errors for the distance and line-of-sight velocity to each data point to generate 1000 samples but do not re-bin the data at all. The errors are then estimated as the standard deviation of the estimates obtained from each sample. Following MB, we take the W velocity error to be $0.7\,\mathrm{km\,s^{-1}}$. The results of this procedure are shown in Fig. 7.4. The error bars are very small giving the impression we have

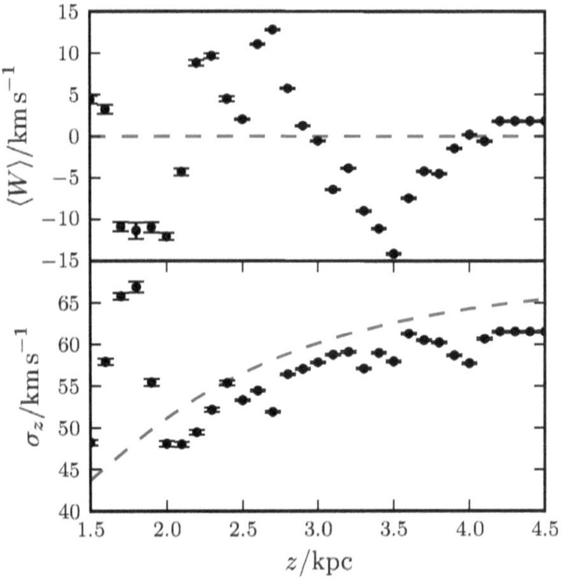

Fig. 7.4 Mean W velocity and W velocity dispersion, σ_z, against height above Galactic plane. The points give the values calculated in each bin using the MB method along with the errors estimated by 1000 samples adding random errors of $0.7\,\mathrm{km\,s^{-1}}$ to the W velocity. The *red dashed lines* show the exact mean velocity and velocity dispersion calculated from the Binney distribution function from which the sample was drawn. Clearly, the error bars do not give a good estimate of the deviation from the true value

very precise results. However, the data are clearly scattered around the true result by amounts much greater than the error bars. This is because we have ignored two much larger sources of error: the errors in the distances moving stars from bin to bin and the Poisson noise.

MB estimate the distance error to be approximately 20 %. As well as a W velocity error, we add a random Gaussian error of 20 % to the z coordinates of the data, re-bin the data and recalculate the mean velocities and dispersions. Repeating this 1000 times, we calculate the errors as the standard deviations of the estimates. These results are shown in Fig. 7.5. The observational errors are now much larger and the results are consistent with the truth within the errors. We have not yet made any estimate for the Poisson noise of the estimate, but it seems that, as the data are consistent with the truth, the observational errors are of the same order as the Poisson noise.

Lines marking one standard deviation are shown in Fig. 7.5. Adding errors in z to the data shifts stars from bin to bin. With such a large distance error, we have many samples in which higher velocity stars have been pulled down to lower Galactic heights and lower velocity stars are displaced to greater heights. This has the effect of flattening the velocity distribution when averaged over the many samples.

7.5.2 Poisson Noise and Systematics

With such a small sample, it is difficult to disentangle the Poisson noise from systematic errors arising from the MB procedure. However, as we have direct access to

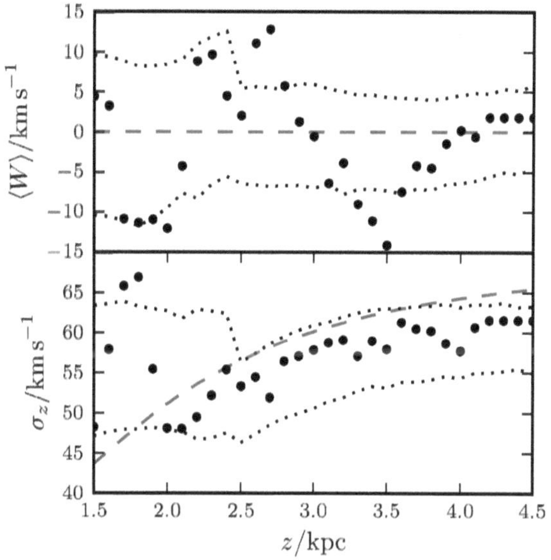

Fig. 7.5 Mean W velocity and W velocity dispersion, σ_z, against height above Galactic plane for the sample of 412 stars. The points give the values calculated in each bin using the MB method. The *dotted lines* show the one standard deviation limits. These were estimated by calculating the standard deviations of the calculated values for 1000 samples formed by adding random errors of $0.7\,\mathrm{km\,s^{-1}}$ to the W velocity and a 20 % error on the distance with re-binning. The *red dashed lines* show the exact mean velocity and velocity dispersion calculated from the Binney distribution function from which the sample was drawn

the distribution function, we can estimate the Poisson noise by repeatedly drawing samples from the distribution function and evaluating the observables for each sample. Therefore, we draw 100 samples of 412 stars and repeat the above procedure for each sample. We then estimate the average sample mean and dispersion in each bin along with the Poisson error in both quantities by calculating the standard deviations. The results are shown in Fig. 7.6.

The calculation of the mean W velocity is entirely consistent with being zero as required, but the recovery of the W velocity dispersion curve is less successful. For low Galactic heights, we are overestimating the velocity dispersion, whilst, for larger Galactic height, we are slightly underestimating the dispersion.

For low z, we are ignoring all data for which $|W| < 30\,\mathrm{km\,s^{-1}}$ when fitting a straight line to the data. This means we give more weight to data with higher W velocities and so the distribution seems broader than it actually is. The probability plot is particularly sensitive at the wings. If we consider an ordered data set that is drawn from a known underlying Gaussian distribution, we can assign each a value of c_i by the method outlined in Sect. 7.2. If we add a single point that is smaller than all the other data points but still drawn from the underlying distribution, the probability of the new point lying above the line with correct mean and standard deviation, but still lower than its neighbouring data point, is equal to the probability of it lying

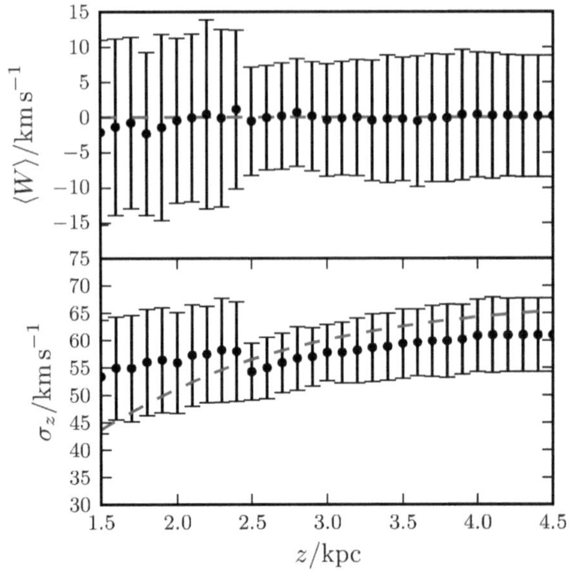

Fig. 7.6 MB method: Average mean W velocity and W velocity dispersion, σ_z, against height above Galactic plane for the 100 samples of 412 stars. The points give the average values calculated in each bin using the MB method. The error bars show the standard deviation of the calculated values in each bin. The *red dashed lines* show the velocity moments of the thick disc calculated using the underlying distribution function. The dispersion is overestimated at low z and slightly underestimated at high z leading to an underestimated gradient

beneath the line. As there is a much larger range of values below the line than above, the estimated gradient in this region will in general be overestimated. We need to use a sufficient number of stars to perform the linear fit in order to reduce this effect.

At high z, we have very few stars in the sample, so, in order to fill the bin with enough stars, we must include stars at lower z. In general, these stars have smaller velocities so the resulting velocity dispersion for the bin is reduced. A very minor effect may also be due to removing stars with $|W| > 150\,\mathrm{km\,s^{-1}}$ to avoid halo contaminants, so the distribution is unfairly weighted by low-velocity stars and the dispersion is underestimated. From Fig. 7.6, we can perform a simple linear fit to the (z, σ_z) plot to find that the data points imply a gradient of $\mathrm{d}\sigma_z/\mathrm{d}z = 2.25\,\mathrm{km\,s^{-1}\,kpc^{-1}}$. A linear fit to the true dispersion curve gives a gradient of $\mathrm{d}\sigma_z/\mathrm{d}z = 6.82\,\mathrm{km\,s^{-1}\,kpc^{-1}}$, so the MB method underestimates the gradient by a factor of three.

7.6 Comparison with Other Work

MB state that their gradient of σ_z with z is shallower than previous authors' work. The sample studied by MB is a subset of the sample studied by Girard et al. (2006; hereafter G06). These authors found a gradient of the V velocity dispersion a factor of two higher than MB, which MB claim is due to G06 not removing thin disc and halo contaminants in their analysis. We conclude by following a method inspired by the method used by G06 to test whether their results are more secure than those reported by MB. The G06 method is very similar to that used by MB. The authors have a sample of approximately 1200 stars. They split each data point into 100 subunits to account for the distance error and form bins of 100 subunits to estimate the U and V velocity dispersion. The probability plot method is used to estimate the dispersion, but crucially only the central 80 % of the data is used in the linear fit and the central region is not excluded for any of the bins. We follow a similar, but simpler, method on each of the 100 samples of 412 stars. We do not split the data points into subunits. We use bins with centres spaced by 0.1 kpc of variable width such that they each contain 32 data points as we have only a third of the number of data points in the G06 sample. We use only the central 80 % of the data in each bin for the linear fit in the probability plot method. Figure 7.7 shows the result of this experiment.

The estimate of the velocity dispersion provided by the G06 method is more reliable than the MB method. At low z, the dispersion is now marginally underestimated

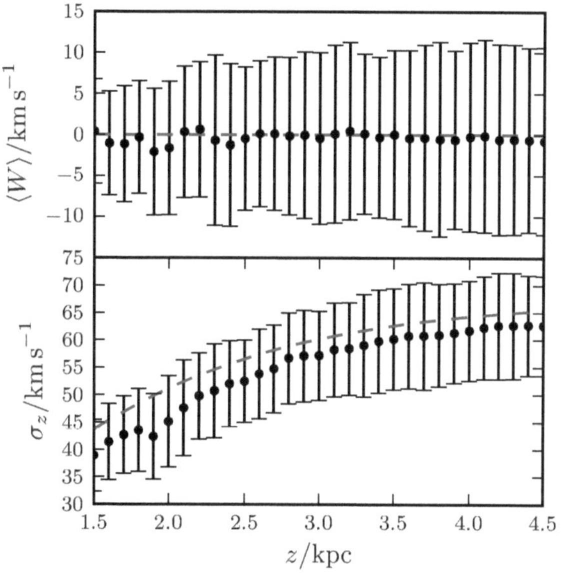

Fig. 7.7 Girard et al. method: average mean W velocity and W velocity dispersion, σ_z, against height above Galactic plane for the 100 samples of 412 stars. The points give the average values calculated in each bin using a method inspired by Girard et al. (2006) outlined in Sect. 7.6. The error bars show the standard deviation of the calculated values in each bin. The *red dashed lines* show the velocity moments of the thick disc calculated using the underlying distribution function. The dispersion is slightly underestimated due to the neglected thin disc contamination and preferentially sampling stars from lower Galactic heights, but in general the recovery of the truth is more successful than using the MB method. In particular the calculated gradient in σ_z is approximately a factor of three larger

which is to be expected due to the thin disc contamination. However, even with fewer stars in the bin, the error in the dispersion at low z is smaller than the equivalent errors in the MB method. This is a clear reflection of the dangers of only using the wings of the distribution to calculate the dispersion. At high z, the problem of preferentially sampling stars at lower heights seems to also have been slightly reduced as the bin size is small enough for the dispersion to be calculated using only local stars. A simple linear fit to the (z, σ_z) plot gives a gradient of $\mathrm{d}\sigma_z/\mathrm{d}z = 7.87\,\mathrm{km\,s^{-1}\,kpc^{-1}}$.

7.7 Conclusions

We have drawn a sample of 412 stars from the distribution function of Binney (2012b). The sample was chosen to replicate the sample presented by MB. We performed the same procedure as MB to extract the mean vertical velocity and vertical velocity dispersion of the thick disc as a function of Galactic height and compared it to the known moments of the vertical velocity of the thick disc of the underlying distribution. We find that the variation of the dispersion with z is far noisier than that found by MB. There are three sources of errors: Poisson errors, distance errors and velocity errors. We show that the observational errors in the velocities, which MB claim as the total error, cannot account for the deviation. Distance errors moving stars from bin to bin and the Poisson noise have a comparable effect on the dispersion estimate, which is much larger than effect of the velocity errors. Both of these sources of errors are neglected by MB.

A large number of samples reveals that the method systematically overestimates the dispersion at low z and underestimates it at high z. The two effects combined lead to a flatter curve of dispersion against z. We recalculated the mean velocity and velocity dispersion using a method inspired by G06, which reveals the causes of these two effects: at low z, only fitting the wings of the distribution to remove thin disc contaminants makes the distribution appear broader, and at high z, large bin sizes preferentially sample stars at lower Galactic height that, in general, have a lower velocity. The G06 method produces a much better fit to the expected velocity dispersion and there is approximately a factor of three discrepancy in the gradient of the dispersion as a function of z between the results of the MB and G06 methods. This discrepancy is not a result of more precise measurements or analysis but purely a result of systematics in the data analysis introduced by MB. We have not touched upon the values given by MB for the U and V dispersions, but similar effects are expected to occur.

The results presented here should serve as a useful demonstration of the expected errors and potential biases that arise when using a method similar to the MB method. We have demonstrated the need to understand the errors and systematics of methods that are applied to observational data, and that pseudo-samples from realistic Galaxy distribution functions are a useful tool in this respect. The effect that the biases and errors demonstrated in this section have on the dark matter mass estimates (Moni Bidin et al. 2012b; Bovy and Tremaine 2012) is beyond the scope of this chapter. However, the results of this study are relevant to both these mass estimate determinations and, more generally, to the understanding of Galactic disc kinematics.

References

Aumer M, Binney JJ (2009) MNRAS 397:1286
Binney J (2010) MNRAS 401:2318
Binney J (2012a) MNRAS 426:1324
Binney J (2012b) MNRAS 426:1328
Binney J et al (2014) MNRAS 439:1231
Binney J, McMillan P (2011) MNRAS 413:1889
Bochanski JJ, Munn JA, Hawley SL, West AA, Covey KR, Schneider DP (2007) AJ 134:2418
Bovy J et al (2012) ApJ 759:131
Bovy J, Tremaine S (2012) ApJ 756:89
Dehnen W, Binney J (1998) MNRAS 294:429
Garbari S, Liu C, Read JI, Lake G (2012) MNRAS 425:1445
Gilmore G, Reid N (1983) MNRAS 202:1025
Girard TM, Korchagin VI, Casetti-Dinescu DI, van Altena WF, López CE, Monet DG (2006) AJ 132:1768
Kuijken K, Gilmore G (1989) MNRAS 239:605
Moni Bidin C, Carraro G, Méndez RA et al (2012a) ApJ 747:101
Moni Bidin C, Carraro G, Méndez RA, Smith R (2012b) ApJ 751:30
Oort JH (1932) Bull. Astron. Inst. Neth. 6:249
Piffl T et al (2014) ArXiv e-prints
Robin AC, Reylé C, Derrière S, Picaud S (2003) A & A 409:523
Sanders J (2012) MNRAS 425:2228
Sharma S, Bland-Hawthorn J, Johnston KV, Binney J (2011) ApJ 730:3
Zhang L, Rix H-W, van de Ven G, Bovy J, Liu C, Zhao G (2013) ApJ 772:108

Chapter 8
Extended Distribution Functions for the Galactic Disc

8.1 Introduction

In the previous chapter, we have seen the importance of fully dynamical distribution functions for the Galaxy. Binney (2010) constructed appropriate action-based distribution functions for the Galactic disc and Binney (2012b) showed they provided good fits to the Geneva-Copenhagen Survey data. In Binney et al. (2014), these distribution functions were used to predict the RAVE data. The distribution functions provide a physically-motivated model of the Galaxy with which we can test methods that are applied to data. In the previous chapter, we presented these distribution functions and used them to inspect a recent method used to extract from data the velocity dispersions of a population.

These distribution functions are purely dynamical—the thin disc is divided by age but, without chemical information, this acts purely a parametrisation. Such distribution functions have limited use when modelling real data sets. All data sets are subject to some selection function such that the population we observe is not a fair sample of the entire Galactic population. In particular, all surveys have a selection in magnitude, which affects the resulting age and metallicity distributions. It is well known that stars of different chemical properties, such as age and metallicity, have very different kinematic properties. For instance, scattering mechanisms in the Galactic disc cause older metal-poor populations to have larger velocity dispersions than their younger metal-rich counterparts. These chemo-dynamical correlations are important, as it means any selection in age or metallicity is also a selection in kinematics. Therefore, any distribution function that does not include metallicity information is of limited utility.

Additionally, the correlations between chemistry and kinematics reveal information regarding the history and evolution of the Galaxy. The correlations between these observables in the data help constrain, for instance, the magnitude of the radial migration in the disc (Sellwood and Binney 2002; Schönrich and Binney 2009) as well as the heating mechanisms that thicken the discs. One area of recent interest is the metallicity correlations with Galactocentric radius and Galactic height (e.g. Lee

J. Sanders, *Dynamics of the Milky Way*, Springer Theses,
DOI 10.1007/978-3-319-18772-3_8

et al. 2011; Boeche et al. 2013). Due to the differing selection functions for each survey, the reported gradients are different, and it is difficult to disentangle whether the differences are due purely to differing selection volumes or metallicity-scale discrepancies between the surveys. A full DF with chemical information helps to disentangle these effects.

The Galaxy is described by a single potential, whereas the distribution function can be composed of many distinct components each associated with a distinct stellar population. For dynamical modelling, we can choose to divide up the Galaxy using any stellar properties except the actions e.g. metallicity, age, mass, chemical abundances etc. Attempting to identify a population of stars from kinematics alone is flawed as populations mix in phase-space due to dynamical processes. However, the chemical composition of a star is fixed, and may be used to tag stellar populations (Mitschang et al. 2013, 2014). Each population must be described by a distribution function that produces a consistent velocity and density structure when viewed in the correct potential. As each population must live in the same potential, this puts a strong constraint on the Galactic potential (e.g. Walker and Peñarrubia 2011). The Galactic disc is conventionally divided into two populations: the thin disc and thick disc. These populations overlap in velocity space, but are believed to have distinct ages. The prospect for obtaining sufficiently accurate ages in the near term are poor. However, the addition of chemistry provides more information about the separation of these two populations. Bovy et al. (2012) split up the SEGUE G dwarf sample into 'mono-abundance populations'—bins in $[\alpha/\mathrm{Fe}]$ and $[\mathrm{Fe}/\mathrm{H}]$. They argued that each population had a simple spatial and velocity structure, and Ting et al. (2013) showed that such populations could be modelled using the distribution functions of Binney. Bovy and Rix (2013) used these distribution functions to constrain the mass profile of the disc by requiring that each mono-abundance population produced consistent dynamics.

In this chapter, we present an approach to extending the distribution functions from Binney (2012b) to include metallicity information. We include the expected correlations between the actions, age and metallicity inspired by the models of Schönrich and Binney (2009). These models can be used to correctly account for the selection effects of a survey, to uncover the chemo-dynamical correlations that reveal information on the Galactic history and evolution, and as a basis on which to constrain the Galactic potential. This approach is different to that taken by Bovy et al. (2012) as the properties of each stellar population are chosen to vary smoothly with analytic dependence on the metallicity, rather than considering discrete metallicity bins. In Sect. 8.2, we present these extended distribution functions. In Sect. 8.3, we discuss the data that will be relevant for using our extended distribution functions. In Sect. 8.4, we discuss selection functions and how one can model the kinematics of a data set without explicitly modelling the selection function. In Sect. 8.5, we go on to fit the parameters of our extended distribution function to the GCS data. In Sect. 8.6, we construct mock catalogues for the GCS and SEGUE G dwarfs from the extended distribution functions.

8.2 Model

In this section, we present and discuss the extended distribution function (EDF). An EDF is the joint distribution function of the phase-space coordinates $(\boldsymbol{x}, \boldsymbol{v})$, and any additional properties of each star, such as ([Fe/H], [α/Fe], T_{eff}, $\log g, \ldots$). Here we discuss extending the usual phase-space distribution function (DF) with the metallicity of the stars, [M/H] (for our purposes we will assume zero α-enhancement for the stars, and so use [M/H] and [Fe/H] interchangeably). This EDF then forms our chemo-dynamical model of the Galaxy. We use as our starting point the DF from Binney (2012b) discussed in the previous chapter. Our EDF is inspired by the models of Schönrich and Binney (2009), who fully model the joint chemical and dynamical evolution of the Galactic disc. Our aim here is to model effectively the chemical evolution part of the Schönrich and Binney (2009) model. In our EDF, the metallicity of stars born at a given location and time are described by the metallicity of the ISM. We begin by discussing how we can effectively model the metallicity of the interstellar medium (ISM).

8.2.1 ISM Metallicity

We begin by assuming that the metallicity of the ISM is solely a function of radius at all times. Following the fuller chemical evolution model of Schönrich and Binney (2009), we choose to model the metallicity of the ISM as a function of age and radius as

$$[\text{Fe/H}](R,\tau) = F(R,\tau) \equiv F(R) + [F(R) - F_m]\Big[\tanh\Big(\frac{\tau_m - \tau}{\tau_F}\Big) - 1\Big]. \quad (8.1)$$

This is an analytic form that approximates the fuller chemical evolution model of Schönrich and Binney (2009). The model assumes that the entire Galaxy began with a metallicity of F_m a time $\tau_m = 12\,\text{Gyr}$ ago. The parameter τ_F controls the rate of enrichment at early times, and, as $\tanh(\tau_m/\tau_F) \approx 1$ for the choice of parameters used later, the metallicity in the disc with radius at the present time is approximately $F(R)$. We adopt a current metallicity-radius relation given by

$$[\text{Fe/H}](R,0) \approx F(R) \equiv F_m \tanh\Big\{\frac{F_R(R - R_F)}{F_m}\Big\}, \quad (8.2)$$

with F_R and R_F as constants. There is an approximately linear metallicity gradient with radius at radii near the Sun governed by F_R, with the parameter R_F controlling the metallicity of stars born at the solar position today. The tanh in the formula for $F(R)$ stops the metallicity decreasing with time for the outermost radii such that the metallicity remains approximately constant with time at large radii.

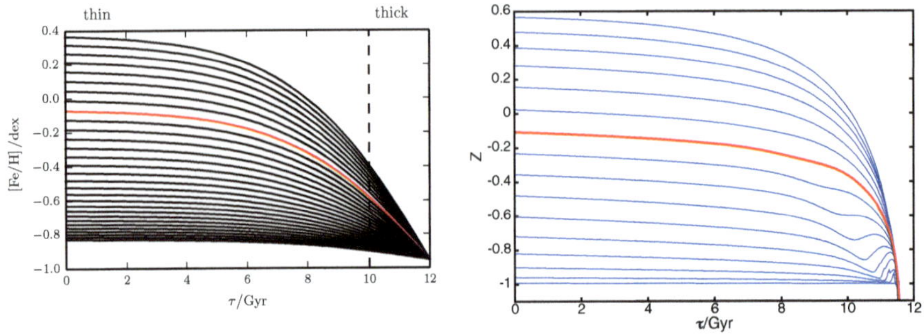

Fig. 8.1 Metallicity against age for our EDF (*left*) and the Schönrich and Binney (2009) model (*right*, reproduced with permission from Ralph Schönrich). On the left, each line shows the metallicity against age for birth radii linearly spaced by 1 kpc between the uppermost line, corresponding to $R = 0$ kpc, and the lowest line, corresponding to $R = 30$ kpc. On the right, the lines are separated by 1.25 kpc

In Fig. 8.1, we plot the metallicity against age for a series of birth radii for the choice of parameters used later, along with the equivalent plot from Schönrich and Binney (2009). In the models of Schönrich and Binney (2009), the innermost radii are enriched with metals most rapidly, metals are advected inwards by flows in the disc and the outer radii are diluted by infalling metal-poor gas. These authors also use a particularly steep metallicity gradient with radius of $F_R \approx -0.082$ dex/kpc.

8.2.2 Extended Distribution Function

With these analytic metallicity relationships, we can extend the DF of Eq. (7.11). The DF is a function of the actions, $\boldsymbol{J}$, such that $f(\boldsymbol{x}, \boldsymbol{v}) \equiv f(\boldsymbol{J}(\boldsymbol{x}, \boldsymbol{v}))$, where we find the actions given $(\boldsymbol{x}, \boldsymbol{v})$ using the algorithm of Binney (2012a) described in Chap. 2. For speed, we use an interpolation scheme to find the actions detailed in Binney (2012a). We decompose our DF into thin and thick disc components as

$$f(\boldsymbol{J}) = (1 - \mathcal{F}) f_{\text{thin}}(\boldsymbol{J}) + \mathcal{F} f_{\text{thick}}(\boldsymbol{J}), \tag{8.3}$$

where

$$f_i(\boldsymbol{J}) = \int \mathrm{d}\tau\ \Sigma_i(\tau) f_{i,\sigma_R(\tau),\sigma_z(\tau)}(\boldsymbol{J}), \quad i = (\text{thin, thick}), \tag{8.4}$$

The star-formation rates, Σ_i, are given by

$$\Sigma_{\text{thin}}(\tau) = \begin{cases} \dfrac{\mathrm{e}^{\tau/\tau_f}}{\tau_f(\mathrm{e}^{\tau_T/\tau_f} - 1)} & \text{if } \tau \leq \tau_T, \\ 0 & \text{otherwise,} \end{cases} \quad \text{and } \Sigma_{\text{thick}}(\tau) = \begin{cases} \dfrac{1}{\tau_m - \tau_T} & \text{if } \tau_T \leq \tau \leq \tau_m, \\ 0 & \text{otherwise.} \end{cases} \tag{8.5}$$

The velocity dispersion parameters for the thin disc are age-dependent and follow the form given in Eq. (7.11). Here we set $\beta_R = 0.33$ and $\beta_z = 0.4$. For the thick disc, the velocity dispersion parameters only depend upon angular momentum and are constant with age. We drop the σ subscripts in what follows for clarity, and instead write $f_{i,\sigma_R(\tau),\sigma_z(\tau)}(\boldsymbol{J})$ as $f_i(\boldsymbol{J},\tau)$ (here the age argument for the thick disc only controls the star-formation rate, which is zero for $\tau < \tau_T$).

We extend the distribution function in Eq. (8.4) by writing

$$f_i(\boldsymbol{J},[\mathrm{Fe/H}]) = \int \mathrm{d}\tau\,\Sigma_i(\tau) f_i(\boldsymbol{J},\tau)\delta([\mathrm{Fe/H}] - F(R_c,\tau)). \tag{8.6}$$

This equation assumes all stars were born cold in the disc at some radius $R = R_\mathrm{c}$ with some metallicity [Fe/H], which together define a unique age. The stars are then heated in time, whilst their guiding radii remain fixed. However, we know that churning plays an important role in the evolution of the Galaxy (Sellwood and Binney 2002; Schönrich and Binney 2009). Around corotation of a rotating perturbing potential, stars may change their angular momentum without being significantly heated. This phenomenon brings metal-rich stars born in the inner Galaxy and metal-poor stars born in the outer Galaxy to the solar radius. Therefore, a star's current angular momentum, L_z, will, in general, be different from its birth angular momentum, L_z', which we assume is drawn from a Gaussian distribution with dispersion σ_L that grows with time:

$$\sigma_L(\tau) = \sigma_{L0}\left(\frac{\tau}{\tau_m}\right)^{\gamma_T}. \tag{8.7}$$

We assume a true random walk and set $\gamma_T = 0.5$.

With churning included, the EDF becomes

$$f_i(\boldsymbol{J},[\mathrm{Fe/H}]) = \int \mathrm{d}L_z' \int \mathrm{d}\tau\,\Sigma_i(\tau)\mathcal{N}(L_z')\frac{\mathrm{e}^{-(L_z-L_z')^2/2\sigma_L^2}}{\sqrt{2\pi\sigma_L^2}} f_i(\boldsymbol{J}',\tau)\delta([\mathrm{Fe/H}] - F(R_\mathrm{c}',\tau)), \tag{8.8}$$

where $\boldsymbol{J}' \equiv (J_R, L_z', J_z)$, $R_\mathrm{c}' \equiv R_\mathrm{c}(L_z')$ and $\mathcal{N}(L_z')$ is a normalization factor given by

$$\mathcal{N}(L_z') = 2\left[1 + \mathrm{erf}\left(\frac{L_z'}{\sqrt{2}\sigma_L}\right)\right]^{-1}, \tag{8.9}$$

which ensures $(2\pi)^3 \int \mathrm{d}L_z'\,\mathrm{d}\tau\,\mathrm{d}[\mathrm{Fe/H}]\,\mathrm{d}^3\boldsymbol{J}\,f = 1$ as shown in Appendix H. The error function appears in $\mathcal{N}$ as the angular momentum is constrained to be positive. $\mathcal{N}$ has very little impact at the solar position for the value of σ_L chosen later.

An important consequence of the introduction of the convolution of the DF with a Gaussian in L_z is that the EDF is no longer exponential in guiding-centre radius. The basic DF was only approximately exponential in radius due to the velocity dispersions in R and z. After deriving the best-fitting parameters later, we will check whether the disc is exponential.

There is a small complication here due to the simplistic approach to the modelling we have taken. In the DF approach from Binney (2010), the dynamical processes (e.g. heating, migration etc.) are all included implicitly in the DF equation. However, we have now identified the migration of stars and included it explicitly in the EDF equation as it helps us to include the metallicity dependence. For each birth angular momentum, we are evaluating the DF at the birth actions, which will be different from the current actions. This means that the heating mechanisms included implicitly in the DF will behave as if the star had spent its entire lifetime at its birth radius. In reality, the total heating will be some combination of heating from the different environments experienced throughout its lifetime.

For a particle undergoing a random walk, the most probable path between L_z and L_z' is linear such that, on average, a star at the Sun has experienced the mean of the heating events at the angular momentum passed through. However, the number of migrated stars must be determined by the star-formation rate at its birth actions. Therefore, we opt for a compromise: the 'density' part of the DF ($\mathrm{e}^{-R_\mathrm{c}/R_\mathrm{d}}$) is evaluated at the birth actions, whilst the 'heating' part ($\mathrm{e}^{-\kappa J_R/\sigma_R^2}\mathrm{e}^{-\nu J_z/\sigma_z^2}$) is evaluated at the current actions. Solway et al. (2012) showed that under radial migration a star's vertical action is conserved. Therefore, in the absence of any heating, the vertical action of a star observed today should be identical to that at birth. Additionally, Minchev et al. (2012) showed from a numerical simulation that the final velocity dispersions of stars that had migrated into a given radial bin matched the final velocity dispersions of those stars that had spent their entire lifetime in that radial bin. This gives us confidence that our approximation is a valid one.

8.2.3 Full DF Evolution with a 3D Action-Space Kernel

The most rigorous procedure for constructing an extended distribution function would be to assume a form for the distribution function for each population at birth (e.g. cold exponential discs), and evolve these with a 3D action-space kernel. The general solution to the Fokker-Planck equation is given by

$$f(\boldsymbol{J},\tau) = \int \mathrm{d}^3\boldsymbol{J}'\, T(\boldsymbol{J},\boldsymbol{J}',\tau) f_0(\boldsymbol{J}',\tau), \tag{8.10}$$

where $T(\boldsymbol{J},\boldsymbol{J}',\tau)$ is the probability of a star of age τ scattering from $\boldsymbol{J}$ to $\boldsymbol{J}'$ in time τ, and $f_0(\boldsymbol{J}',\tau)$ is the DF of a population of age τ at birth. Here we have in essence followed this procedure but with several simplifying assumptions to make the models faster to calculate. Our action-space kernel only depends on the age, initial action and final action, and not the series of actions that a given star could pass through. Each population is born as a δ-function in J_R' and J_z' and an exponential in R_c' such that

$$f_0(\boldsymbol{J}',\tau) = \Sigma(\tau)\delta(J_R')\delta(J_z')\frac{2\Omega(L_z')R_c'}{\kappa^2(L_z')R_d^2(\tau)}\mathrm{e}^{-R_c'/R_d(\tau)}. \tag{8.11}$$

We assume $R_d(\tau) = R_d$—inside-out formation would require a scale-length that decreases with age. The kernel for the population of age τ is given by

$$T(\boldsymbol{J}, \boldsymbol{J}', \tau) = \frac{\kappa(L_z)}{\sigma_R^2(L_z, \tau)} \mathrm{e}^{-\kappa(L_z)(J_R - J_R')/\sigma_R^2(L_z)} \frac{\mathcal{N}(L_z')}{\sqrt{2\pi\sigma_L^2(\tau)}} \mathrm{e}^{-(L_z - L_z')^2/2\sigma_L^2(\tau)}$$
$$\times \frac{\nu(L_z)}{\sigma_z^2(L_z)} \mathrm{e}^{-\nu(L_z)(J_z - J_z')/\sigma_z^2(L_z,\tau)}. \tag{8.12}$$

The J_R and J_z kernels depend upon the final angular momentum. We can perform the integral over both J_R' and J_z' using the δ-functions, and the resulting distribution function for the population of age τ is the one given in Eq. (7.4). Note here that the choice of kernel T should conserve total z-component of the angular momentum. Our choice does not, as it causes the disc to broaden. Schönrich and Binney (2009) make the transition probability depend upon the product of the mass of adjacent annuli, such that T depends upon f, and explicitly conserves angular-momentum. Whether there are more general, physically-motivated choices of angular-momentum kernel is unclear.

8.2.4 Performing the Integrals

To evaluate Eq. (8.8), we need to evaluate a double integral over τ and L_z'. Fortunately, one of these integrals is trivial as the integrand contains a δ-function. When we use the δ-function to evaluate the integral, we obtain derivatives of F with respect to either τ or L_z' in the denominator of the integrand. The choice of whether to perform the τ or L_z' integral first depends upon the properties of these derivatives over the integration range. We note that $\partial F/\partial R_c(\tau = \tau_m) = 0$, and $\partial F/\partial\tau(\tau = 0) \approx 0$. Therefore, for the thin disc we use the δ-function to perform the L_z' integral and for the thick disc we use it to perform the τ integral. For the thin disc, we obtain

$$f_{\text{thin}}(\boldsymbol{J}, [\text{Fe/H}]) = \int_0^{\tau_T} \mathrm{d}\tau \, \frac{\Sigma_{\text{thin}}(\tau) f_{\text{thin}}(\boldsymbol{J}', \tau)}{|\partial F/\partial R_c||\partial R_c/\partial L_z|} \frac{\mathrm{e}^{-(L_z - L_z')^2/2\sigma_L^2}}{\sqrt{2\pi\sigma_L^2}}, \tag{8.13}$$

where L_z' is given by $F(R_c(L_z'), \tau) = [\text{Fe/H}]$, which may be inverted analytically. Additionally, we have that

$$\frac{\partial F}{\partial R_c}(R, \tau) = -F_R \operatorname{sech}^2\Big(\frac{F_R(R - R_F)}{F_m}\Big) \tanh\left(\frac{\tau_m - \tau}{\tau_F}\right),$$
$$\frac{\partial R_c}{\partial L_z} = \frac{2\Omega}{R_c \kappa^2}. \tag{8.14}$$

For the thick disc, we have

$$f_{\rm thick}(\boldsymbol{J}, [{\rm Fe/H}]) = \int_0^\infty {\rm d}L'_z\, \Sigma_{\rm thick}(\tau) \frac{{\rm e}^{-(L_z-L'_z)^2/2\sigma_L^2}}{\sqrt{2\pi\sigma_L^2}} \frac{f_{\rm thick}(\boldsymbol{J}')}{|\partial F/\partial\tau|}, \qquad (8.15)$$

where

$$\frac{\partial F}{\partial \tau} = \frac{1}{\tau_F}(F(R) - F_m)\,{\rm sech}^2\Big(\frac{\tau_m - \tau}{\tau_F}\Big). \qquad (8.16)$$

For convenience, we limit the integration range to $\pm 3\sigma_{L0}$ and perform the integral over R'_c. These two 1D integrals are then performed numerically using a 5-point Gaussian quadrature scheme.

8.2.5 *Halo* EDF

One practical problem with the above EDF is that any star that falls outside the allowed range in [Fe/H] (e.g. $[{\rm Fe/H}] < F_m$) is deemed unphysical by the model. This problem can be solved by the inclusion of a halo distribution function. The data we are considering are not very sensitive to the halo, but its inclusion allows us to assign any 'unphysical' star to the halo.

We construct a simple action-based distribution function for the halo of the form (Posti et al. 2015)

$$f_{\rm halo}(\boldsymbol{J}) = \frac{k_{\rm halo}}{(J_0 + J_R + 0.68|L_z| + 0.7J_z)^3}. \qquad (8.17)$$

This model has a simple power-law density profile with a core that is specified by the parameter J_0. We choose $J_0 = 180\,{\rm km\,s^{-1}\,kpc}$. This model has a density profile $\rho \propto r^{-3}$ (i.e. NFW Navarro et al. 1996) outside a scale radius of $r \approx 5\,{\rm kpc}$, and $\rho \approx$ const. inside, and has velocity dispersions at the Sun of $\sigma_U \approx \sigma_W \approx 130\,{\rm km\,s^{-1}}$ (Brown et al. 2010). The factors multiplying $|L_z|$ and J_z are approximately Ω_ϕ/Ω_R and Ω_z/Ω_R at the solar position such that the halo model is approximately isotropic. In addition to this action-based part, we include a simple Gaussian in metallicity such that our halo EDF is given by

$$f_{\rm halo}(\boldsymbol{J}, [{\rm Fe/H}]) = f_{\rm halo}(\boldsymbol{J}) \frac{{\rm e}^{-([{\rm Fe/H}]-F_h)/2\sigma_F^2}}{\sqrt{2\pi\sigma_F^2}}. \qquad (8.18)$$

We set the mean metallicity as $F_h = -1.5$ dex and the width of the metallicity distribution function as $\sigma_F = 0.5$ dex. We assume all stars in the halo are of age 12 Gyr. In what follows, the weight of the halo, $k_{\rm halo}$, is allowed to vary, but we expect that it will be such that the halo contributes ~0.1 % in the solar neighbourhood (Jurić et al. 2008).

8.3 Data

Here we present and discuss the data that will prove useful for fitting and comparing with our model. For working with the extended distribution functions, we require 7D data (6D phase-space coordinates and the metallicity [M/H]). We use 7D data from the GCS and SEGUE survey, complemented by the stellar density data from Gilmore and Reid (1983). Additionally, we use the peculiar solar velocity from Schönrich et al. (2010), we place the Sun at $R_0 = 8$ kpc and $z_0 = 0.014$ kpc (Binney et al. 1997), and we use the adjusted Dehnen and Binney (1998) potential detailed in the previous chapter.

8.3.1 Geneva-Copenhagen Survey

The Geneva-Copenhagen Survey (GCS) (Nordström et al. 2004) is a sample of 16682 nearby F and G stars extending out to ~200 pc. Through a combination of $uvby\beta$ photometry, radial velocity, Hipparcos parallax and proper motion observations, the catalogue provides a view of the chemo-dynamical structure of the solar neighbourhood. We use the most recent re-analysis of the survey from Casagrande et al. (2011). These authors used the infrared flux method (IRFM) to produce more consistent effective temperature and metallicity scales. This re-analysis found that the stars were on average 0.1 dex more metal rich than in previous analyses. We use all stars in the catalogue with proper motions and that were flagged by Casagrande et al. (2011) as having reliable metallicity determinations. This reduces the data set to 12723 stars.

The local nature of the GCS means that it is dominated by thin disc stars, and the influence of the thick disc is subtle (Binney 2012a). Due to the accuracy of the Hipparcos parallaxes, the GCS provides us with accurate velocity histograms for the solar neighbourhood. The accuracy of the parallaxes has led to the discovery of substructure in the solar neighbourhood (Dehnen 1998). Both the v_R and v_ϕ velocity distributions are rich with substructure (see Chap. 2). In particular, the peak of the v_ϕ distribution is due to the Hyades moving group, and the flat-top of the v_R distribution is due to the Hyades and Sirius moving groups. The v_z distribution is free of substructure (Dehnen 1998). The presence of this substructure is important when attempting to fit a zeroth order model as it makes the comparison of model and data more difficult to interpret.

The 7D data we use for each star are $(l, b, \varpi, v_{||}, \boldsymbol{\mu}, \mathrm{[Fe/H]})$, along with the corresponding errors, where ϖ is the parallax and the other symbols have their usual meanings. We adopt the reported errors in $(l, b, \varpi, v_{||}, \boldsymbol{\mu}$, and following Casagrande et al. (2011) we use $\sigma_{[\mathrm{Fe/H}]} = 0.12$ dex for all stars.

8.3.2 SEGUE G dwarfs

The Sloan Extension for Galactic Understanding and Exploration (SEGUE) (Yanny et al. 2009) is a low-resolution spectroscopic survey of stars fainter than 14th magnitude, complemented by *ugriz* photometry. As such, it provides a view of the outer parts of the disc, dominated by the thick disc, and the stellar halo of the Galaxy, and so complements the more local GCS sample. The SEGUE data are available as part of SDSS DR10 (Ahn et al. 2014). These data were reduced using an improved SEGUE Stellar Parameter Pipeline (SSPP) (Smolinski et al. 2011), which, like the latest GCS re-analysis, used the IRFM to produce more consistent effective temperatures. However, this did not significantly affect the obtained metallicities.

Here we use all SEGUE data that satisfy the SEGUE target selection criteria for G dwarfs. A SEGUE program 'G dwarf' is defined as lying in the colour-magnitude range $14 < r < 20.2$ and $0.48 < g - r < 0.55$. These selected stars produced valid parameter estimates from the SSPP, are fainter than $r < 15$ to ensure completeness at the bright end, and were not flagged as noisy or with a temperature mismatch. In addition, we also impose a cut in surface gravity ($\log g \geq 4.2$) to ensure we are only observing dwarf stars, remove all stars with SNR< 15, remove those stars in fields with $E(B - V) \geq 0.3$ from the Schlegel et al. (1998) extinction maps, and remove those stars with no measured radial velocity or proper motions. The final sample contains 18,575 stars.

We estimate the distances using the method presented in Schlesinger et al. (2012). The majority of the stars are from the outer disc, so we expect them to be old. We, therefore, assume all stars have an age of 10 Gyr. Using the 10 Gyr YREC isochrone provided by An et al. (2009), we first bracket the provided metallicity for each star with two isochrones. For each isochrone, we find the closest entry to the star's reported $(g - r)$ colour. The *ugriz* magnitudes are found by linearly interpolating between the two entries in each isochrone. The distance s is determined by averaging the five estimates from each of the extinction-corrected *ugriz* bands. We make no consideration of the errors in the colours, magnitudes and metallicities. A Bayesian distance-estimation algorithm, such as that presented by Burnett and Binney (2010), would be preferable. However, for dwarf stars, we expect the cruder estimates to be equally valid. Schlesinger et al. estimate the errors in the obtained distances using the method presented here. They estimate that there is a random distance uncertainty of 18 % for stars with $[\mathrm{Fe/H}] > -0.5$ dex, and 8 % for more metal-poor stars. Additionally, there are systematic distance uncertainties arising from the single-age assumption (expected to produce a -3 % shift for the metal-rich stars), the presence of undetected binaries and errors in the isochrones.

The 7D data we use for each star are $(l, b, s, v_{||}, \boldsymbol{\mu}, [\mathrm{Fe/H}])$, and associated errors.

8.3.3 Gilmore-Reid Density Curve

Gilmore and Reid (1983) measured the stellar density as a function of distance away from the Galactic plane, by observing a sample of K dwarfs towards the South Galactic Pole. This was the first study to indicate the existence of a thick disc.

8.4 Selection Functions

Before comparing our model to data, we must understand the selection effects of a survey. Here we will discuss how we include the selection function in our modelling approach, the selection functions for the GCS and the SEGUE survey, and how we can potentially avoid explicitly using the selection function. The selection function of a survey is the probability of a star being in the catalogue given its properties. The selection is nearly always done on the basis of the *observed* properties of a star. If we denote S as meaning 'in the survey', the probability of the data, D, given the model, M, and given it is in the survey, S, is given by

$$p(D|M, S) = \frac{p(S|D)p(D|M)}{p(S|M)}, \tag{8.19}$$

where we call $p(S|D)$ the selection function, $p(D|M)$ the distribution function, and $p(S|M)$ is a normalising constant that only comes into play when fitting the model to the data.

Often stars are selected on the basis of their colours and magnitudes. In the case of spectroscopic surveys, e.g. SEGUE and RAVE, the colours and magnitudes from a large photometric survey are used to choose stars that are then spectroscopically observed. To relate colours and magnitudes to the physically interesting stellar properties, e.g. age and metallicity, we must engage with isochrones. An isochrone gives the properties of a population of stars of age, τ, and metallicity, [Fe/H], each born with some initial mass, m. We use 19 BaSTI isochrones (Pietrinferni et al. 2004) spaced by $\sim$0.25 Gyr for $\tau < 2$ Gyr and 1 Gyr for $\tau > 2$ Gyr of 12 different metallicities listed in Table 8.1. We assume that all populations of fixed metallicity and age were born with a universal initial mass function (IMF), $\xi(m)$. We adopt the Kroupa et al. (1993) IMF given by

$$\xi(m) \propto \begin{cases} 0.035m^{-1.3} & \text{if } 0.08 \leq m < 0.5 \\ 0.019m^{-2.2} & \text{if } 0.5 \leq m < 1.0 \\ 0.019m^{-2.7} & \text{if } m \geq 1.0. \end{cases} \tag{8.20}$$

Table 8.1 Metallicities of the BaSTI isochrones used

Z	Y	[Fe/H]
0.00001	0.245	−3.27
0.0001	0.245	−2.27
0.0003	0.245	−1.79
0.0006	0.246	−1.49
0.001	0.246	−1.27
0.002	0.248	−0.96
0.004	0.251	−0.66
0.008	0.256	−0.35
0.01	0.259	−0.25
0.0198	0.2734	0.06
0.03	0.288	0.26
0.04	0.303	0.40

Here m is the mass of the star in units of the solar mass. With this choice, we can write down our full distribution function as

$$f(\boldsymbol{x},\boldsymbol{v},[\mathrm{Fe/H}],\tau,m)=f(\boldsymbol{x},\boldsymbol{v},[\mathrm{Fe/H}],\tau)\xi(m)=\int \mathrm{d}L_z'\, f(\boldsymbol{J},[\mathrm{Fe/H}],\tau,L_z')\xi(m). \tag{8.21}$$

For later purposes, the selection functions considered will be in colour, apparent magnitude, l and b. A combination of m, [Fe/H] and τ uniquely determine a colour and an absolute magnitude from the isochrones. Coupled with a distance, the absolute magnitude uniquely defines an apparent magnitude. Therefore, we can consider a selection in colour and magnitude as a selection in mass, metallicity, age and distance. We will assume that the form of this selection function is independent of the line of sight (l, b) and the line of sight only affects the scaling of the selection function i.e. the completeness. For later purposes, we will use the reported l and b to construct catalogues so the completeness is not an issue.

If we want to determine the distribution of the arguments of our distribution function, $\boldsymbol{X} = (\boldsymbol{x}, \boldsymbol{v}, \tau, [\mathrm{Fe/H}], L_z')$ with a selection of this form we write

$$p(\boldsymbol{X}|M,S) \propto \int \mathrm{d}m\, p(S|s,m,\tau,[\mathrm{Fe/H}])\xi(m)f(\boldsymbol{X}) = p(S|s,\tau,[\mathrm{Fe/H}])f(\boldsymbol{X}), \tag{8.22}$$

where $p(S|s,\tau,[\mathrm{Fe/H}]) = \int \mathrm{d}m\, p(S|s,m,\tau,[\mathrm{Fe/H}])\xi(m)$. $p(S|s,\tau,[\mathrm{Fe/H}])$ can be calculated independent of the dynamical model. This is useful as we must engage with the isochrones only once. The resulting pre-tabulation can then be interpolated for any choice of $(\tau, s, [\mathrm{Fe/H}])$ and any call outside the grid uses the nearest grid point. This formulation is also useful as we can still use the results from the previous section but with f replaced by $p(S|s,\tau,[\mathrm{Fe/H}])f$.

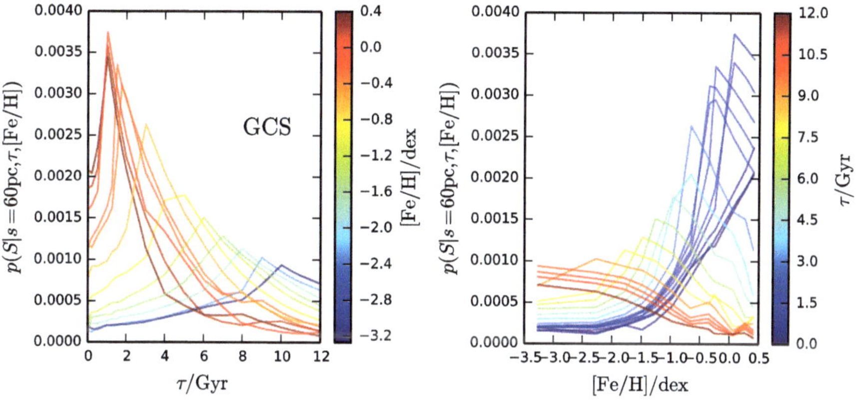

Fig. 8.2 GCS selection function for a star at 60 pc. The *left panel* shows the selection in age coloured by metallicity, and vice versa for the *right panel*

8.4.1 GCS Selection Function

For the GCS, we use the selection function in Schönrich and Binney (2009). This selection function was approximately constructed using the selection rules from Nordström et al. (2004) and comparison with the target catalogues. The result is a simple selection function in Strömgren colour $(b - y)$ and apparent magnitude, v. Following Schönrich and Binney (2009), we cut the isochrones at the bottom of the red giant branch reported in the BaSTI isochrones. This cut is not strict enough as it still includes many horizontal branch isochrone points not observed in the GCS. Therefore, we also cut all isochrone points with $M_y < 1$ and $M_y < -62.5(\log T_{\rm eff} - 3.78)$ to approximately reproduce the edge of the sample observed by Casagrande et al. (2011). For each isochrone, we form a grid in the logarithm of distance between a minimum and maximum value when the selection function falls to zero. At each distance s, we find $\int \mathrm{d}m\; p(S_{GCS}|s, m, \tau, [\mathrm{Fe/H}])\xi(m)$ for each isochrone. Thus, we construct a 3D grid that may be interpolated for any choice of $(s, \tau, [\mathrm{Fe/H}])$. The resulting selection function is shown in Fig. 8.2 for a star located at $s = 60\,$pc. We see that it peaks at around 2 Gyr, where the majority of stars in GCS lie (Casagrande et al. 2011).

8.4.2 SEGUE Selection Function

In Bovy et al. (2012), it is shown that the SEGUE G dwarf selection is uniform in $(g - r)$ and a near step function in r. The position of this step depends upon the plate P, in particular whether it is a bright or faint plate, and has the functional form

$$p(S|r, g - r, P) = \frac{W_P}{2}\left[1 - \tanh\left(\frac{r - r_{\rm cut} + 0.1}{\exp(-3)}\right)\right] \tag{8.23}$$

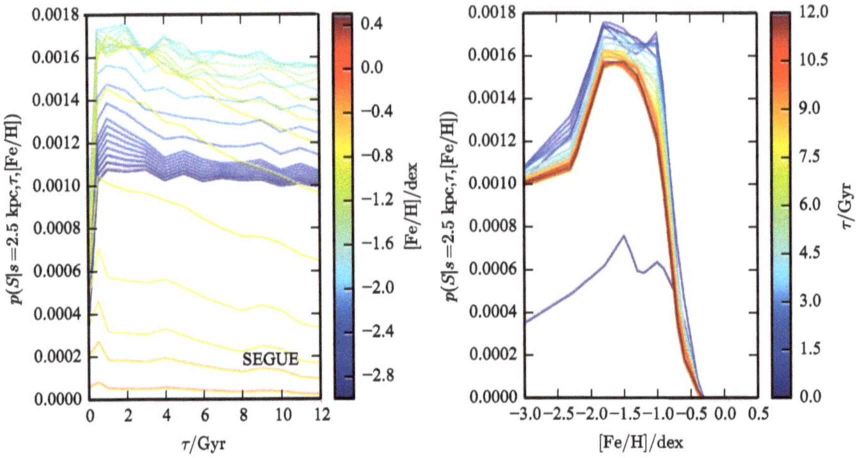

Fig. 8.3 SEGUE selection function for a star at 2.5 pc for the faint plate #1881. The *left panel* shows the selection in age coloured by metallicity, and vice versa for the *right panel*

where r_{cut} depends upon the plate, and W_P is the proportion of spectroscopic objects to number of targets in SDSS. The selection function is set to zero outside the r magnitude interval [14.5, 17.8] for bright plates and [17.8, 20.2] for faint plates. We use the publicly available code from Bovy et al. (2012) to find the location of r_{cut} for each plate by comparison to SDSS. In Fig. 8.3, we show the selection function for a star at 2.5 kpc observed in the faint plate #1881. We see that it is approximately flat with age and falls to zero for $[\mathrm{Fe/H}] > -0.3$ dex. Even without a physically-motivated model i.e. one in which the stars at high altitude are metal-poor, the SEGUE selection function is such that metal-poor stars are preferentially selected.

8.4.3 Using Minimal Knowledge of the Selection Function

We have detailed appropriate selection functions for the GCS and the SEGUE survey based on the targeting strategy for these surveys. However, often the selection function of a survey can be very complex and depend on many historical and social factors. In these cases, it may be difficult to accurately construct the selection function to allow for an appropriate model and data comparison. However, there is still an appropriate route to take which we will discuss here.

Say, from a survey, we have a complete set of observables for each star denoted $\boldsymbol{X}$. If we know that the survey is constructed by selecting in a subset of these observables, denoted $\boldsymbol{y} \subset \boldsymbol{X}$, then there is a subset of observables that have not been explicitly selected in, which we denote $\boldsymbol{x} = \boldsymbol{X} - \boldsymbol{y}$. For instance, we may know that the selection was performed in colour but not in velocity. Therefore, the velocities are free from an explicit selection but they are implicitly biased by the selection due to any relationship in the model between velocity and colour, e.g. bluer stars are younger and so have a lower velocity dispersion, hence we preferentially choose smaller velocities given a colour cut that selects blue stars.

In this case, we can still proceed by considering the conditional probability $p(\boldsymbol{x}|\boldsymbol{y}, S)$. We write

$$p(\boldsymbol{x}|\boldsymbol{y}, S) = \frac{p(\boldsymbol{x}, \boldsymbol{y}, S)}{p(\boldsymbol{y}, S)} = \frac{p(S|\boldsymbol{x}, \boldsymbol{y})p(\boldsymbol{X})}{p(S|\boldsymbol{y})p(\boldsymbol{y})} = \frac{p(S|\boldsymbol{x}, \boldsymbol{y})p(\boldsymbol{X})}{p(S|\boldsymbol{y}) \int \mathrm{d}^n\boldsymbol{x}\; p(\boldsymbol{y}, \boldsymbol{x})}. \tag{8.24}$$

We know that the selection is only in the observables $\boldsymbol{y}$ so $p(S|\boldsymbol{x}, \boldsymbol{y}) = p(S|\boldsymbol{y})$. Therefore, we find that

$$p(\boldsymbol{x}|\boldsymbol{y}, S) = \frac{p(\boldsymbol{X})}{\int \mathrm{d}^n\boldsymbol{x}\; p(\boldsymbol{y}, \boldsymbol{x})} = p(\boldsymbol{x}|\boldsymbol{y}). \tag{8.25}$$

The conditional probability of the observables $\boldsymbol{x}$ given the other observables $\boldsymbol{y}$ and the fact this star is in the selection S is just the conditional probability of the observables $\boldsymbol{x}$ given the other observables $\boldsymbol{y}$. We do not need to know the explicit form for the selection function, just that we have selected only in $\boldsymbol{y}$.

This argument is the underpinning for modelling the disc using only the velocity histograms (Binney 2012a; Binney et al. 2014). For a survey such as RAVE, the selection is not performed on the line-of-sight velocities or proper motions. Therefore, given a set of observables $\boldsymbol{y}$, we can sample a set of velocities from the model from which we can construct histograms to compare with the data. However, we must be sure we use *all* the selected observables $\boldsymbol{y}$. By not using all the selected observables, we will not restrict ourselves to stars of certain ages, which will affect the velocity distributions.

This route around the problem of the selection function seems useful, as we have avoided explicitly engaging with the selection function. It is particularly useful when constructing mock catalogues for surveys. However, the disadvantage is that we end up not using all of the available information to the full. For instance, consider the case of the RAVE velocity sampling. With a full dynamical distribution function, the power comes from the link between the spatial and velocity distributions, which are tied to the potential. For a fixed potential, we may be able to find a velocity distribution that matches the data, but the spatial distribution will then not match the data unless we have found the true potential. If we consider only velocity information, we lose this power and can only rely on spatial gradients in the velocity distribution for any constraint. Therefore, for constraining the potential, it seems necessary to use all the available information, which means engaging with the selection function.

The second problem with the above approach is that it has increased the computation we need to do when fitting the data. If we use Eq. (8.19) to fit the data, we must calculate the denominator $p(S)$ once to an appropriate precision such that, when multiplied by the number of data points, the noise does not dominate our posterior probability (McMillan and Binney 2013). However, this is a single computation for each considered model. When we do not explicitly use the selection function, as in

Eq. (8.25), we must compute the denominator $\int \mathrm{d}^n \boldsymbol{x}\, p(\boldsymbol{y}, \boldsymbol{x})$ for each datum. One way of approaching this is to tabulate the integral on a grid in $\boldsymbol{y}$ and interpolate. However, the dimensionality of the grid can be large, and it is challenging to reduce interpolation errors such that the noise does not dominate the posterior probability.

In conclusion, not explicitly using the selection function seems an attractive route, but we are not using all the available information to the full, and the increase in computation for a given model can be large. However, in certain cases it may be the only way to proceed. For the rest of the chapter, we will consider the selection function explicitly.

8.5 Choice of Parameters

We now turn to fitting our EDF to the discussed data. To choose the parameters of our model, we use the GCS and Gilmore-Reid data, and vary the 15 parameters:

$$\text{Thin: } R_{d,thin}, R_{\sigma,thin}, \sigma_{R0,thin}, \sigma_{z0,thin}; \tag{8.26}$$

$$\text{Thick: } R_{d,thick}, R_{\sigma,thick}, \sigma_{R0,thick}, \sigma_{z0,thick}, \mathcal{F}; \tag{8.27}$$

$$\text{Metallicity: } \sigma_L, \tau_F, F_R, F_m, R_F; \tag{8.28}$$

$$\text{Halo: } k_{\text{halo}}. \tag{8.29}$$

We seek to maximise $p(D|S_{GCS}, M)$ given by

$$p(D|S_{GCS}, M) = \prod_i p(l_i, b_i, \varpi_i, v_{||i}, \boldsymbol{\mu}_i, [\text{Fe/H}]_i|S_{GCS}), \tag{8.30}$$

where

$$\begin{aligned} p(l, b, \varpi, v_{||}, \boldsymbol{\mu}, [\text{Fe/H}]|S_{GCS}, M) = {} & \frac{1}{p(S_{GCS})} \int \mathrm{d}^5\boldsymbol{g}'\, G^5(\boldsymbol{g} - \boldsymbol{g}', \boldsymbol{\sigma}_g) \\ & \times s'^6 \cos b \int \mathrm{d}\tau\, p(S_{GCS}|s', \tau, [\text{Fe/H}]') \\ & \times f(\boldsymbol{x}', \boldsymbol{v}', \tau, [\text{Fe/H}]'), \end{aligned} \tag{8.31}$$

where G^5 is a 5D Gaussian to give the convolution of the observables $\boldsymbol{g} = (\varpi, [\text{Fe/H}], v_{||}, \boldsymbol{\mu})$ with the errors $\boldsymbol{\sigma}_g$ and the primed quantities are functions of $\boldsymbol{g}'$. $p(S_{GCS}|s', \tau, [\text{Fe/H}]')$ is the selection function as detailed in Sect. 8.4.1. We perform the integral over the errors ($\boldsymbol{g}'$) using Monte Carlo integration. We use a fixed sampling of 100 per star to remove numerical noise (McMillan and Binney 2013) and it allows for pre-computation of the actions. If we were to resample the Monte Carlo sampling points for each set of parameters of our distribution function, we would be very sensitive to the noise in the integration. By using a fixed sampling, we have removed this numerical noise such that the log-likelihood should vary more smoothly as we traverse the parameter space.

We assume that the completeness along each line-of-sight is the same such that

$$p(S_{GCS}|M) = \int \mathrm{d}l\,\mathrm{d}b\,\mathrm{d}s\,s^2\cos b \int \mathrm{d}^3\boldsymbol{v}\,\mathrm{d}[\mathrm{Fe/H}]\,\mathrm{d}\tau\,\mathrm{d}L'_z\,p(S|s,\tau,[\mathrm{Fe/H}]) \times f(\boldsymbol{x},\boldsymbol{v},\tau,[\mathrm{Fe/H}],L'_z). \tag{8.32}$$

We perform this integral using the VEGAS algorithm implemented in the CUBA package Hahn (2005).

We also use the Gilmore-Reid data in the fits. The log-likelihood of the Gilmore-Reid data is given by

$$\log \mathscr{L}_{\mathrm{GR}} = \sum_z \left|\frac{\log_{10}[\rho_{\mathrm{GR}}(z)/\rho_{\mathrm{DF}}(z)]}{\sigma_{\mathrm{GR}}(z)}\right|^2, \tag{8.33}$$

where ρ_{GR} is the density profile from Gilmore and Reid (1983), σ_{GR} are the errors in $\log_{10}(\rho_{\mathrm{GR}})$ and ρ_{DF} is the density profile calculated using the EDF integrated over all metallicities. The quantity we seek to maximise is

$$\log \mathscr{L} = \log p(D_{GCS}|S_{GCS}, M) + \chi \log \mathscr{L}_{\mathrm{GR}}. \tag{8.34}$$

We perform this procedure using the Nelder-Mead multi-dimensional minimization routine (Nelder and Mead 1965) implemented in the Gnu Science Library (Galassi et al. 2009). χ is some weight which we set to 10. This is to encourage the fitting procedure to take the rather few Gilmore-Reid data points seriously. The results of this procedure are shown in Table 8.2.

Table 8.2 EDF Parameters: parameters found from the fitting procedure presented in Sect. 8.5 and used to produce mock catalogues in Sect. 8.6

Thick	R_{d}/ kpc	2.9662
	R_σ/ kpc	5.82241
	σ_{R0}/ km s^{-1}	49.1744
	σ_{z0}/ km s^{-1}	52.1151
Thin	R_{d}/ kpc	2.55303
	R_σ/ kpc	9.25603
	σ_{R0}/ km s^{-1}	45.9825
	σ_{z0}/ km s^{-1}	28.4704
Other	F_R/dex kpc^{-1}	−0.0576
	F_m/dex	−0.951
	$\mathcal{F}$	0.1945
	γ_T	0.5
	σ_L/100 kpc km s^{-1}	9.96
	τ_F/Gyr	4.48
	R_F/kpc	6.94
	k_{halo}	1.43×10^{-3}

We note that whilst this procedure is statistically sound, we know the model will not perfectly match the data because (i) we are using a fixed potential, (ii) our model ignores substructure like that seen in the GCS velocity distribution, and (iii) the Nelder-Mead algorithm is likely to select a local minimum, particularly for high-dimensional problems. For our purposes, we want a model that is a good, but not statistically optimal, representation of the data. The initial choice of parameters was chosen by eye, so the results reflect some of this personal judgement.

In Appendix I, we present the results of a much fuller MCMC search of the parameters. From the full MCMC procedure, we find that the scale-length of the thin disc ($R_d = 2.9\,\text{kpc} \pm 0.2\,\text{kpc}$) is longer than that of the thick disc ($R_d = 2.3\,\text{kpc} \pm 0.3\,\text{kpc}$) (as in Bovy et al. 2012). The velocity-dispersion scale-length of the thin disc is found to be $R_\sigma = (9.5 \pm 0.5)\,\text{kpc}$ and for the thick disc $R_\sigma = (4.8 \pm 0.2)\,\text{kpc}$. The velocity dispersions were well constrained as $\sigma_{R0} = (47.3 \pm 0.5)\,\text{km s}^{-1}$ and $\sigma_{z0} = (31.5 \pm 0.4)\,\text{km s}^{-1}$ for the thin disc and $\sigma_{R0} = (57 \pm 3)\,\text{km s}^{-1}$ and $\sigma_{z0} = (48 \pm 2)\,\text{km s}^{-1}$ for the thick disc. Also, the thick disc fraction seems well pinned down as $\mathcal{F} = (0.19 \pm 0.01)$. However, as we will discuss later, the metallicity distribution results are not favourable, so we have opted for the local minimum solution from the Nelder-Mead algorithm.

8.6 Results

We now show results from the best-fitting model found using the procedure in Sect. 8.5. We begin by inspecting the global properties of the model. In Fig. 8.4, we show the global R and z profiles of the full model and of the individual components. We see that the discs have a strong core that may in part be due to the radial migration prescription. At the solar radius, the total radial profile is approximately exponential with a scale-length of 3.7 kpc. The radial migration prescription has broadened the discs slightly but maintained the exponential profile. Figure 8.5 shows the Gilmore-Reid density curve is well matched, although the density at low z is slightly overestimated.

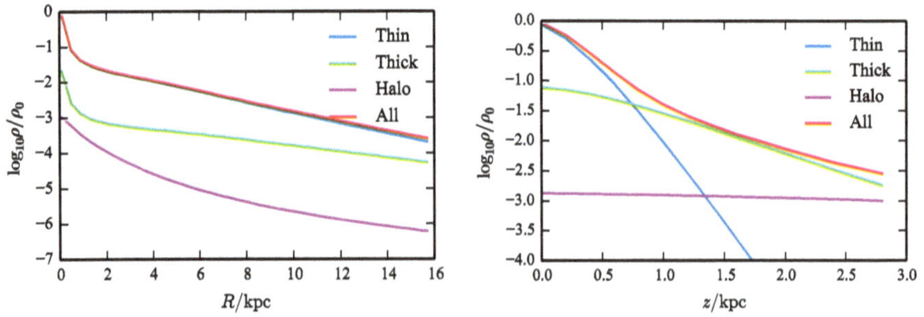

Fig. 8.4 R (at $z = 0$) and z (at $R = R_0$) density profiles for the full extended distribution function

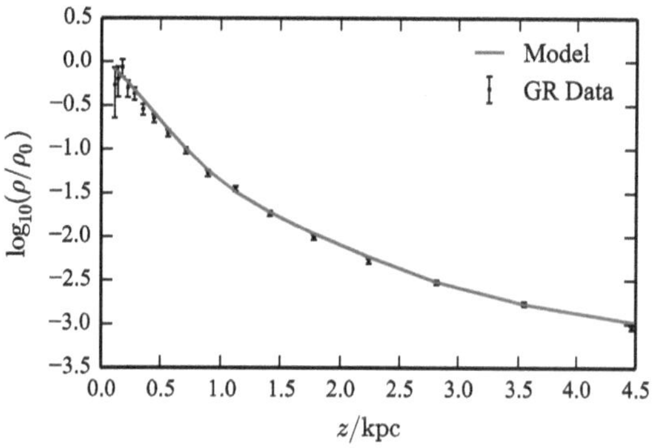

Fig. 8.5 Full vertical density profile at the solar radius along with the data from Gilmore and Reid (1983)

8.6.1 Sampling Mock Catalogues

To compare our model and data, we sample mock catalogues from the model. At the reported l and b of each star we sample distance, metallicity and velocities. We use a simple rejection-sampling technique using a uniform distribution in distance and Gaussians in metallicity and Galactocentric velocities. We sample from the distributions $p(s, [\mathrm{Fe/H}], v_R, v_\phi, v_z|l_i, b_i, S_k)$ (where i denotes the datum and k the survey), convert the quantities to Galactic coordinates and scatter by the reported error in the datum. For this procedure to be valid, we require the errors in the observables to be independent of s, [Fe/H], v_R, v_ϕ and v_z. In the histograms that follow, we use black points for the data, whilst the coloured points correspond to mock catalogues drawn from the EDF.

8.6.2 Apparent Magnitude Cut

Before comparing to actual data, we perform a brief experiment to demonstrate the importance of selection functions and including metallicity information in the DF. We construct a sample of 10,000 stars along the line-of-sight $l = 0$, $b = \pi/4$ with the selection function given by

$$p(S|V) = 1 \text{ if } V < 8 \text{ else } 0. \tag{8.35}$$

Such a selection function corresponds to a magnitude-limited sample. The resulting metallicity and velocity distributions are shown in red in Fig. 8.6. We now turn the selection function off (set $p(S|s, \tau, [\mathrm{Fe/H}]) = 1$) and resample the velocities given l, b, distance and metallicity. The resulting distributions are also shown in Fig. 8.6 in blue. This procedure is performed to highlight the effects of a selection function on the velocity distributions. If there are no correlations between the kinematics and chemical properties of the stars, the velocity distributions of the two mock catalogues

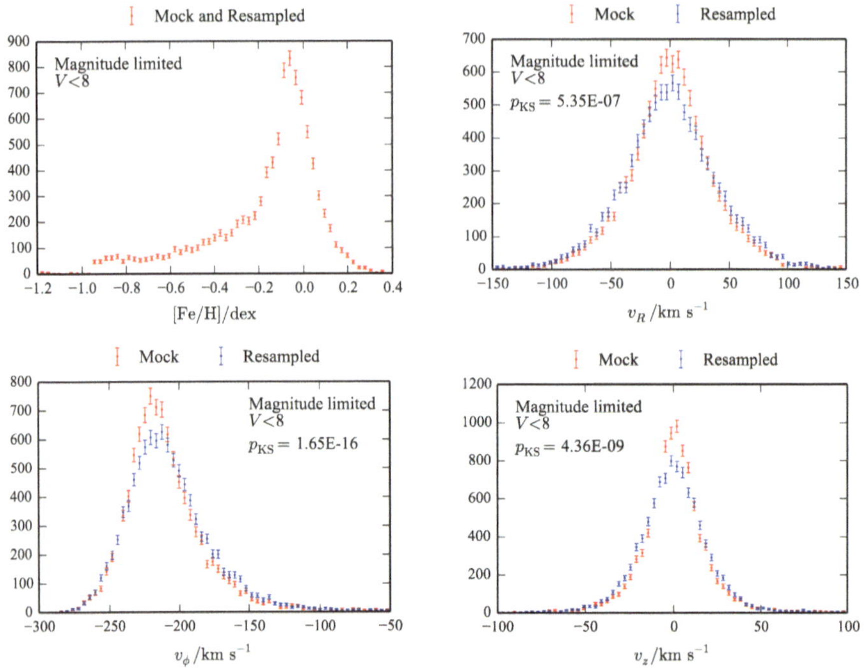

Fig. 8.6 Mock catalogues for a magnitude-limited survey. The *red points* gives the mock catalogue constructed by sampling distances, metallicities and velocities along the line-of-sight $l = 0, b = \pi/4$ for a magnitude-complete sample out to $V = 8$. The *blue points* give the mock catalogue formed by resampling the velocities given the metallicities and distances of the first mock catalogue with the selection function switched off. In each plot, we show the Kolmogorov-Smirnov probabilities, $p_{\rm KS}$, that the two samples were drawn from the same distribution

should be identical. However, in our model we have included chemo-dynamical correlations, and we see there are small differences between the two samples. The Kolmogorov-Smirnov probabilities that the two samples are drawn from the same distribution are $\lesssim 10^{-7}$.

When we select in apparent magnitude, we are implicitly selecting in age. The magnitude-limited sample has more young stars, and hence narrower velocity distributions. Although we have not explicitly selected in kinematics, the chemo-dynamical correlations in the model cause an implicit selection in the kinematics. When we turn off the selection function, we are not biased in age and the resulting velocity distributions are broader. The adopted selection function is probably the simplest selection function, and real surveys use much more complicated selection functions. This indicates that the inclusion of these implicit selection effects are crucial for the modelling of any data set, and that DFs must be extended to include metallicity information to take account of these subtle biases.

We performed this experiment with a fainter cut of $V = 10$ and found that the difference between the two velocity distributions was negligible. This fainter cut decreases the relative proportion of nearby young stars and increases the proportion of distant old stars. The resulting age distribution of the sample is thus more representative of that of the underlying population. As such the resulting velocity distributions are more representative.

8.6.3 GCS

We now construct a mock catalogue for the GCS. We use the selection function given in Sect. 8.4.1 and sample new distances, Galactocentric velocities and metallicities given the l and b values for each star in the GCS catalogue. After we have drawn a sample distance, metallicity and Galactocentric velocities, we calculate the proper motions and line-of-sight velocities and scatter the observables by the reported errors for each star. This procedure neglects any correlation between the errors and distance, metallicity or velocity. The most prominent correlation for the GCS sample is the parallax error with the V magnitude (Pearson correlation coefficient between $\log V$ and $\log \sigma_{\varpi} \approx 0.6$). However, the parallax error does not seem to correlate as well with the parallax (correlation between $\log \varpi$ and $\log \sigma_{\varpi} \approx -0.2$), so we neglect this effect and simply use the reported parallax errors for each star.

In Fig. 8.7, we show the distance distribution of the data and mock sample. The data distribution is offset from the mock distribution such that we have too few distant stars and too many close stars. Over the small volume that the GCS probes the distance distribution is a far stronger function of the selection function than it is of the underlying model. Therefore, the differences in the data and model distance distributions indicate that the selection function is not optimal. Note that due to our method of assigning observational errors to each star, we anticipate that the low distance stars will have, on average, been assigned higher distance errors than expected, whilst the opposite is true for the high distance stars. From the correlation coefficients calculated previously, we expect this effect is small. However, it would produce a signal of this kind.

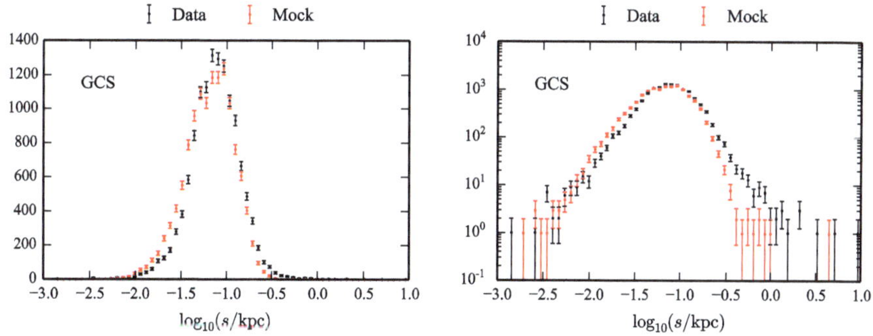

Fig. 8.7 GCS distance distributions: *black* shows the data, *red* the mock catalogue

In Fig. 8.8, we show the metallicity and velocity distributions.[1] The model (red points) fails to produce the metal-rich wing of the metallicity distributions. It is not clear whether the metallicity error estimation for these extremal stars is reliable. However, assuming they are valid measurements, our model has too few high-metallicity stars. We have chosen to evaluate the velocity dispersion parameters at the current angular momentum. However, as noted earlier, this is an approximation and, in fact, the heating is due to a series of environments through which the star migrates. Stars from the inner disc will have been heated more than we have allowed, which increases the number of high-action stars and high-metallicity stars at the solar position.

Another more important problem with the EDF is the restrictive nature of our model for the ISM as a function of radius. The full MCMC procedure detailed in Appendix I shows that the data favours a very large σ_L, whilst F_R (the gradient of the ISM metallicity with radius) is well constrained by the low-action GCS stars as $F_R = (-0.055 \pm 0.005)$ dex/kpc. Our model for the ISM assumes an approximately linear gradient with radius from the solar radius to the Galactic centre, such that the metallicity at the Galactic centre in the model is $\sim$0.38 dex. Therefore, there are essentially no stars in our EDF with metallicities greater than this (the halo DF has small but non-zero weight at high metallicities). Our model believes that all of the data with metallicities higher than this are due to observational errors. With small F_R, models with a very large σ_L are favoured as these have more metal-rich stars in the solar neighbourhood. However, such models are disfavoured on physical grounds as larger σ_L broadens the disc such that the scale-length becomes large and the disc loses its exponential profile. The GCS data is not sensitive to these effects due to its local nature.

The situation should be remedied by making the ISM metallicity model more flexible. We require the metallicity to increase non-linearly towards the Galactic centre (for instance, we could use an exponential). Note that Schönrich and Binney (2009) used a very steep metallicity gradient in the ISM, so this problem was avoided. Clearly, with a simplistic model for the ISM, there is some tension between the gradient today and the number of metal-rich stars observed locally.

The velocity distributions in Fig. 8.8 are fitted well by the model, particularly the v_R and v_z distributions. The model v_ϕ distribution fails to match the peak of the data distribution, but this peak is due to the Hyades stream, a non-equilibrium feature that we are not concerned with matching. Figure 8.9 shows the mock catalogue weighted by the thin-disc, thick-disc and halo membership probabilities. We see that, as expected, the thin disc dominates. For $[\mathrm{Fe/H}] \lesssim 0.6$ dex and $v_z \gtrsim 50\,\mathrm{km\,s^{-1}}$, the thick disc dominates.

In each of the plots in Fig. 8.8, we also show a mock catalogue generated by sampling new metallicities and velocities at the solar position with the selection function turned off (blue points). It is believed that the GCS velocity histograms are fair samples of the local velocity distributions. If this is the case, the velocity histograms for the two mock samples should be similar. Note that we have scattered

[1] We choose positive radial velocity, v_R, to be away from the Galactic Centre and positive azimuthal velocity, v_ϕ, and hence positive angular momentum, L_z, to be in the direction of Galactic rotation.

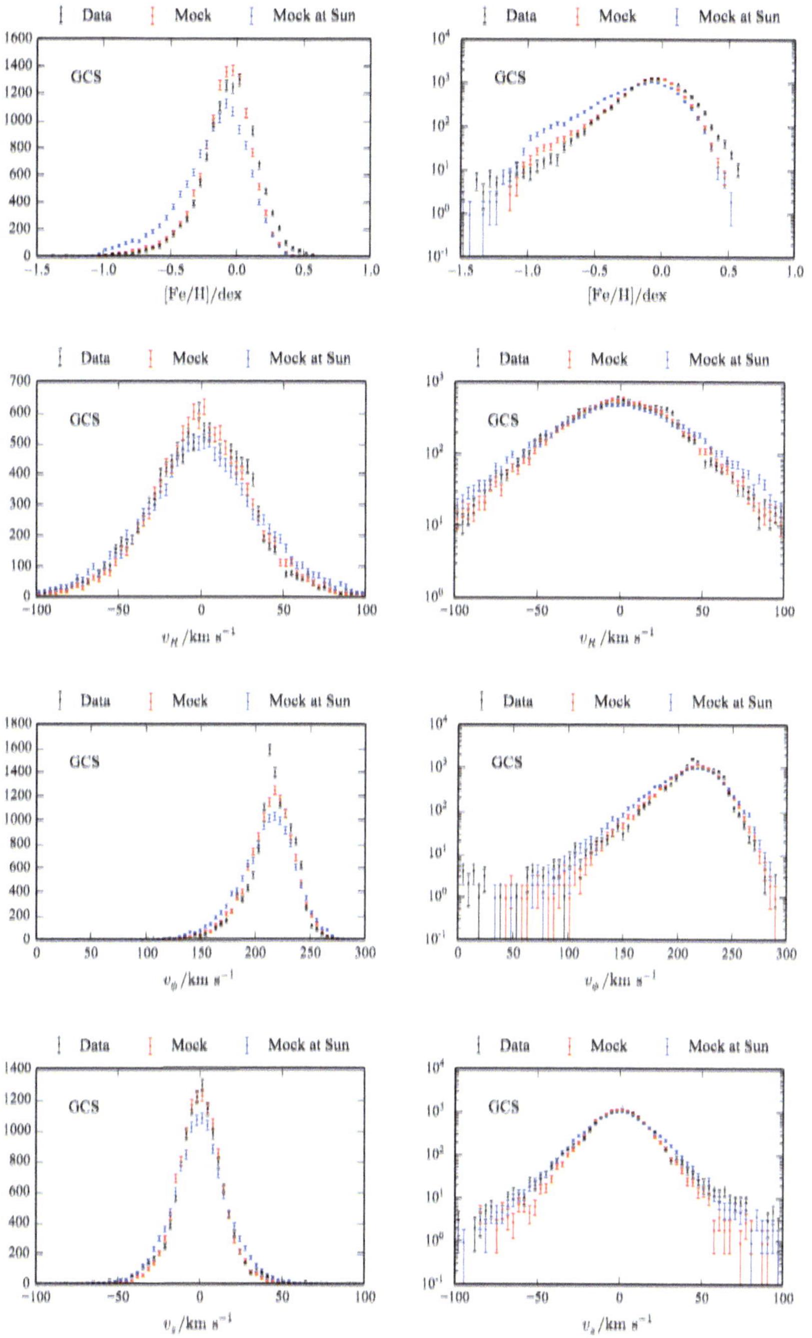

Fig. 8.8 GCS metallicity and velocity distributions with a linear scale (*left*) and logarithmic (*right*): the top row shows the metallicity distribution, second row the v_R distribution, third row the v_ϕ distribution and the final row the v_z distribution. The *black* shows the data, *red* the mock catalogue, and *blue* if we only sample at the Sun

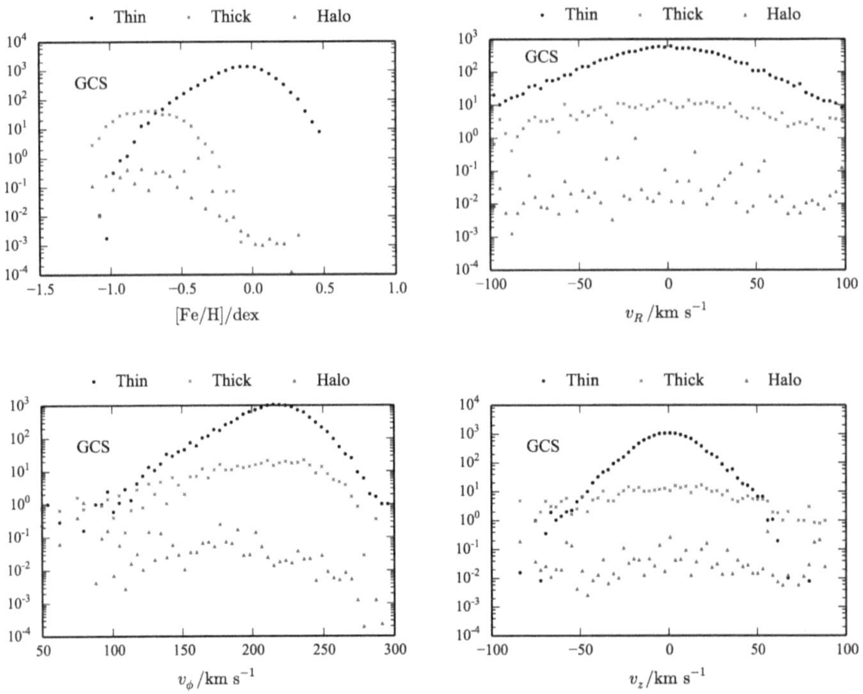

Fig. 8.9 GCS weighted metallicity and velocity distributions: The model histograms are weighted by the probability of being a member of the thin disc (*black circles*), the thick disc (*red crosses*) and the halo (*blue triangles*). The top left shows the metallicity distribution, top right the v_R distribution, bottom left the v_ϕ distribution and the *bottom right* the v_z distribution

the sampled metallicities by the reported errors but not the velocities. The v_R and v_z distributions of this mock sample are broader than the true GCS mock sample. This is in part due to not including velocity errors in this second sample, but also we have increased the number of high-age stars by turning off the selection function. This is also seen in the metallicity distribution, which has a broader metal-poor wing.

We find a best-fitting vertical velocity dispersion parameter for the thin disc of $\sigma_{z,\mathrm{thin}} \approx 30\,\mathrm{km\,s^{-1}}$. Other fits to the Geneva-Copenhagen Survey data produce lower velocity dispersion parameters for the thin disc (e.g. Chap. 7 uses $\sigma_{z,\mathrm{thin}} \approx 20\,\mathrm{km\,s^{-1}}$). The reason for our higher $\sigma_{z,\mathrm{thin}}$ is the inclusion of a selection function. The GCS preferentially selected low age stars (∼2 Gyr, see Sect. 8.4.1), so the model is weighted towards younger, colder populations. The velocity dispersion of the GCS sample is smaller than that of the full population. This can be seen by comparing the two mock catalogues. From both the experiment of the previous section with a magnitude-limited survey and the experiment shown here with the Geneva-Copenhagen Survey, we see the importance of the selection function. Although the

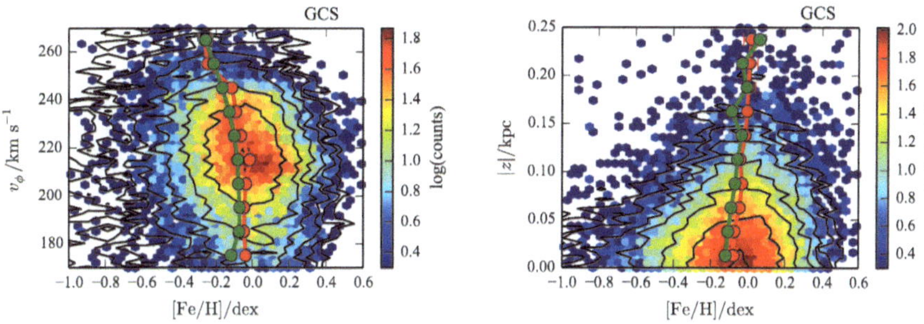

Fig. 8.10 2D histograms of metallicity [Fe/H] and azimuthal velocity v_ϕ, and [Fe/H] against absolute height above the plane $|z|$—the *coloured* histogram shows the data and the *black* contours are for the mock catalogue. The *red* and *green lines* give the mean [Fe/H] in equal-width bins centred on the *dots* for the data and model

velocities are not explicitly used to select the stars, there is an implicit velocity selection due to the relationship between the velocity and the hidden parameter, age.

Figure 8.10 shows the 2D histograms of [Fe/H]-v_ϕ and [Fe/H]-$|z|$. The data and models match well in these planes, and the means of v_ϕ and $|z|$ binned in metallicity are well recovered with a small systematic of ~0.05 dex, due to the lack of metal-rich stars.

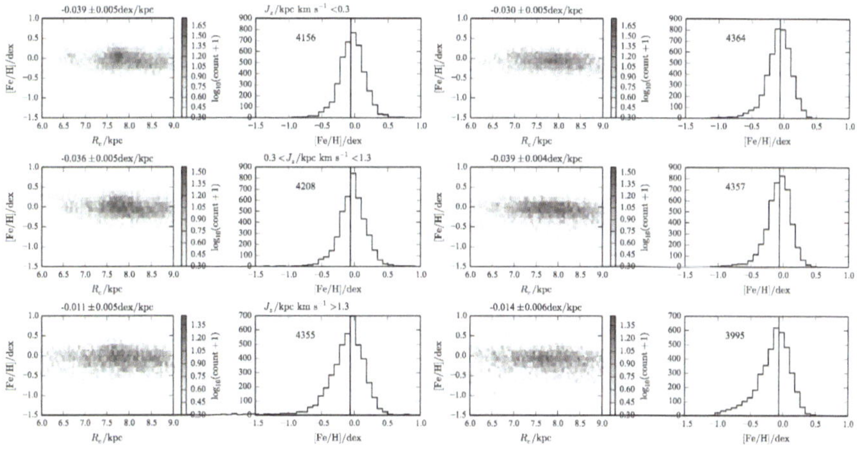

Fig. 8.11 2D histograms of metallicity [Fe/H] and guiding-centre radius R_c and 1D [Fe/H] histograms for three bins in J_z (*top row* $J_z < 0.3\,\mathrm{km\,s^{-1}\,kpc}$, *middle row* $0.3\,\mathrm{km\,s^{-1}\,kpc} < J_z < 1.3\,\mathrm{km\,s^{-1}\,kpc}$, and *bottom row* $J_z > 1.3\,\mathrm{km\,s^{-1}\,kpc}$). The left two plots show the GCS data whilst the right two show the mock GCS catalogue. We show the gradient d[Fe/H]/dr_c for the samples in each bin excluding [Fe/H] < -1 dex and only using data with $6\,\mathrm{kpc} < R_c < 9\,\mathrm{kpc}$ above the relevant plot, and the number of stars in each bin in the second and fourth column panels. The *vertical line* shows the location of the peak of the data metallicity distribution in the lowest action bin

Figure 8.11 shows the gradients of the metallicity with respect to the guiding-centre radius and the metallicity distributions in three bins in vertical action. In all three action bins, the gradient is well recovered. Additionally, we find that the metallicity distributions for all three action bins are well matched. Note that the peak of the metallicity distributions remains fixed with increasing vertical action for both the data and the model.

8.6.4 SEGUE G dwarfs

Using our extended distribution function from the previous section, we now construct a mock SEGUE G dwarf catalogue. For each star in the G dwarf sample, we take the l and b values and draw a new distance, metallicity and Galactocentric velocities, convert to the observables and scatter by the reported errors. We found that the halo weight, k_{halo}, was too large, so we reduced k_{halo} by a factor of six. The GCS does not constrain the halo properties well, and the strongest constraint on k_{halo} comes from the Gilmore-Reid data. By making this alteration, the higher-z densities are underestimated. However, the degree to which we can trust these high-z densities is unclear. Additionally, the problem could be that our models are not appropriate. The halo distribution is intentionally very crude, and there is the suggestion from Binney et al. (2014) that the J_z distribution for the thick disc is not appropriate. Therefore, there is plenty of scope for adjusting the models, and it is clear that the model we have presented here is not optimal.

We plot the resulting data and model distance distributions in Fig. 8.12. Despite using only the local GCS sample and the Gilmore-Reid density curve to fit our parameters, we see that the resulting distance distribution of our model matches the SEGUE data well. Additionally, we have used different isochrones to find the distances to the G dwarfs and then to construct our model. This indicates the errors in the distances obtained from the isochrones are accurate.

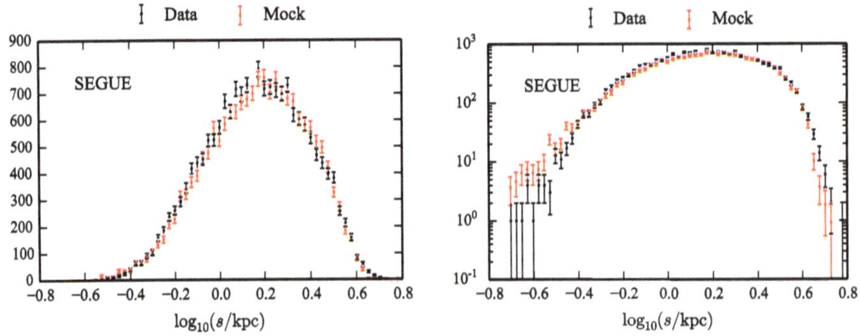

Fig. 8.12 SEGUE distance distributions: *black* shows the data, *red* the mock catalogue

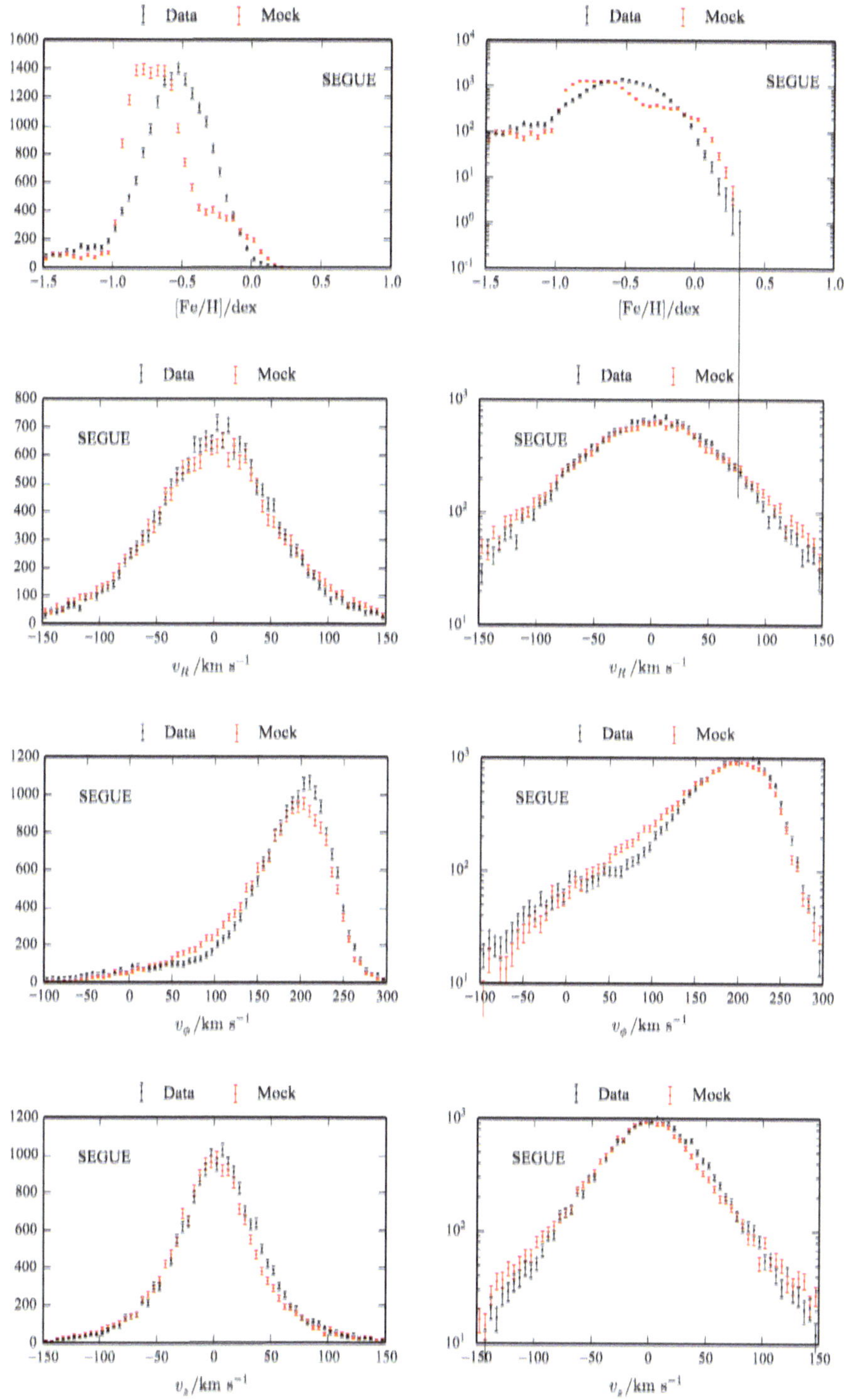
Data
Mock
SEGUE
[Fe/H]/dex
v_R/km s^{-1}
v_ϕ/km s^{-1}
v_z/km s^{-1}

◀**Fig. 8.13** SEGUE metallicity and velocity distributions with a linear scale (*left*) and logarithmic (*right*): the top row shows the metallicity distribution, second row the v_R distribution, third row the v_ϕ distribution and the final row the v_z distribution. The *black* shows the data and *red* the mock catalogue

In Fig. 8.13, we plot the metallicity and velocity distributions of both the data and the model. The velocity distributions are a good match—the v_R distribution matches very well; the mock v_ϕ distribution fails to match the data peak and is slightly broader than the data, which is perhaps due to a sub-optimal potential; the v_z distribution broadly matches the data but the mean of the data distribution is clearly offset from zero. There has been much in the literature recently associating mean vertical velocity shifts with modes in the disc (e.g. Widrow et al. 2012; Williams et al. 2013). However, this shift in the peak could be equally well accounted for by systematic distance errors or zero-point errors in the SDSS proper motions.

The metallicity distribution in Fig. 8.13 is a very poor match to the data. We see that our metallicity distribution peaks at a much lower metallicity than the data. This

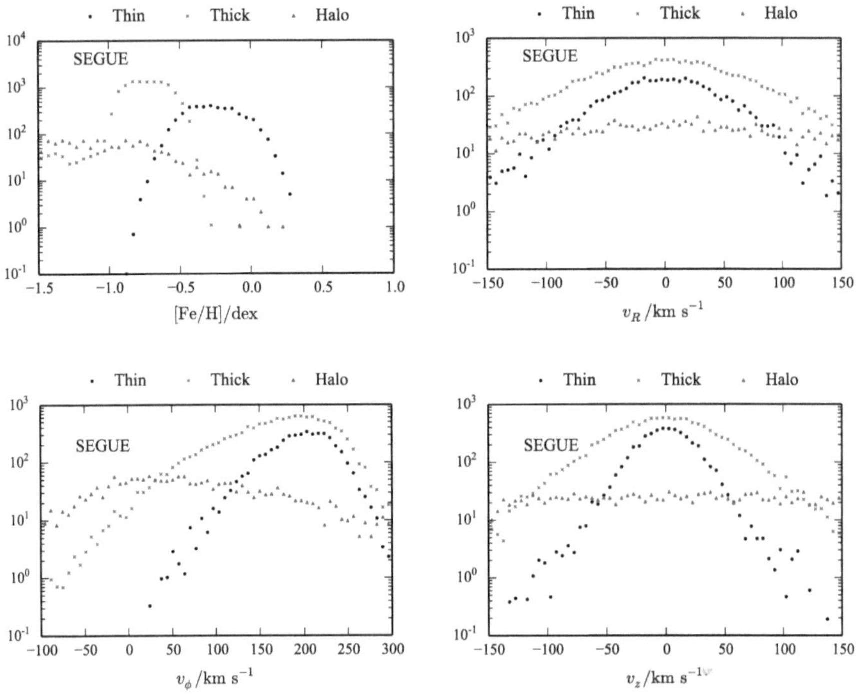

Fig. 8.14 SEGUE weighted metallicity and velocity distributions: The model histograms are weighted by the probability of being a member of the thin disc (*black circles*), the thick disc (*red crosses*) and the halo (*blue triangles*). The *top left* shows the metallicity distribution, *top right* the v_R distribution, *bottom left* the v_ϕ distribution and the *bottom right* the v_z distribution

discrepancy can be amended by adjusting the star-formation rate in the thick disc in our model. This is entirely valid as currently we have adopted an overly simplistic uniform star-formation rate. The other option for resolving the discrepancy is to revise the metallicity estimates from the SEGUE pipeline. Schönrich and Bergemann (2014) argue that there is a bias in the SEGUE stellar parameter pipeline that causes an artificial build-up of stars at [Fe/H] ≈ -0.5 dex.

In Fig. 8.14, we show the metallicity and velocity distributions weighted by the probability of being a member of the thin disc, thick disc or halo. The thick disc is dominant with the thin disc forming an important contribution to the metallicity distributions at metallicities [Fe/H] $\gtrsim -0.5$ dex.

In Fig. 8.15, we show the 2D histograms of metallicity and v_ϕ, and metallicity and $|z|$. The broad trends are reproduced, but there is a clear offset in the means of ~ 0.1 dex for intermediate z ($1\,\text{kpc} < |z| < 2\,\text{kpc}$) and high v_ϕ ($v_\phi > 180\,\text{km s}^{-1}$). Finally, in Fig. 8.16 we show the 2D metallicity—guiding-centre-radius histograms separated by vertical action. As noted in Lee et al. (2011), the α-enhanced stars exhibit a positive gradient of v_ϕ with metallicity. We see that, for all vertical-action bins, the metallicity gradient with guiding-centre radius is positive, unlike for the GCS. Our model fails to match this. This is due to the simplicity of our model for the thick disc. We have assumed the thick disc is a single homogeneous quasi-isothermal. However, much like the thin disc, it should be made a superposition of quasi-isothermals. To account for the positive gradient, we require inside-out formation such that the scale-length of the thick disc is a decreasing function of age. The populations migrating from inside are dominated by metal-poor stars, whilst those from outside are dominated by metal-rich stars. Additionally, we could make the velocity dispersion parameters or star-formation rate a function of age for the thick disc.

Also, in Fig. 8.16, we show the metallicity histograms in each vertical-action bin. The high-action bin provides a good match to the data, but, for the low-action bin, the data are mono-modal, whilst we produce a flat-topped, near bimodal distribution

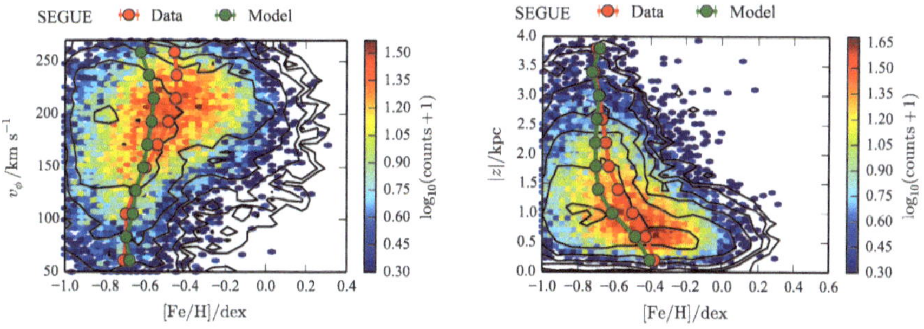

Fig. 8.15 2D histograms of metallicity [Fe/H] and azimuthal velocity v_ϕ, and [Fe/H] against absolute height above the plane $|z|$ for the SEGUE data and mock sample—the *coloured* histogram shows the data and the *black* contours are for the mock catalogue. The *red* and *green lines* give the mean [Fe/H] in equal-width bins centred on the *dots* for the data and model

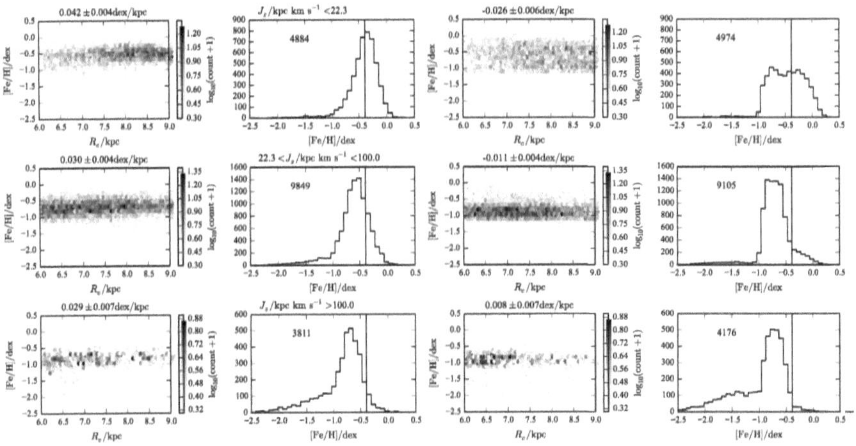

Fig. 8.16 2D histograms of metallicity [Fe/H] and guiding-centre radius R_c and 1D [Fe/H] histograms for three bins in J_z (*top row* $J_z < 22.3\,\mathrm{km\,s^{-1}\,kpc}$, *middle row* $22.3\,\mathrm{km\,s^{-1}\,kpc} < J_z < 100\,\mathrm{km\,s^{-1}\,kpc}$, and *bottom row* $J_z > 100\,\mathrm{km\,s^{-1}\,kpc}$). The left two plots show the SEGUE G dwarf data, whilst the right two show the mock SEGUE G dwarf catalogue. We show the gradient $\mathrm{d[Fe/H]}/\mathrm{d}R_c$ for the samples in each bin excluding $\mathrm{[Fe/H]} < -1$ dex above the relevant plot, and the number of stars in each bin in the second and fourth column panels. The *vertical line* shows the location of the peak of the data metallicity distribution in the lowest action bin

due to the competing contributions from the thin and thick discs. This could also be remedied by making the thick disc a superposition of populations.

8.7 Conclusions

We have presented a simple extension of the action-based distribution functions presented in Binney (2012a) to include metallicity information. Inspired by the full chemo-dynamical evolution models of Schönrich and Binney (2009), we included analytically the relationship between metallicity and dynamics. The model explicitly includes a radial migration prescription that causes the solar metallicity distribution to broaden due to stars migrating from the inner and outer disc. This is the first time an action-based DF has been extended to include metallicity and compared to data.

All surveys are subject to selection functions in magnitude, and hence metallicity. The kinematics depend upon the metallicity, so, in general, any selection in metallicity affects the velocity distributions. Therefore, any distribution function that does not include metallicity information has limited use for modelling large Galactic data sets. Additionally, the correlations between metallicity and kinematics reveal information about the history of the Galaxy. The metallicity indicates where a star was born, which, when combined with its current kinematics, provides information on the history of a star in the Galaxy. Finally, extended dynamical distribution functions are a useful platform on which to constrain the Galactic potential, as each unique chemical population has different dynamics that must be consistent with the same Galactic potential. Our presented EDFs are an important step towards fully modelling current and future Galactic data.

We fitted the parameters of our extended distribution function to the Geneva-Copenhagen survey (GCS) and the Gilmore and Reid (1983) stellar density curve. From the resulting model, we sampled a mock GCS catalogue and compared to the data. Our model reproduced many features of the data, but failed to produce the required number of high metallicity stars. This is believed to be due to the simplistic model of the metallicity of the ISM adopted.

The EDF allows us to test the effects of selections in magnitude and colour on the kinematics of the observed sample. To observe the effects of a simple selection function on the kinematics, we constructed a magnitude-limited sample of stars along a single line-of-sight and compared the resulting velocity distributions of the sample to a catalogue formed by neglecting the implicit selection in velocities that results from the selection function. We found that the sample which neglected the selection function had slightly broader velocity distributions as there are more old stars included in the sample than in the selected sample. Even with the simplest selection function, one must take into account the selection function to truly model the dynamics.

The GCS velocity distributions are believed to be a fair sample of the local velocity distributions. To test this, we drew a sample of stars from our EDF that are all at the solar position, and are not subject to any colour-magnitude selection. We found that this sample had broader velocity distributions than the full colour-magnitude-selected sample, due to the hidden selection in age. Finally, we constructed a mock SEGUE G dwarf sample and compared to the data. The data and model provided a good match with the velocity distributions, but not such a good match with the metallicity distribution. This could be due to inconsistent metallicity scales between the GCS and SEGUE, or point to the need for a more refined model to fit the wealth of data. Additionally, our model failed to reproduce the positive gradients of [Fe/H] with R_c observed for the SEGUE G dwarfs.

8.7.1 Future Work

The models presented in this chapter should prove extremely powerful for modelling data from current and future Galactic surveys. We now discuss the direction this work should be developed in the future, and what we can hope to achieve with the EDFs.

1. Initially, we should make adjustments to the model that seem necessary in light of the results of the chapter. Here we have treated the thick disc as a single homogeneous quasi-isothermal, but, like with the thin disc, the thick disc should be made a superposition of quasi isothermals. This seems necessary in light of the SEGUE G dwarf data, which require a positive gradient of [Fe/H] with

guiding-centre radius. This requires the thick disc to be composed of a series of EDFs with decreasing scale-length. We have also seen the importance of correctly modelling the selection function for understanding of the thin disc. Additionally, we have seen that our model of the ISM with radius at the current time is not appropriate, so it is necessary make the model more flexible to account for a changing gradient with Galactocentric radius.
2. The next step for the extended distribution function modelling is to fit the model to the RAVE (Kordopatis et al. 2013) and upcoming Gaia-ESO (Gilmore et al. 2012) data. Both surveys provide $\sim 10^5$ stars at intermediate distances from the Galactic plane, so nicely complement the thin-disc-dominated view from the GCS and the thick-disc dominated view from the SEGUE survey. Binney et al. (2014) has shown that the DFs provide a good account of the RAVE data, whilst Piffl et al. (2014) used the RAVE stars to constrain the Galactic potential. These analyses essentially assumed the selection function of the RAVE survey is uniform in age, and they did not use the metallicity information. The EDF models should be used to repeat this analysis to simultaneously constrain the Galactic potential and uncover signatures of evolutionary processes in the Galaxy, such as radial migration. Gaia-ESO will provide higher-resolution spectra than RAVE, so will produce more accurate metallicities, as well as being able to see fainter stars. The constraints from these two surveys should be highly complementary.
3. Additionally, the EDF should be extended further to account for the $[\alpha/\mathrm{Fe}]$ abundances of the stars. $[\alpha/\mathrm{Fe}]$ is a crucial quantity for measuring the age of a star, and the ($[\alpha/\mathrm{Fe}]$, [Fe/H]) plane is a powerful space to disentangle different Galactic populations (Schönrich and Binney 2009; Bovy et al. 2012).
4. As discussed in Sect. 8.2, the extended distribution function presented here is only an approximation to a more physically rigorous extended distribution function that explicitly follows the full action-space diffusion. Construction of such models is possible through either a Monte-Carlo route or by solving the Fokker-Planck equation on a discrete action-space grid. Both of these routes require action-space kernels, which may be produced by theoretical calculations or extracted from full N-body simulations.

References

Ahn CP et al (2014) ApJS 211:17
An D et al (2009) ApJ 700:523
Binney J (2010) MNRAS 401:2318
Binney J (2012a) MNRAS 426:1324
Binney J (2012b) MNRAS 426:1328
Binney J et al (2014) MNRAS 439:1231
Binney J, Gerhard O, Spergel D (1997) MNRAS 288:365
Boeche C et al (2013) A & A 559:A59
Bovy J, Rix H-W (2013) ApJ 779:115
Bovy J, Rix H-W, Liu C, Hogg DW, Beers TC, Lee YS (2012) ApJ 753:148

Brown WR, Geller MJ, Kenyon SJ, Diaferio A (2010) AJ 139:59
Burnett B, Binney J (2010) MNRAS 407:339
Casagrande L, Schönrich R, Asplund M, Cassisi S, Ramírez I, Meléndez J, Bensby T, Feltzing S (2011) A & A 530:A138
Dehnen W (1998) AJ 115:2384
Dehnen W, Binney J (1998) MNRAS 294:429
Galassi M, Davies J, Theiler J, Gough B, Jungman G (2009) GNU scientific library—reference manual, for GSL version 1.12, 3rd edn. Network Theory Ltd, pp 1–573
Gilmore G et al (2012) Messenger 147:25
Gilmore G, Reid N (1983) MNRAS 202:1025
Hahn T (2005) Comput Phys Commun 168:78
Jurić M et al (2008) ApJ 673:864
Kordopatis G et al (2013) AJ 146:134
Kroupa P, Tout CA, Gilmore G (1993) MNRAS 262:545
Lee YS et al (2011) ApJ 738:187
McMillan PJ, Binney JJ (2013) MNRAS 433:1411
Minchev I, Famaey B, Quillen AC, Dehnen W, Martig M, Siebert A (2012) A & A 548:A127
Mitschang AW, De Silva G, Sharma S, Zucker DB (2013) MNRAS 428:2321
Mitschang AW, De Silva G, Zucker DB, Anguiano B, Bensby T, Feltzing S (2014) MNRAS 438:2753
Navarro JF, Frenk CS, White SDM (1996) ApJ 462:563
Nelder JA, Mead R (1965) Comput J 7:308
Nordström B et al (2004) A & A 418:989
Pietrinferni A, Cassisi S, Salaris M, Castelli F (2004) ApJ 612:168
Piffl T et al (2014) ArXiv e-prints
Posti L, Binney J, Nipoti C, Ciotti L (2015) MNRAS 447:3060
Schlegel DJ, Finkbeiner DP, Davis M (1998) ApJ 500:525
Schlesinger KJ et al (2012) ApJ 761:160
Schönrich R, Bergemann M (2014) MNRAS 443:698
Schönrich R, Binney J (2009) MNRAS 396:203
Schönrich R, Binney J, Dehnen W (2010) MNRAS 403:1829
Sellwood JA, Binney JJ (2002) MNRAS 336:785
Smolinski JP et al (2011) AJ 141:89
Solway M, Sellwood JA, Schönrich R (2012) MNRAS 422:1363
Ting Y-S, Rix H-W, Bovy J, van de Ven G (2013) MNRAS 434:652
Walker MG, Peñarrubia J (2011) ApJ 742:20
Widrow LM, Gardner S, Yanny B, Dodelson S, Chen H-Y (2012) ApJL 750:L41
Williams MEK et al (2013) MNRAS 436:101
Yanny B et al (2009) AJ 137:4377

Chapter 9
Conclusions

9.1 Overview

Milky Way science is blossoming. The recent launch of the *Gaia* satellite has heralded a new era in the study of the Milky Way and presents exciting prospects for the field of galactic astronomy. *Gaia* will take an unprecedented stereoscopic census of one billion stars in our Galaxy. The huge quantity of data will enable us to view the present structure of the Galaxy, investigate the formation and history of the Galaxy, understand the structure of the dark matter that envelopes the Milky Way and test theories of galaxy formation.

The Milky Way is a highly complex object with many dynamical and chemical processes continuously occurring. As with any complex physical system, an understanding can be gained only through simplification. This takes the form of modelling that focusses on the key physics of the problem and neglects any processes that are of secondary importance. In the era of *Gaia*, the union of sophisticated models and large quantities of data are necessary for extracting all possible information from the data, and advancing our understanding of the Milky Way and all galaxies.

To this end, appropriate methods must be developed to synthesise data from several surveys and extract the important information. In this thesis, we have presented a variety of methods for exploiting data with sophisticated modelling techniques. We briefly detail the conclusions of each chapter below.

9.1.1 Angle-Action Estimation in a General Axisymmetric Potential

Angle-action coordinates are known to be very useful variables in Galactic dynamics. However, we can calculate these variables exactly only in a very limited set of potentials. In Chap. 2, we presented algorithms for estimating the angle-action variables in general axisymmetric potentials. The first, and most thoroughly

J. Sanders, *Dynamics of the Milky Way*, Springer Theses,
DOI 10.1007/978-3-319-18772-3_9

explored method proceeds by first fitting the region an orbit probes with a Stäckel potential, and then estimating the angles and actions in the true potential as those in the fitted Stäckel potential. We investigated the accuracy of the method for a range of tori in a realistic Galactic potential, and we presented an application of the method by inspecting the Geneva-Copenhagen survey in angle-action space. We presented another new method for finding the actions based on the adiabatic approximation (Binney 2010; Schönrich and Binney 2012) but assuming the orbit is separable in spheroidal coordinates. We compared the accuracy of the action estimates and the computation time for these two methods as well as the axisymmetric Stäckel fudge from Binney (2012a) and the adiabatic approximation from Schönrich and Binney (2012). Finally, we showed how these approximate non-convergent methods can be combined with more accurate results from torus construction (McMillan and Binney 2008) to produce an efficient, and powerful, iterative scheme for finding the actions.

9.1.2 Actions, Angles and Frequencies from Numerically Integrated Orbits

When the potential is axisymmetric, the routines from Chap. 2 are suitable for finding actions and angles. For a more general potential e.g. triaxial, such as those from an N-body simulation, we must develop more general algorithms to find the angles and actions from a series of phase-space points. In Chap. 3, we presented a method that proceeds by constructing a generating function to take us from an analytic torus to our target torus, and solves for the components of this generating function using a series of time samples from an orbit integration. We demonstrated the accuracy of the method by comparing to known actions for a triaxial Stäckel potential, and presented a brief application of the method by inspecting a typical surface of constant energy in a triaxial potential of astrophysical interest (Law and Majewski 2010). We closed the chapter with a short discussion of the impact of irregular orbits on the method.

9.1.3 Action Estimation Using a Triaxial Stäckel Approximation

Finding the Fourier components of a generating function from a numerically integrated orbit is a general, but slow, method for finding the actions from an initial phase-space point. When constructing distribution functions for triaxial components of the Galaxy or external galaxies, we require fast algorithms for finding the actions such that we can rapidly calculate integrals over the distribution function. In Chap. 4, we presented an extension to the method of Binney (2012a) for estimating the actions using a Stäckel approximation. The method uses only the initial phase-space point to find estimates for the two classical integrals in an integrable triaxial potential. These estimates are based on the equivalent equations for the triaxial Stäckel potential. We inspected the accuracy of this method for estimating actions in a triaxial Navarro-Frenk-White potential, and showed that the accuracy obtained for the loop orbits, which only probe a limited region of the potential, is significantly better than the accuracy obtained for the box orbits, which reach down to the centre of the potential. We went on to construct triaxial stellar systems from specified DFs $f(\boldsymbol{J})$ in Sect. 4.6, and we showed the mass of these models is well recovered using the triaxial Stäckel approximation. We demonstrated that, notwithstanding the errors in individual actions, both a radially-biased model and a tangentially-biased model satisfy the Jeans equations to good accuracy. When computing moments, individual errors largely cancel out during integration over velocities.

9.1.4 Stream-Orbit Misalignment

We presented a discussion of stream formation in Chap. 5 and, in particular, discussed the misalignment between streams and orbits. The formation of tidal streams is simple in angle-action space, which provides a natural platform on which to quantify the misalignment. We discussed the expected misalignment between a stream and progenitor orbit for a range of potentials of interest in Galactic dynamics. Specifically, we focussed on the expected misalignment for the known streams. By introducing a two-parameter family of realistic Galactic potentials, we demonstrated that assuming these streams delineate orbits can lead to order one errors in the halo flattening and halo-to-disc force ratio at the Sun. We presented a discussion of the dependence of these results on the progenitor mass, and demonstrated that the misalignment is mass-independent for the range of masses of observed streams. Hence, orbit-fitting does not yield better constraints on the potential if one uses narrower, lower-mass streams.

9.1.5 Stream Modelling in Angle-Frequency Space

In Chap. 6, we embarked on constructing an algorithm for constraining the Galactic potential without assuming the stream delineates an orbit. We used the angle-action formalism of stream formation to show that the angle-frequency structure of the stream is very simple. In the correct potential, the angle structure of the stream should align with the frequency structure, such that we can use the misalignment between the angle and frequency structure as a measure of the fit of the potential. We showed how this algorithm functioned by recovering the parameters of a simple two-parameter logarithmic potential from a GD-1-like stream simulation. We went on to investigate how the algorithm performed when we include observational errors from current and future surveys. We found that observational errors lead to biases in our method, as observational errors do not translate into simple error distributions in angle-frequency space. This is the disadvantage of taking the data into the model space. To fix these problems, we developed the approach further into a fully probabilistic model that can fully handle observational errors, missing data and outliers. We used the insights of the expected angle-frequency structure of the stream to construct a model in the space of observables. We showed that the probabilistic model could constrain the parameters of the potential from a simulation for error-free data, data with outliers and data with small observational errors.

9.1.6 Determining the Velocity Dispersion of the Thick Disc

In Chap. 7, we attempted to recover the mean vertical velocity and vertical velocity dispersion as a function of the distance from the Galactic plane for a sample drawn from a realistic Galaxy distribution function by following the method presented in Moni Bidin et al. (2012). We found that, for the sample size used, the observational error in the velocities is much smaller than the Poisson noise while Moni Bidin et al. assumed that observational errors dominated. We repeated the analysis on a large number of samples to estimate the contribution of the Poisson noise and to uncover any systematics. We found that the dispersion is systematically overestimated at low Galactic distances from the plane and slightly underestimated at high Galactic distances, leading to an underestimate of the gradient of the dispersion with Galactic height. The causes of the systematics were revealed by repeating the calculation using a method inspired by Girard et al. (2006). This method recovered the expected dispersion much more successfully and in particular yields a gradient of the dispersion with Galactic height that is approximately three times that found using the method presented by Moni Bidin et al.

9.1.7 Extended Distribution Functions for the Galactic Disc

In Chap. 8, we presented a simple extension to the fully dynamical action-based distribution functions of Binney (2012b) to functions that analytically include metallicity information and the expected correlations between metallicity and kinematics inspired by the models of Schönrich and Binney (2009). This extension to include metallicity is essential for correctly modelling the selection effects of any survey, as well as providing vital constraints on the history of the Galaxy. We discussed the inclusion of the selection function when modelling the data from a survey, and showed how we could avoid explicitly using the selection function when we know what observables the data are selected in. We fitted our extended distribution functions to the Geneva-Copenhagen Survey data, and demonstrated the effects of a simple selection function on the kinematics of a sample drawn along a single line of sight. We showed that a simple magnitude-limited survey produced small biases in the resultant velocity distributions due to the hidden selection in age. To close, we presented mock catalogues for the Geneva-Copenhagen Survey and a G dwarf sample from the SEGUE survey, and compared with the data. We discussed how the models could be improved to give a better account of the data. In particular, we found that the required metallicity gradient in the ISM today is in tension with the number of high-metallicity stars in the solar neighbourhood, assuming a linear radial metallicity gradient from the Sun to the Galactic centre. Also, we found that our assumption that the thick disc is a single homogeneous quasi-isothermal population was not appropriate to model the SEGUE data.

9.2 Context

To conclude, we will now set the work of this thesis in a broader context. The goal of a Milky Way scientist is to understand how the Galaxy operates as a machine. This involves understanding at first how the Galaxy is presently structured, and then how it formed and evolved to be in this configuration. Here we will describe the prospects for these grand goals with reference to the work in this thesis.

9.2.1 Galactic Structure

Measuring the distribution of dark matter in the Galaxy is a key aim of Milky Way science. It is believed that the Milky Way, like all galaxies, is enveloped in a dark halo, but its exact properties are still very much uncertain. The local dark matter density is still debated, whilst the shape of the halo is a crucial parameter for which we have very little information. Very little is known about the nature of dark matter, although multiple particle physics experiments have attempted to detect these elusive particles. Astrophysics seems the most immediate route to advancing our understanding of dark matter. Dark matter can currently only be detected through its impact on the dynamics of gas (Begeman 1989; e.g. the rotation curve of NGC3198), photons (e.g. gravitational lensing) and stars. The enormous number of stellar tracers available from current data, as well as those promised from future missions such as *Gaia*, means the last of these probes is being pushed to the fore, and we should be able to pin down the distribution of the dark matter in the Milky Way from the dynamics of these tracers.

In this thesis, we have described two complementary modelling techniques that should prove useful in this respect. Extended distribution function modelling of the Galactic disc will provide us with improved constraints of the local dark matter density as we use the requirement that each distinct chemical population must have consistent dynamics in the same potential. Additionally, the inclusion of chemistry in the distribution function is essential for the modelling of the selection function of any survey, which is necessary for extracting unbiased dynamics from a trace population. The modelling of tidal streams will provide more global constraints on, in particular, the shape of the dark matter halo. More and more tidal streams are being discovered and so provide a very thorough probe of the gravitational field of a large volume of the Galactic halo. Unbiased, probabilistic methods for the measurement of Galactic properties from tidal streams are key for the maximal exploitation of the data, as well as for combining with other constraints on the Galactic potential. Our work on the computation of angle-action variables in both axisymmetric and triaxial potentials should prove valuable in these studies.

Over the coming years, the combined effort of these modelling approaches should provide strong constraints on the dark matter halo of our Galaxy and hence give insights into the structure of dark matter halos throughout the Universe. With these insights, we will take a step closer to understanding the nature of dark matter.

9.2.2 Galactic History

Galactic dynamics helps us to understand the present structure of the Galaxy. In order to understand how the Galaxy came to be in this state, we must broaden our models and approaches. Stars in the Galaxy experience a series of environments throughout their lives as they are scattered by gas clouds and spiral arms. Over time, the stars gradually lose memory of their birth properties as these dynamical processes scatter the stars in action space. Given a snapshot of the Galaxy, how then do we infer the history of the Galaxy? Fortunately, the internal properties of stars retain memory of the environment in which they were born such that the present chemo-dynamical correlations reveal information regarding the history of the Galaxy. These correlations are best uncovered through the fitting of an appropriate model, such as the extended distribution functions presented in this thesis, to current data. Many surveys are now viewing the Galaxy in a rich multi-dimensional chemo-dynamical space. Constraints on the historical dynamical processes in the Galaxy should be extracted by fitting these models to this wealth of data.

The dynamical processes in the Galaxy are known to scatter stars in action space. However, the effect of say a spiral wave on the action distribution of a stellar population is still unclear. Inspection of N-body simulations provides a promising way to determine the appropriate action-space diffusion coefficients generated by realistically complex spiral structure. In this thesis, we have detailed many methods for finding actions which are highly applicable to an N-body simulation. From a series of controlled N-body simulations, the diffusion coefficients for a series of dynamical processes can be measured using the methods outlined in this thesis. These must then be used to refine the extended distribution functions.

The combination of an understanding of the Galactic present and the Galactic past will enable us to piece together the history and evolution of the Milky Way. Insights from the Milky Way can then be paired with those from studies of external galaxies to draw conclusions regarding the evolution and structure of *all* galaxies, and produce a coherent picture of the history of structure in the Universe.

9.3 Future Work

At the end of each chapter, we have detailed further work that could be pursued in light of the work in the chapter. Here we summarise what should be done next with regards to the work in this thesis.

1. One route to understanding our Galaxy is through N-body modelling. With an N-body model, the 'stars' have no errors associated with their phase-space coordinates and we know the potential. Therefore, we are in a position to calculate the angle-action structure of the N-body model. Pontzen and Governato (2013) did this for an N-body dark matter halo under the simplification that the halo is assumed to be spherical, and found that the particles followed a simple distribution in action space. With the techniques for finding angle-action variables presented in this thesis, a next step would be to project an N-body model into angle-action space. Additionally, measuring the change in actions for the particles in the model provides crucial information on processes such as radial migration driven by non-axisymmetric features such as the bar and spiral arms (Solway et al. 2012).
2. The discs of our Galaxy seem to have simple distribution functions in action space (Binney 2010, 2012b; Binney et al. 2014). The next step is to build action-based distribution functions for the other components of the Galaxy, such as the bar, bulge and halo. These features may well be triaxial. Following the work in Chaps. 3 and 4 a next step would be to construct triaxial action-based distribution functions for these components and fit to data such as that from the ARGOS survey (Freeman et al. 2013).
3. The approach to stream modelling presented in Chap. 6 seems promising but in its current state is computationally expensive. Constructing the stream models using the torus machine McMillan and Binney (2008) seems a more promising way to go. Additionally, in Chap. 6 we have only inspected a single stream model and shown that our modelling approach is valid. Before being fully applied to data, the modelling approach should be further validated by inspecting a range of stream models in a range of potentials. For instance, we should create a much broader suite of progenitor models that could have richer internal properties such as rotation, and launch these progenitors on eccentric orbits in a realistic Galactic potential. The results of Chap. 5 show that the impact of the Galactic disc is significant, so inspecting low-J_z stream models will be interesting.
4. After the above two points have been fulfilled, we will be in a position to use the modelling approach in anger on the data. There are many streams with good data, in particular GD-1 (Grillmair and Dionatos 2006; Koposov et al. 2010), and many more will be discovered by *Gaia*. Simultaneous application of the method to many streams should provide improved constraints on the Galactic potential.
5. The next step for the extended distribution function modelling is to fit the model to the SEGUE, RAVE and upcoming Gaia-ESO data. We have already seen that attempting to fit precisely the very local GCS sample is a challenge, despite the flexibility in our model. This points towards interesting correlations that have not

been accounted for in our model. In Chap. 8, we have described various alterations to the extended distribution functions that should be implemented immediately. With data that covers a larger volume of the Galaxy, our model can be further refined and tested. When performing this procedure, we should simultaneously constrain the Galactic potential.
6. Additionally, the EDF should be extended further to account for the [α/Fe] abundances of the stars. [α/Fe] is a crucial quantity for measuring the age of a star, and the ([α/Fe], [Fe/H]) plane is a powerful space to disentangle different Galactic populations (Schönrich et al. 2010; Bovy et al. 2012).

If this plan of future work is fulfilled before the arrival of the *Gaia* data, we will be in a very strong position for fully exploiting the new *Gaia* data, and further advancing our understanding of the Milky Way.

References

Begeman KG (1989) A A 223:47
Binney J (2010) MNRAS 401:2318
Binney J (2012a) MNRAS 426:1324
Binney J (2012b) MNRAS 426:1328
Binney J et al (2014) MNRAS 439:1231
Bovy J et al (2012) ApJ 759:131
Freeman K et al (2013) MNRAS 428:3660
Girard TM, Korchagin VI, Casetti-Dinescu DI, van Altena WF, López CE, Monet DG (2006) AJ 132:1768
Grillmair CJ, Dionatos O (2006) ApJL 643:L17
Koposov SE, Rix H-W, Hogg DW (2010) ApJ 712:260
Law DR, Majewski SR (2010) ApJ 714:229
McMillan PJ, Binney JJ (2008) MNRAS 390:429
Moni Bidin C, Carraro G, Méndez RA (2012) ApJ 747:101
Pontzen A, Governato F (2013) MNRAS 430:121
Schönrich R, Binney J (2009) MNRAS 396:203
Schönrich R, Binney J (2012) MNRAS 419:1546
Schönrich R, Binney J, Dehnen W (2010) MNRAS 403:1829
Solway M, Sellwood JA, Schönrich R (2012) MNRAS 422:1363

Appendix A
Computing the Angle-Action Variables in a Axisymmetric Stäckel Potential

In this appendix, we give expressions for the angle-action variables in an axisymmetric Stäckel potential, which are used in Chap. 2. The approach taken here, as well as the majority of the formulae, have been taken from Eyre (2010). Following on from Eq. (2.12), the action J_τ is given by an integral over a full oscillation in τ. A full oscillation in λ involves integrating twice over the interval (λ_0, λ_1). λ_0 and λ_1 are the roots of p_λ, which may be found by Brent's method using Eq. (2.10). There is a complication when calculating J_ν due to the definition of ν. ν is only uniquely defined for $z \geqslant 0$ such that a full oscillation in ν corresponds to half an oscillation in z. Therefore we calculate J_ν by integrating four times over the interval (ν_0, ν_1), where $\nu_0 = c^2$, as all orbits cross the $z = 0$ plane, and ν_1 is the root of p_ν found by Brent's method. The actions are given explicitly by

$$J_\lambda = \frac{1}{\pi}\int_{\lambda_0}^{\lambda_1} p_\lambda \mathrm{d}\lambda, \quad J_\nu = \frac{2}{\pi}\int_{\nu_0}^{\nu_1} p_\nu \mathrm{d}\nu. \tag{A.1}$$

In order to calculate the angle coordinates, we use the generating function, W, defined as

$$W(\tau, \phi, E, L_z, I_3) = W_\phi + \sum_{\tau=\lambda,\nu} W_\tau = \int_0^\phi L_z \mathrm{d}\phi' + \sum_{\tau=\lambda,\nu}\int_{\tau_0}^{\tau} p_{\tau'}\mathrm{d}\tau'. \tag{A.2}$$

This generating function defines the canonical transformation between the canonical coordinates $(\tau, \phi, p_\tau, L_z)$ and $(J_\tau, L_z, \theta_\tau, \theta_\phi)$. The angles are now computed as

$$\theta_\tau = \frac{\partial W}{\partial J_\tau} = \frac{\partial W}{\partial E}\frac{\partial E}{\partial J_\tau} + \frac{\partial W}{\partial L_z}\frac{\partial L_z}{\partial J_\tau} + \frac{\partial W}{\partial I_3}\frac{\partial I_3}{\partial J_\tau}, \tag{A.3}$$

for $\tau = \lambda, \nu$. The derivatives of the classical integrals with respect to the actions may be found by inverting the 3-by-3 matrix of the derivatives of the actions with respect to the classical integrals. These derivatives are simpler to calculate as they

J. Sanders, *Dynamics of the Milky Way*, Springer Theses,
DOI 10.1007/978-3-319-18772-3

follow from Eq. (A.1) and the definition of p_τ from Eq. (2.10):

$$
\begin{aligned}
\frac{\partial J_\lambda}{\partial E} &= \frac{1}{4\pi}\int_{\lambda_0}^{\lambda_1} \frac{\mathrm{d}\lambda}{(\lambda - a^2)p_\lambda},\\
\frac{\partial J_\lambda}{\partial L_z} &= -\frac{L_z}{4\pi}\int_{\lambda_0}^{\lambda_1} \frac{\mathrm{d}\lambda}{(\lambda - a^2)^2 p_\lambda},\\
\frac{\partial J_\lambda}{\partial I_3} &= -\frac{1}{4\pi}\int_{\lambda_0}^{\lambda_1} \frac{\mathrm{d}\lambda}{(\lambda - a^2)(\lambda - c^2)p_\lambda},\\
\frac{\partial J_\nu}{\partial E} &= \frac{1}{2\pi}\int_{\nu_0}^{\nu_1} \frac{\mathrm{d}\nu}{(\nu - a^2)p_\nu},\\
\frac{\partial J_\nu}{\partial L_z} &= -\frac{L_z}{2\pi}\int_{\nu_0}^{\nu_1} \frac{\mathrm{d}\nu}{(\nu - a^2)^2 p_\nu},\\
\frac{\partial J_\nu}{\partial I_3} &= -\frac{1}{2\pi}\int_{\nu_0}^{\nu_1} \frac{\mathrm{d}\nu}{(\nu - a^2)(\nu - c^2)p_\nu}.
\end{aligned}
\tag{A.4}
$$

The derivatives of the generating function with respect to the classical integrals may be calculated in the same spirit as

$$
\begin{aligned}
\frac{\partial W}{\partial E} &= \sum_{\tau=\lambda,\nu} \frac{1}{4}\int_{\tau_0}^{\tau} \frac{\mathrm{d}\tau'}{(\tau' - a^2)p_{\tau'}}\\
\frac{\partial W}{\partial L_z} &= \phi - \sum_{\tau=\lambda,\nu} \frac{L_z}{4}\int_{\tau_0}^{\tau} \frac{\mathrm{d}\tau'}{(\tau' - a^2)^2 p_{\tau'}}\\
\frac{\partial W}{\partial I_3} &= -\sum_{\tau=\lambda,\nu} \frac{1}{4}\int_{\tau_0}^{\tau} \frac{\mathrm{d}\tau'}{(\tau' - a^2)(\tau' - c^2)p_{\tau'}}.
\end{aligned}
\tag{A.5}
$$

We note that the second equation is simply the angle conjugate to L_z, θ_ϕ.

With the scheme given above there is a degeneracy in θ_ν between points in the orbit at $\pm z$. This is simply resolved by adding 2π to θ_ν if $z < 0$, and dividing θ_z by 2.

As $p^2(\tau)$ vanishes at the endpoints of many of these integrals, we want to avoid evaluating the integrands at the endpoints. We do this by performing a change of variables and estimating the integral using a Gauss-Legendre quadrature scheme. Here we will outline the procedure for calculating J_λ but the same principle follows for the rest of the integrals. We perform a change of variables to

$$\lambda = \hat{\lambda}\sin\vartheta + \bar{\lambda};\ \bar{\lambda} = \frac{1}{2}(\lambda_0 + \lambda_1);\ \hat{\lambda} = \frac{1}{2}(\lambda_1 - \lambda_0), \tag{A.6}$$

such that the integral is now over $\vartheta = (-\frac{\pi}{2}, \frac{\pi}{2})$:

$$J_\lambda = \frac{1}{\pi}\int_{-\pi/2}^{\pi/2} \hat{\lambda}\cos\vartheta\, p(\lambda(\vartheta))\mathrm{d}\vartheta. \tag{A.7}$$

This integral can now be computed numerically using a 10-point Gaussian-Legendre quadrature scheme.

The frequencies, $\mathbf{\Omega}$, are related to the derivatives of the actions with respect to the integrals of the motion by

$$\begin{aligned} \Omega_R &= \frac{1}{\Gamma}\frac{\partial J_z}{\partial I_3}, \\ \Omega_\phi &= \frac{1}{\Gamma}\left(\frac{\partial J_z}{\partial L_z}\frac{\partial J_R}{\partial I_3} - \frac{\partial J_z}{\partial I_3}\frac{\partial J_R}{\partial L_z}\right), \\ \Omega_z &= -\frac{1}{\Gamma}\frac{\partial J_R}{\partial I_3}, \end{aligned} \tag{A.8}$$

where

$$\Gamma = \left(\frac{\partial J_R}{\partial E}\frac{\partial J_z}{\partial I_3} - \frac{\partial J_R}{\partial I_3}\frac{\partial J_z}{\partial E}\right). \tag{A.9}$$

Reference

Eyre A (2010) D.Phil thesis. University of Oxford, Oxford

Appendix B
Derivation of Best-Fitting Stäckel Potential Functions

In this appendix, we give a derivation of the local best-fitting Stäckel potential to a general potential used in Chap. 2. To find the best-fitting Stäckel potential, we must minimise Eq. (2.17) with respect to the function f. It is useful to consider minimisation with respect to the two parts of the function, $f(\lambda)$ and $f(\nu)$. This yields

$$\begin{aligned}\int_{\lambda_-}^{\lambda_+} \mathrm{d}\lambda\, \Lambda(\lambda)[\chi(\lambda,\nu) - f(\lambda) + f(\nu)] = 0,\\ \int_{\nu_-}^{\nu_+} \mathrm{d}\nu\, N(\nu)[\chi(\lambda,\nu) - f(\lambda) + f(\nu)] = 0.\end{aligned} \tag{B.1}$$

Rearranging each of these, and noting that $\Lambda(\lambda)$ and $N(\nu)$ are normalized over the integration range, we find

$$\begin{aligned}f(\lambda) &= \bar{\chi}(\lambda) + \int_{\nu_-}^{\nu_+} \mathrm{d}\nu\, N(\nu) f(\nu),\\ f(\nu) &= -\bar{\chi}(\nu) + \int_{\lambda_-}^{\lambda_+} \mathrm{d}\lambda\, \Lambda(\lambda) f(\lambda),\end{aligned} \tag{B.2}$$

where the definition of $\bar{\chi}(\tau)$ is given in Eq. (2.20). Substitution of the expression for $f(\nu)$ into the expression for $f(\lambda)$ we find

$$f(\lambda) = \bar{\chi}(\lambda) - \bar{\bar{\chi}} + \int_{\lambda_-}^{\lambda_+} \mathrm{d}\lambda\, \Lambda(\lambda) f(\lambda), \tag{B.3}$$

where $\bar{\bar{\chi}}$ is defined in Eq. (2.20). This equation along with Eq. (2.17) implies that

$$\int_{\lambda_-}^{\lambda_+} \mathrm{d}\lambda\, \Lambda(\lambda) f(\lambda) - \int_{\nu_-}^{\nu_+} \mathrm{d}\nu\, N(\nu) f(\nu) - \bar{\bar{\chi}}. \tag{B.4}$$

J. Sanders, *Dynamics of the Milky Way*, Springer Theses,
DOI 10.1007/978-3-319-18772-3

As we are only constraining the difference $[f(\lambda) - f(\nu)]$ we are free to choose the values of these integrals as long as their difference equals $\bar{\bar{\chi}}$. We opt for the symmetric choice

$$f(\lambda) = \bar{\chi}(\lambda) - \frac{1}{2}\bar{\bar{\chi}}, \quad f(\nu) = -\bar{\chi}(\nu) + \frac{1}{2}\bar{\bar{\chi}}. \tag{B.5}$$

Appendix C
Angle-Actions in the Isochrone Potential

In this appendix, we give expressions for the angle-actions in the isochrone potential useful for Chap. 3. We first convert our phase-space point into spherical polar coordinates given by $(r, \phi, \vartheta, v_r, v_\phi, v_\vartheta)$. The three actions are the radial actions, J_r, the angular momentum, L, and the vertical action, $L - |L_z|$, where L_z is the angular momentum about the z-axis. If E is the orbital energy, the radial action is given by

$$J_r = \frac{1}{\sqrt{-2E}} + \frac{1}{2}(L + \sqrt{L^2 + 4GMb}). \tag{C.1}$$

The expressions for the angles are more involved. We define

$$\begin{aligned} c &\equiv \frac{GM}{-2E} - b, \\ e &\equiv \sqrt{1 - \frac{L^2}{GMc}(1 + b/c)}, \\ \eta &= \arctan\left(\frac{r v_r}{\sqrt{-2E}(b + c - \sqrt{b^2 + r^2})}\right). \end{aligned} \tag{C.2}$$

The radial angle, θ_r, is then given by

$$\theta_r = \eta - \frac{ec \sin \eta}{c + b}. \tag{C.3}$$

Additionally, the frequencies, $\partial H / \partial \boldsymbol{J}$, are given by

$$\begin{aligned} \Omega_r &= \frac{(-2E)^{3/2}}{GM}, \\ \Omega_\phi &= \frac{1}{2}\Omega_r\left(1 + \frac{L}{\sqrt{L^2 + 4GMb}}\right). \end{aligned} \tag{C.4}$$

J. Sanders, *Dynamics of the Milky Way*, Springer Theses,
DOI 10.1007/978-3-319-18772-3

Note $\Omega_\phi = \partial H/\partial L_z = 0$. To find θ_ϕ and θ_z, we further define

$$\begin{aligned} \psi &\equiv \arctan\left(\frac{L\cos\vartheta}{-v_\vartheta r\sin\vartheta}\right) \\ a &\equiv \sqrt{\frac{1+e}{1-e}}, \\ a_p &\equiv \sqrt{\frac{1+e+2b/c}{1-e+2b/c}}, \end{aligned} \tag{C.5}$$

such that

$$\theta_z = \psi + \frac{\Omega_\phi\theta_r}{\Omega_R} - \arctan(a\tan(\eta/2)) - \frac{\arctan(a_p\tan(\eta/2))}{\sqrt{1+4GMbL^{-2}}}. \tag{C.6}$$

Finally, we define

$$\sin u \equiv \frac{L_z\tan\vartheta}{L\sqrt{1-(L_z/L)^2}}, \tag{C.7}$$

such that

$$\theta_\phi = \phi - u + \mathrm{sign}(L_z)\theta_z. \tag{C.8}$$

In both the expression for θ_ϕ and θ_z, we must be careful in defining the branch cuts of the trigonometric functions to ensure the angles increase continuously from 0 to 2π. We define

$$\arctan(p\tan q) = \begin{cases} \arctan(p\tan q) & \text{if } -\frac{\pi}{4} < q < \frac{\pi}{4}, \\ \frac{\pi}{2} - \arctan\left(p\tan\left[\frac{\pi}{2} - q\right]\right) & \text{if } q < -\frac{\pi}{4}, \\ -\frac{\pi}{2} + \arctan\left(p\tan\left[\frac{\pi}{2} + q\right]\right) & \text{if } q > \frac{\pi}{4}. \end{cases} \tag{C.9}$$

Appendix D
Symmetries

In Sect. 3.2.2, we asserted that for a time-reversible Hamiltonian the Fourier components of the generating function, $S_{\boldsymbol{n}}$, are real. However, it must also be true that there is a point on the target torus where $\dot{\boldsymbol{J}} = \boldsymbol{0}$. McGill and Binney (1990) show that this is true if the toy potential is an isochrone and the target Hamiltonian is axisymmetric. Additionally, they demonstrated that, when the potential is symmetric about the plane $z = 0$, Fourier components of the generating function with odd n_z vanish. Here we repeat these arguments extended to the 3D triaxial case. Let's first consider the loop orbits. Suppose we have a target Hamiltonian of the form

$$H(r, \phi, \vartheta) = \frac{1}{2}p_r^2 + \frac{p_\phi^2}{2r^2\sin^2\vartheta} + \frac{p_\vartheta^2}{2r^2} + \Phi(r, \phi, \vartheta), \tag{D.1}$$

where (r, ϕ, ϑ) are standard spherical polar coordinates. The equations of motion for the toy actions are

$$\begin{aligned}\dot{J}_i &= -\frac{\partial H}{\partial \theta_i} \\ &= \left(\frac{p_\vartheta^2}{r^3} + \frac{p_\phi^2}{r^3\sin^2\vartheta} - \frac{\partial\Phi}{\partial r}\right)\frac{\partial r}{\partial\theta_i} + \left(\frac{p_\phi^2\cos\vartheta}{r^2\sin^3\vartheta} - \frac{\partial\Phi}{\partial\vartheta}\right)\frac{\partial\vartheta}{\partial\theta_i} \\ &\quad - \frac{\partial\Phi}{\partial\phi}\frac{\partial\phi}{\partial\theta_i} - p_r\frac{\partial p_r}{\partial\theta_i} - \frac{p_\vartheta}{r^2}\frac{\partial p_\vartheta}{\partial\theta_i} - \frac{p_\phi}{r^2\sin^2\vartheta}\frac{\partial p_\phi}{\partial\theta_i}.\end{aligned} \tag{D.2}$$

Now let us consider the point $\boldsymbol{\theta} = (0, 0, \pi/2)$: at this point the particle is at pericentre, at a maximum in its vertical oscillation and at $\phi = 0$. At this point, we have that

$$\frac{\partial r}{\partial\theta_i} = \frac{\partial\vartheta}{\partial\theta_i} = p_r = p_\vartheta = \frac{\partial p_\phi}{\partial\theta_i} = 0, \tag{D.3}$$

J. Sanders, *Dynamics of the Milky Way*, Springer Theses,
DOI 10.1007/978-3-319-18772-3

so

$$\dot{J}_i = -\frac{\partial\Phi}{\partial\phi}\frac{\partial\phi}{\partial\theta_i}. \tag{D.4}$$

In a triaxial potential, $x = 0$ is a symmetry plane of the potential so $\partial\Phi/\partial\phi|_{\phi=0} = 0$ and $\dot{J}_i = 0$. This is the requirement introduced in Sect. 3.2.2 for the Fourier components of the generating function to be real. Now let's consider the point $\boldsymbol{\theta} = (0, 0, 0)$. Here the particle is at pericentre, crossing the $z = 0$ plane, and at $\phi = 0$. At this point we have

$$\frac{\partial r}{\partial\theta_i} = \cos\vartheta = p_r = \frac{\partial p_\vartheta}{\partial\theta_i} = \frac{\partial p_\phi}{\partial\theta_i} = 0, \tag{D.5}$$

so

$$\dot{J}_i = -\frac{\partial\Phi}{\partial\phi}\frac{\partial\phi}{\partial\theta_i} - \frac{\partial\Phi}{\partial\vartheta}\frac{\partial\vartheta}{\partial\theta_i}. \tag{D.6}$$

As we saw before, the first term is zero as $x = 0$ is a symmetry plane of the potential. The second term is also zero as $z = 0$ is also a symmetry plane. By a similar argument at $\boldsymbol{\theta} = (0, \pi/2, 0)$, $\partial\Phi/\partial\phi|_{\phi=\pi/2} = 0$ as $y = 0$ is a symmetry plane of the potential.

We calculate $\dot{\boldsymbol{J}}$ from Eq. (3.8) as

$$\dot{\boldsymbol{J}} = \sum_{\boldsymbol{n}\in\mathbf{N}} 2\boldsymbol{n}\left(\imath\boldsymbol{n}\cdot\dot{\boldsymbol{\theta}}\right)S_{\boldsymbol{n}}(\boldsymbol{J}')\sin\boldsymbol{n}\cdot\boldsymbol{\theta} \tag{D.7}$$

At the point $\boldsymbol{\theta} = (0, 0, \pi/2)$, we know $\dot{\boldsymbol{J}} = \mathbf{0}$, so we require $\sin\pi n_3/2 = 0$ such that n_3 must be even. Similarly, we know $\dot{\boldsymbol{J}} = \mathbf{0}$ at $\boldsymbol{\theta} = (0, \pi/2, 0)$ so n_2 is restricted to even values. However, n_1 can take any integer value.

Now let us consider the box orbits. We have a target Hamiltonian of the form

$$H = \frac{1}{2}\sum_i p_i^2 + \Phi(x, y, z) \tag{D.8}$$

where $p_i = (p_x, p_y, p_z)$ and the equations of motion for the toy actions are

$$\dot{J}_i = -\sum_j \frac{\partial\Phi}{\partial x_j}\frac{\partial x_j}{\partial\theta_i} - p_j\frac{\partial p_j}{\partial\theta_i} \tag{D.9}$$

Consider the point $\boldsymbol{\theta} = (0, 0, 0)$. Here the orbit is turning in all three coordinates so $\boldsymbol{p} = \mathbf{0}$ and $\partial\boldsymbol{x}/\partial\theta_i = \mathbf{0}$ so $\dot{\boldsymbol{J}} = \mathbf{0}$ as required in Sect. 3.2.2. Now let's consider the point $\boldsymbol{\theta} = (\pi/2, 0, 0)$. Here the orbit is turning in y and z and is passing through the $x = 0$ plane at which point $\partial p_x/\partial\theta_i = 0$ as p_x is at a maximum. Therefore, we have

$$\dot{J}_i = -\frac{\partial\Phi}{\partial x}\frac{\partial x}{\partial\theta_i}. \tag{D.10}$$

For a triaxial potential aligned with our choice of Cartesian axes, $x = 0$ is a symmetry plane so $\partial\Phi/\partial x|_{x=0} = 0$. Therefore $\dot{\boldsymbol{J}} = \mathbf{0}$ here and by similar arguments to the loop orbit case we are restricted to even n_1. We can employ the same arguments by considering the stationary points $\boldsymbol{\theta} = (0, \pi/2, 0)$ and $\boldsymbol{\theta} = (0, 0, \pi/2)$ to show that n_2 and n_3 must be even.

Reference

McGill C, Binney J (1990) MNRAS 244, 634

Appendix E
Angles and Frequencies from Stäckel Fudge

With the framework presented in Sect. 4.3, we are also in a position to find the angles, $\boldsymbol{\theta}$, and frequencies, $\boldsymbol{\Omega}$. Following de Zeeuw (1985), we write

$$\begin{aligned}\frac{\partial E}{\partial E}&=1=\sum_{\tau=\lambda,\mu,\nu}\Omega_\tau\frac{\partial J_\tau}{\partial E},\\ \frac{\partial E}{\partial a}&=0=\sum_{\tau=\lambda,\mu,\nu}\Omega_\tau\frac{\partial J_\tau}{\partial a},\\ \frac{\partial E}{\partial b}&=0=\sum_{\tau=\lambda,\mu,\nu}\Omega_\tau\frac{\partial J_\tau}{\partial b}.\end{aligned}\tag{E.1}$$

Inversion of these equations gives, for instance,

$$\Omega_\lambda=\frac{1}{\Gamma}\frac{\partial(J_\mu,J_\nu)}{\partial(a,b)}\text{ where }\Gamma=\frac{\partial(J_\lambda,J_\mu,J_\nu)}{\partial(E,a,b)},\tag{E.2}$$

and Ω_μ and Ω_ν are given by cyclic permutation of $\{\lambda,\mu,\nu\}$. To find the derivatives of J_τ with respect to the integrals, we differentiate Eq. (4.13) under the integral sign at constant τ. From Eq. (4.17), we know $p_\tau(\tau,E,A_\tau,B_\tau)$. We note that

$$\begin{aligned}\frac{\partial}{\partial a}\bigg|_\tau&=\frac{\partial A_\tau}{\partial a}\bigg|_\tau\frac{\partial}{\partial A_\tau}\bigg|_\tau=\frac{\partial}{\partial A_\tau}\bigg|_\tau,\\ \frac{\partial}{\partial b}\bigg|_\tau&=\frac{\partial B_\tau}{\partial b}\bigg|_\tau\frac{\partial}{\partial B_\tau}\bigg|_\tau=\frac{\partial}{\partial B_\tau}\bigg|_\tau,\end{aligned}\tag{E.3}$$

as $A_\tau=a-C_\tau$ and $B_\tau=b+D_\tau$ where C_τ and D_τ are independent of τ. The required derivatives are

J. Sanders, *Dynamics of the Milky Way*, Springer Theses,
DOI 10.1007/978-3-319-18772-3

$$\begin{aligned}
\left.\frac{\partial p_\tau}{\partial E}\right|_\tau &= \frac{\tau^2}{4p_\tau(\tau+\alpha)(\tau+\beta)(\tau+\gamma)},\\
\left.\frac{\partial p_\tau}{\partial a}\right|_\tau &= -\frac{1}{\tau}\left.\frac{\partial p_\tau}{\partial E}\right|_\tau,\\
\left.\frac{\partial p_\tau}{\partial b}\right|_\tau &= \frac{1}{\tau^2}\left.\frac{\partial p_\tau}{\partial E}\right|_\tau.
\end{aligned} \tag{E.4}$$

Note that p_τ can vanish at the limits of integration. The change of variables

$$\tau = \hat{\tau}\sin\vartheta + \bar{\tau};\ \bar{\tau} = \frac{1}{2}(\tau_- + \tau_+);\ \hat{\tau} = \frac{1}{2}(\tau_+ - \tau_-)$$

causes the integrand to go smoothly to zero at the limits. To find the angles, we use the generating function, $W(\lambda, \mu, \nu, J_\lambda, J_\mu, J_\nu)$, given by

$$W = \sum_{\tau=\lambda,\mu,\nu} W_\tau = \sum_{\tau=\lambda,\mu,\nu} \int_{\tau_-}^{\tau} \mathrm{d}\tau' p'_\tau + \mathcal{F}_\tau(p_\tau, \boldsymbol{x}) \int_{\tau_-}^{\tau_+} \mathrm{d}\tau' |p'_\tau|. \tag{E.5}$$

$\mathcal{F}_\tau$ are factors included to remove the degeneracy in the τ coordinates such that θ_τ covers the full range 0–2π over one oscillation in the Cartesian coordinates. These factors can be written in the form

$$\begin{aligned}
\mathcal{F}_\lambda(p_\lambda, \boldsymbol{x}) &= \Pi(\lambda_- + \alpha)\Theta(-x) + \Theta(-p_\lambda),\\
\mathcal{F}_\mu(p_\mu, \boldsymbol{x}) &= \Pi(\mu_- + \beta)[\Theta(-y) + \Theta(-p_\mu)]\\
&\quad + \Pi(\nu_+ + \beta)\Pi(\mu_+ + \alpha)\left[\frac{1}{2} + \Theta(-x)\right]\\
&\quad + \Pi(\nu_+ + \beta)\Pi(\lambda_- + \alpha)\Theta(-p_\mu),\\
\mathcal{F}_\nu(p_\nu, \boldsymbol{x}) &= \Theta(-z) + \Theta(-p_\nu),
\end{aligned} \tag{E.6}$$

where Θ is the Heaviside step function and Π is one when its argument is zero and zero otherwise. The Π function in $\mathcal{F}_\lambda$ takes care of the cases when the orbit is a box or inner long-axis loop. The Π functions in $\mathcal{F}_\mu$ take care of the cases when the orbit is a short-axis loop or a box, an outer long-axis loop, and an inner long-axis loop respectively. The angles are given by

$$\theta_\tau = \frac{\partial W}{\partial J_\tau} = \sum_{I=E,a,b} \frac{\partial W}{\partial I}\frac{\partial I}{\partial J_\tau}. \tag{E.7}$$

The first term on the right is, up to factors, the indefinite integral of the derivatives of J_τ with respect to the integrals found previously, whilst the second term is found from inverting these derivatives. We have chosen the zero-point of θ_τ to correspond to $\tau = \tau_-$, $p_\tau > 0$ and $\dot{x}_i \geq 0$ for all i, except for the outer long-axis loop orbits

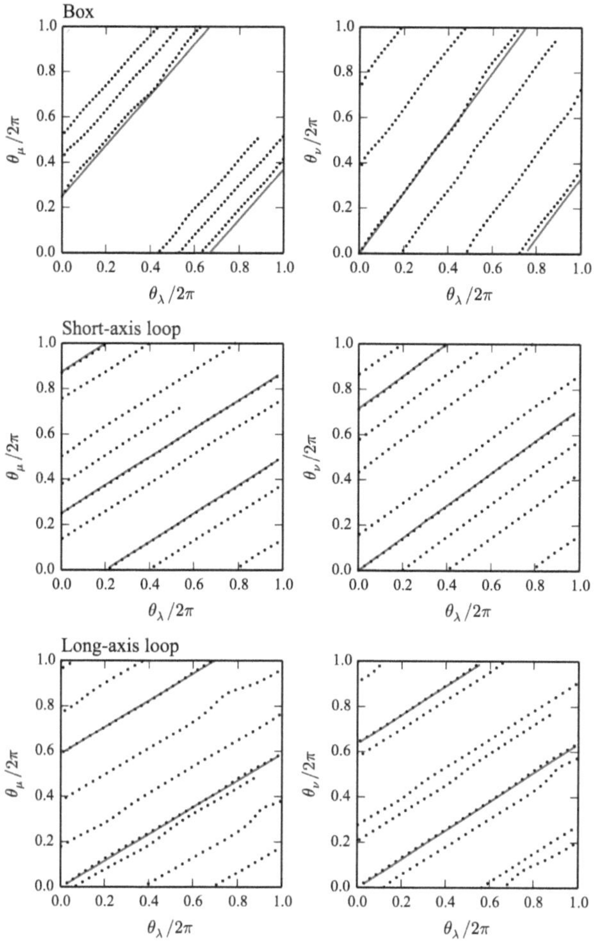

Fig. E.1 Angles calculated using the triaxial Stäckel fudge presented in Chap. 4 for three different orbits in the triaxial NFW potential. The *solid red lines* show the angles calculated from the initial angle estimate and the frequency estimates for approximately one period

which have $\theta_\mu = 0$ at $\mu = -\alpha$, $p_\tau > 0$ and $\dot{x}_i \geq 0$. Note that the angles are the 2π modulus of the θ_τ found from the above scheme.

In Fig. E.1, we show the angles calculated from the Stäckel fudge for the three orbits investigated in Sect. 4.5. We use the automatic choice of Δ_i for the box and short-axis loop orbit, and the choice that minimises the spread in actions for the long-axis loop orbit. The short-axis loop orbit shows the expected straight-line structure in the angle coordinates, whilst for the long-axis loop and box orbits there is clear deviation from this expected straight line. We also show the angles calculated using the initial angle estimate and the median of the frequency estimates along the orbit.

We see that they are well recovered but, after approximately one period, the error in the frequencies is sufficient for these angles to deviate from the angle estimates.

Reference

de Zeeuw T (1985) MNRAS 216, 273

Appendix F
A Family of Two-Parameter Potentials

We wish to construct a suite of realistic Galaxy potentials that are defined by two parameters: the density flattening of the halo, Q, and the ratio k of the force on the Sun due to visible matter and dark matter. This gives us a range of appropriate potentials that we explore to find the best-fitting Galaxy potential. It acts as a prior in our exploration of all possible Galaxy potentials.

For our base model, we adopt the usual multi-component model: a bulge, thick and thin discs, and a dark halo. For each of these components, we use the functional forms discussed by McMillan (2011). The bulge is taken to be a Bissantz-Gerhard model, and we adopt exponential discs for the thick and thin discs. For the dark matter profile, we adopt the NFW profile (Navarro et al. 1996) with a flattening introduced:

$$\rho_h = \frac{\rho_{\mathrm{h},0}}{x(1+x)^2} \text{ where } x = \frac{\sqrt{R^2 + (z/Q)^2}}{r_h}. \tag{F.1}$$

This introduces the first of our two parameters, the halo flattening Q. The second is defined as

$$k \equiv \frac{1}{N}\frac{g_{\mathrm{disc}}(R_0, z_0)}{g_{\mathrm{halo}}(R_0, z_0)} \tag{F.2}$$

where g_i is the magnitude of the gravitational force on the Sun due to the ith component, and the normalisation N is chosen such that $k = 1$ for the 'best' potential from McMillan (2011). The model with $(k, Q) = (1, 1)$ corresponds exactly to McMillan's best potential. We take $(R_0, z_0) = (8.29, 0.0)$ kpc.

We would like all these potentials to satisfy the observational constraints that have been collected and listed by McMillan (2011). These include maser observations and terminal velocity curves. However, the most important constraint is the circular speed at the solar position which is largely constrained by the motion of Sgr A* (Reid and Brunthaler 2004). Therefore, we only adjust the parameters until the circular speed at the solar position is correct, which McMillan found to be $v_c = 239.1\,\mathrm{km\,s^{-1}}$.

J. Sanders, *Dynamics of the Milky Way*, Springer Theses,
DOI 10.1007/978-3-319-18772-3

F.1 Procedure

For a given pair of the parameters (k_s, Q_s), we follow this procedure to find a realistic Galactic potential with these parameters:

1. Construct McMillan's best potential corresponding to $(k, Q) = (1, 1)$.
2. Set $Q = Q_s$.
3. Adjust $\rho_{\mathrm{h},0}$ until $k = k_s$.
4. Calculate the circular speed at the solar position in this model, v_{cs}.
5. Scale $\rho_{\mathrm{h},0}$ and Σ_{d} by the same factor $p = (v_c/v_{cs})^2$ so that the circular speed at the solar position is restored to $v_c = 239.1\ \mathrm{km\ s}^{-1}$.

F.2 Tabulation

As the constructed potentials are simply described by two parameters, it is convenient to construct a 2D grid of these potentials in (k, Q) space. We construct the potentials using the above procedure for N values of k and N values of Q. At each of these points in parameter space, we store $\rho_{\mathrm{h},0}$ and Σ_{d} in an $N \times N$ array. This grid may then be linearly interpolated for a given pair of (k, Q). For any call that falls outside the grid range, we use the full procedure outlined above. In Table F.1, we list the parameters of a sample of (k, Q) models, and, in Fig. F.1, these parameters are plotted as a function of k.

Table F.1 Parameters for 64 (k, Q) models: $\rho_{\mathrm{h},0}$ is the central dark matter halo density (in units of $10^6 M_\odot\ \mathrm{kpc}^{-3}$) and $\Sigma_{\mathrm{d,thin}}$ and $\Sigma_{\mathrm{d,thick}}$ are the surface densities of the thin and thick disc (in units of $10^6 M_\odot\ \mathrm{kpc}^{-2}$)

k	Q	$\rho_{\mathrm{h},0}$	$\Sigma_{\mathrm{d,thin}}$	$\Sigma_{\mathrm{d,thick}}$
1	1	8.46	817	209
0.25	0.25	37.4	344	88.3
0.25	0.50	22.0	344	88.3
0.25	0.75	16.8	344	88.3
0.25	1.00	14.3	344	88.3
0.25	1.25	12.7	344	88.3
0.25	1.50	11.7	344	88.3
0.25	1.75	10.9	344	88.3
0.25	2.00	10.4	344	88.3
0.50	0.25	30.5	560	144
0.50	0.50	17.9	560	144
0.50	0.75	13.7	560	144
0.50	1.00	11.6	560	144
0.50	1.25	10.3	560	144

Table F.1 (continued)

k	Q	$\rho_{h,0}$	$\Sigma_{d,thin}$	$\Sigma_{d,thick}$
1	1	8.46	817	209
0.50	1.50	9.49	560	144
0.50	1.75	8.88	560	144
0.50	2.00	8.43	560	144
0.75	0.25	25.7	709	182
0.75	0.50	15.1	709	182
0.75	0.75	11.6	709	182
0.75	1.00	9.78	709	182
0.75	1.25	8.71	709	182
0.75	1.50	8.00	709	182
0.75	1.75	7.49	709	182
0.75	2.00	7.10	709	182
1.00	0.25	22.2	817	209
1.00	0.50	13.0	817	209
1.00	0.75	9.99	817	209
1.00	1.00	8.46	817	209
1.00	1.25	7.53	817	209
1.00	1.50	6.92	817	209
1.00	1.75	6.47	817	209
1.00	2.00	6.14	817	209
1.25	0.25	19.5	899	231
1.25	0.50	11.5	899	231
1.25	0.75	8.80	899	231
1.25	1.00	7.45	899	231
1.25	1.25	6.63	899	231
1.25	1.50	6.09	899	231
1.25	1.75	5.70	899	231
1.25	2.00	5.41	899	231
1.50	0.25	17.5	964	247
1.50	0.50	10.3	964	247
1.50	0.75	7.86	964	247
1.50	1.00	6.65	964	247
1.50	1.25	5.93	964	247
1.50	1.50	5.44	964	247
1.50	1.75	5.09	964	247
1.50	2.00	4.83	964	247
1.75	0.25	15.8	1020	261
1.75	0.50	9.27	1020	261
1.75	0.75	7.10	1020	261
1.75	1.00	6.01	1020	261

(continued)

Table F.1 (continued)

k	Q	$\rho_{h,0}$	$\Sigma_{d,thin}$	$\Sigma_{d,thick}$
1	1	8.46	817	209
1.75	1.25	5.35	1020	261
1.75	1.50	4.92	1020	261
1.75	1.75	4.60	1020	261
1.75	2.00	4.36	1020	261
2.00	0.25	14.4	1060	272
2.00	0.50	8.46	1060	272
2.00	0.75	6.48	1060	272
2.00	1.00	5.48	1060	272
2.00	1.25	4.88	1060	272
2.00	1.50	4.48	1060	272
2.00	1.75	4.20	1060	272
2.00	2.00	3.98	1060	272

The top lines with $(k, Q) = (1, 1)$ are identical to McMillan's best potential

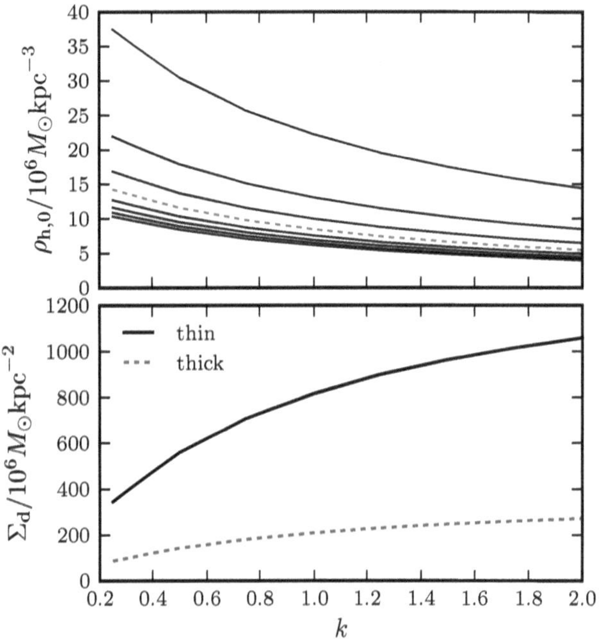

Fig. F.1 $\rho_{h,0}$ and Σ_d as a function of k for the two-parameter (k, Q) potentials. The *top panel* shows lines of constant Q. The lines are spaced by $\Delta Q = 0.25$ with the *uppermost line* showing $Q = 0.25$. The *red dashed line* gives $Q = 1$. The *bottom panel* shows the variation of Σ_d for the thin and thick discs, which has no Q dependence

References

McMillan PJ (2011) MNRAS 414, 2446
Navarro JF, Frenk CS, White SDM (1996) ApJ 462,563
Reid MJ, Brunthaler A (2004) ApJ 616, 872

Appendix G
Finding Angles and Frequencies for Stream Particles

G.1 Estimating I_3

In the algorithm presented in Chap. 2, I_3 was estimated by averaging three estimates along the orbit. I_3 is an integral of the motion in a Stäckel potential so, in Chap. 2, it was desirable to avoid the estimates of I_3 depending on the initial position on the orbit. This produced much smaller errors in the estimated actions so was a favourable scheme for the task presented. However, problems arise around the turning points of the orbit with a poor choice of I_3. A small error in I_3 can move the boundary of the orbit away from the true boundary, so points which are close to the boundary appear further away than they should be. As I_3 is a separation constant, it can be estimated in two different ways, using $\dot{\lambda}$ or $\dot{\nu}$:

$$\begin{aligned} I_3 &\approx (\lambda - c^2)\left(E - \frac{L_z^2}{2(\lambda - a^2)} + \frac{f(\lambda)}{\lambda - c^2} - \frac{\dot{\lambda}^2(\lambda-\nu)^2}{8(\lambda - a^2)(\lambda - c^2)^2}\right), \\ &\approx (\nu - c^2)\left(E - \frac{L_z^2}{2(\nu - a^2)} + \frac{f(\nu)}{\nu - c^2} - \frac{\dot{\nu}^2(\nu-\lambda)^2}{8(\nu - a^2)(\nu - c^2)^2}\right). \end{aligned} \tag{G.1}$$

In a Stäckel potential, these two estimates are identical but, in a general potential, we expect a discrepancy. Using only the $\dot{\lambda}$ estimate leads to errors in θ_z near the turning points and, similarly, using only the $\dot{\nu}$ estimate leads to errors in θ_R near the turning points.

When analysing the stream data, we are primarily interested in the angles and the behaviour of the stream near apsis. Also, we are only interested in relative differences between stars in the stream. Therefore, we use a modified scheme to that presented in Chap. 2. We choose to only estimate I_3 at the given phase-space point. We also calculate two different estimates for I_3 using the $\dot{\lambda}$ and $\dot{\nu}$. The equation of motion for λ uses the $\dot{\lambda}$ estimate and similarly for ν. In this way, we remove noise around the turning points and the stream structure is recovered much more cleanly.

J. Sanders, *Dynamics of the Milky Way*, Springer Theses,
DOI 10.1007/978-3-319-18772-3

Several other small alterations have been made to the algorithm presented in Chap. 2, which improve the performance for the present task. We have increased the number of α-estimates per particle, and increased the size of the potential-fitting region and selected the weight function $\Lambda(\lambda) = 3\lambda^{-4}(\lambda_+^{-3} - \lambda_-^{-3})^{-1}$ to reduce noise in the frequencies.

G.2 Impact of Errors on Suggested Algorithm

The Stäckel-fitting algorithm used to estimate the angles and frequencies introduces errors that systematically depend on the orbital phase. Using the same method to estimate the errors as that presented in Chap. 2, we find that the standard errors in the angles for the orbit used in Sect. 6.3 are $(\sigma_{\theta_R}, \sigma_{\theta_\phi}, \sigma_{\theta_z}) = (9.4, 5.7, 5.4) \times 10^{-3}$ rad. For a single orbit, the frequencies should be constant at all points along the orbit. We estimate the error in the calculated frequencies as the standard deviation of their estimates around the orbit which yields $(\sigma_{\Omega_R}, \sigma_{\Omega_\phi}, \sigma_{\Omega_z}) = (3.7, 4.0, 2.6) \times 10^{-3}\,\mathrm{Gyr}^{-1}$. These errors are much smaller than the size of the distributions shown in Fig. 6.4. The errors in the frequency and angle estimation depend upon the phase of the particle, with the largest error occurring at pericentre.

However, these absolute errors do not give a good indication of the accuracy of the algorithm presented in Chap. 6. As we are measuring the gradients of the frequency distribution and the angle distribution, we are instead concerned with the errors in the differences of the frequencies/angles of the particles in the stream, which are subject to more subtle effects. For simplicity, we limit the discussion to the errors in the frequencies but similar effects are found in the angles. Let us consider two particles in the stream: one in the cluster with frequency $\boldsymbol{\Omega}_0 = (\Omega_{R0}, \Omega_{\phi 0}, \Omega_{z0})$ and another particle that has just been stripped from the progenitor with frequency $\boldsymbol{\Omega} = (\Omega_{R0}+\Delta\Omega_R, \Omega_{\phi 0}+\Delta\Omega_\phi, \Omega_{z0}+\Delta\Omega_z)$. The Stäckel-fitting algorithm produces deviations in the frequencies, given by $\boldsymbol{\sigma}(t, \boldsymbol{\Omega})$, that depend on the orbital phase so fluctuate in time at a rate determined by the frequencies.

At a given time, the difference in the estimated frequencies is given by

$$\begin{aligned} \delta\Omega_i &= [\Omega_i + \sigma_i(t, \boldsymbol{\Omega})] - [\Omega_{0i} + \sigma_i(t, \boldsymbol{\Omega}_0)] \\ &\approx \Delta\Omega_i + 2g_i(\Delta\boldsymbol{\Omega}, t), \end{aligned} \tag{G.2}$$

where $g_i(\Delta\boldsymbol{\Omega}, t)$ is an oscillating function of the frequency difference, and we have dropped the fast-oscillating part. Therefore, there is an error in the difference which oscillates in time at a frequency related to the beat frequency. In Fig. G.1, we plot the azimuthal frequency difference, $\Delta\Omega_\phi$, as a function of time for a single particle from the simulation integrated for a long time. The particle was stripped at the third pericentric passage and then integrated independent of the simulation for many more periods. We observe that the fluctuations in the frequency introduced by the algorithm are producing beats in $\Delta\Omega_\phi$ with a frequency of $\sim$0.3 Gyr^{-1}. We expect that, at early

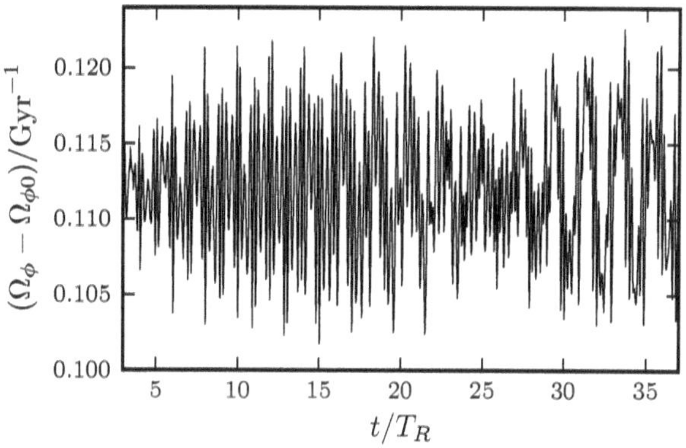

Fig. G.1 The azimuthal frequency difference between the progenitor and a single particle taken from the simulation, and integrated for a long time. For all shown times, the particle is moving freely in the external Galactic potential. The error in the frequency difference exhibits beats at a frequency of ~0.3 Gyr^{-1} characteristic of two particles oscillating at close frequencies drifting in and out of phase. The units give the times of pericentric passage of the progenitor

times ($t \ll 5$ Gyr), the error in the difference will be small as the systematic errors in the progenitor and particle frequency-estimates are in phase and so nearly cancel. At large times ($t \approx 5$ Gyr), the error fluctuations in the particle and progenitor frequencies have drifted out of phase, so the error in the difference is large. The error in $\Delta\Omega_\phi$ reaches a maximum of ~10 %. A consequence of this systematic is that, at early times, the particles with the largest frequency difference will introduce the largest errors, as they were the first to leave the cluster. This explains the broadening of the extremes of the frequency distributions in Fig. 6.4. A similar effect is observed in the angles shown in Fig. 6.2. The largest error in the angles occurs at pericentre (Chap. 2). Around each pericentric passage, the Stäckel-fitting algorithm produces a small blip in the angle difference. This blip increases in magnitude with time as the systematic errors introduced by the algorithm shift out of phase with each other.

The discussion has been limited to the errors introduced when finding the gradient using a single particle and the progenitor. It is more difficult to assess how the linear regression of the angles and frequencies of many particles in the stream is affected, but it is clear that it is systematic and not random, as the errors of neighbouring stream particles are correlated. The errors depend upon the initial cluster conditions which govern $\Delta\boldsymbol{\Omega}$, the time since the cluster started being stripped, and the phase at which the stream is observed. We also expect this source of error to decrease with the mass of the progenitor. Larger progenitor masses produce larger frequency distributions (see Chap. 5), so the errors introduced by the Stäckel-fitting algorithm become less significant.

G.3 Calculating the Hessian

As part of our likelihood in Sect. 5.3, we require the Hessian matrix, $\mathbf{D} = \partial^2 H/\partial \boldsymbol{J}^2$. In a Stäckel potential, this matrix may be found numerically by following the procedure presented in the Appendix of Eyre (2010). This involves finding the second derivatives of the analytic integral expressions for the actions with respect to the integrals of motion: the energy, E, the z-component of the angular momentum, L_z, and the third integral, I_3. The resulting integrals are performed analytically using Gaussian quadrature, but care must be taken due to the divergence of the frequency integrand at the end-points. These considerations are taken care of by introduction of a dummy variable as described in Eyre (2010).

For our purposes, we are using a Stäckel approximation to the true potential (Chap. 2), so we estimate the true Hessian matrix as that calculated in the approximate Stäckel potential. In the true potential, the error in each component of the Hessian matrix is less than 10 %. However, the error in the determinant is larger ($\sim$30 %). As the potentials considered are near-spherical, the determinant of the Hessian is small (it is zero for the spherical case), which arises due to cancelling terms in the calculation. Therefore, small errors in each component can give rise to larger errors in the determinant. We recover the appropriate trends in the determinant of the Hessian matrix with the potential parameters. There is a slight bias in Fig. 6.11 that is probably due to the errors in the Hessian matrix. However, it is not significant. The results shown in this paper demonstrate that the magnitude of the observational errors in the data dominates any systematic errors in estimating the angles, frequencies, or the determinant of the Hessian matrix.

A better, but more time-consuming, estimate of the Hessian matrix may be found using the torus machine (McMillan and Binney 2008) as described in Chap. 5.

Reference

McMillan PJ, Binney JJ (2008) MNRAS 390, 429

Appendix H
EDF Normalization

Here we show that the EDF presented in Chap. 8 is correctly normalized i.e.

$$\int \mathrm{d}^3\boldsymbol{J}\,\mathrm{d}L_z'\,\mathrm{d}[\mathrm{Fe/H}]\,\mathrm{d}\tau\; f(\boldsymbol{J}, L_z', [\mathrm{Fe/H}], \tau) = 1.$$

The full DF is

$$\begin{aligned} &f(J_R, L_z, J_z, L_z', [\mathrm{Fe/H}], \tau)\\ &\quad = \Sigma(\tau)\frac{\mathrm{e}^{-(L_z-L_z')^2/2\sigma_L^2}}{\sqrt{2\pi\sigma_L^2}}\frac{\frac{1}{2}[1+\tanh(L_z/L_0)]}{\frac{1}{2}[1+\mathrm{erf}(L_z'/\sqrt{2}\sigma_L)]}\frac{2\Omega_c(L_z')}{8\pi^3R_d^2\kappa^2(L_z')}\frac{\nu(L_z)\kappa(L_z)}{\sigma_R^2(L_z)\sigma_z^2(L_z)}\times\\ &\quad \mathrm{e}^{-R_c'/R_d}\mathrm{e}^{-\kappa(L_z)J_R/\sigma_R^2(L_z)}\mathrm{e}^{-\nu(L_z)J_z/\sigma_z^2(L_z)}\delta([\mathrm{Fe/H}]-F(R_c',\tau)) \qquad \text{(H.1)}\end{aligned}$$

This EDF assumes that all the heating occurred at the current angular momentum (i.e. σ_R and σ_z are functions of L_z). Note the error function in the denominator, and the arguments of the epicyclic frequencies.

For the thin disc, we have

$$\Sigma(\tau) = \Sigma_{\mathrm{thin}}(\tau) = \frac{(1-\mathcal{F})\mathrm{e}^{\tau/\tau_f}}{\tau_f(\mathrm{e}^{\tau_T/\tau_f}-1)} \qquad \text{(H.2)}$$

and for the thick disc

$$\Sigma(\tau) = \Sigma_{\mathrm{thick}}(\tau) = \frac{\mathcal{F}}{\tau_m - \tau_T}. \qquad \text{(H.3)}$$

We now show that this EDF integrates to unity. To perform the integral, we carry out the following steps.

J. Sanders, *Dynamics of the Milky Way*, Springer Theses,
DOI 10.1007/978-3-319-18772-3

1. Perform integral over [Fe/H]: integrates to one if $R'_c > 0$ and $\tau < \tau_m$.
2. Integrate over J_R, J_z from 0 to ∞: Exponentials produce factors $\frac{\sigma_R^2(L_z)}{\kappa(L_z)}$ and $\frac{\sigma_z^2(L_z)}{\nu(L_z)}$ that cancel with part of the fraction.
3. Integrate over L_z from $-\infty$ to ∞: The only terms that now depend upon L_z are the tanh and the Gaussian. The tanh limits the integration limits to 0 to ∞ such that the integral over the Gaussian is given by the error function term in the denominator so cancels.
4. Change integration variable from L'_z to R'_c: $\frac{\mathrm{d}L'_z}{\mathrm{d}R'_c} = \frac{R'_c\kappa(L'_z)}{2\Omega_c(L'_z)}$. This piece cancels with the appropriate terms in the fraction. Again the integral is from $-\infty$ to ∞ but this time we don't have a tanh to cancel out the negative piece. However, all stars are born in the disc with positive angular momentum. The negative birth angular momenta are forbidden. We are left with the integral

$$\int_0^\infty \mathrm{d}R'_c\, f(\tau, R'_c) = \int_0^\infty \mathrm{d}R'_c\, g(\tau)\frac{R'_c}{8\pi^3 R_d^2}\mathrm{e}^{-R'_c/R_d} \tag{H.4}$$

which integrates to $f(\tau) = g(\tau)/8\pi^3$.
5. Integration over the three angle variables removes the factor $8\pi^3$.
6. Finally, integration over τ gives

$$\int \mathrm{d}\tau\, f(\tau) = \int_0^{\tau_T} \Sigma_{\text{thin}}(\tau) + \int_{\tau_T}^{\tau_m} \Sigma_{\text{thick}}(\tau) = \mathcal{F} + (1 - \mathcal{F}) = 1. \tag{H.5}$$

Appendix I
Markov Chain Monte Carlo EDF Parameter Search

In Sect. 8.5 of Chap. 8, we showed how we chose the parameters of the EDF using the GCS and Gilmore-Reid density data. The parameters were chosen by using the Nelder-Mead downhill simplex minimization routine (Nelder and Mead 1965). Such a procedure risks finding local minima. Here we present the results of a much fuller Markov Chain Monte Carlo (MCMC) search. We use an affine-invariant ensemble sampler implemented in the *emcee* package from Foreman-Mackey et al. (2013) and evaluate the log-likelihood given in Eq. (8.34). We use a group of 128 walkers and choose logarithmic priors for all scale parameters. We limited $\sigma_L < 1800\,\mathrm{kpc\,km\,s^{-1}}$. We show the resulting covariance matrix of the MCMC samples after a sufficient burn-in in Fig. I.1. In Table I.1, we show the recovered parameter estimates assuming the distributions to be uncorrelated Gaussians.

I.1 Discussion

We will now discuss the results of our full MCMC procedure. The majority of the 1D histograms shown in Fig. I.1 are approximately Gaussian. The most notable exceptions to this are the histograms for F_m and σ_L.

We chose to limit $\sigma_L < 1800\,\mathrm{kpc\,km\,s^{-1}}$ and clearly the data require σ_L to be large. With σ_L this large, it means stars are able to migrate $\sim R_0$ in the lifetime of the Galaxy. The low-J_z stars of the GCS constrain $\mathrm{d[Fe/H]}/\mathrm{d}R_\mathrm{c}$ (see Fig. 8.11), and hence F_R such that $F_R \approx -0.055$ dex/ kpc. The GCS contains some metal-rich stars of $\sim$0.5 dex. These stars must migrate from the inner Galaxy but as the metallicity is assumed to grow linearly towards the Galactic centre, these stars cannot physically exist in our model. The solution to this is to increase the gradient of the ISM (e.g. Schonrich and Binney (2009) use a steep gradient of -0.082 dex/ kpc) but this contradicts the low-J_z $\mathrm{d[Fe/H]d}R_\mathrm{c}$ gradient. Another solution is to alter the model such that the gradient is steeper towards the centre. Additionally, the errors in the metallicities for these high metallicity stars could be larger than that reported.

J. Sanders, *Dynamics of the Milky Way*, Springer Theses,
DOI 10.1007/978-3-319-18772-3

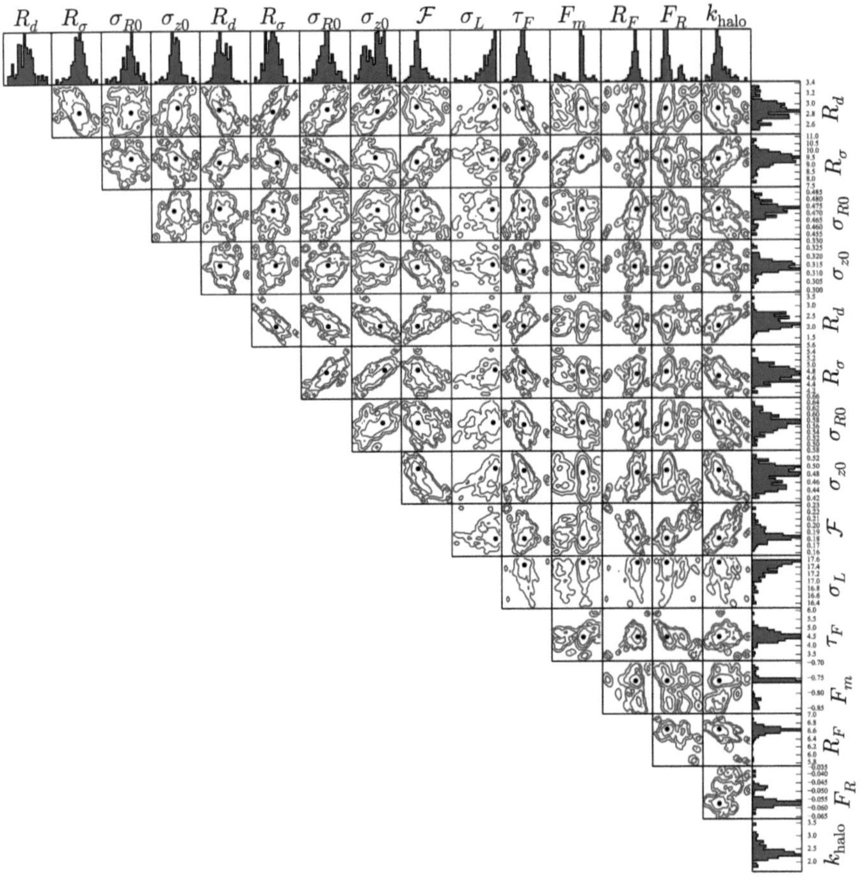

Fig. I.1 Covariance matrix for the MCMC results of the EDF parameter search. The *top row* and *right-most column* show the 1D histograms for each parameter. The other panels show the 2D histograms with contours containing 68, 95 and 99 % of the samples

The F_m distribution is determined by the very few metal-poor stars in the survey. It is therefore no surprise that the recovered parameter estimate is noisy.

We can also see that there are some parameters that are highly correlated. One notable correlation is τ_F with the thin disc scale-length, R_d. Increasing the scale-length flattens the radial profile. We see that τ_F is anti-correlated with R_d. Increasing τ_F flattens the metallicity of the ISM with time such that there are more metal-poor stars and fewer metal-rich stars in the solar neighbourhood now. This can be combated by decreasing the scale-length, which reduces the number of metal-poor stars and boosts the number of metal-rich stars. Additionally, we see that there are strong correlations between R_σ and σ_{R0} or σ_{z0} for the thick disc. Increasing R_σ lowers the number of hot stars coming from inside which must be compensated for by increasing σ_0. Similarly, R_d and R_σ are anti-correlated for the thick disc as

Table I.1 EDF parameter estimates from MCMC procedure on GCS data

Thick	R_d/kpc	2.3 ± 0.3
	R_σ/kpc	4.8 ± 0.2
	σ_{R0}/km s^{-1}	57 ± 3
	σ_{z0}/km s^{-1}	48 ± 2
Thin	R_d/kpc	2.9 ± 0.2
	R_σ/kpc	9.5 ± 0.5
	σ_{R0}/km s^{-1}	47.3 ± 0.5
	σ_{z0}/km s^{-1}	31.5 ± 0.4
Other	$\mathcal{F}$	0.19 ± 0.01
	σ_L/(100 km s^{-1} kpc)	17.31 ± 0.27
	τ_F/Gyr	4.6 ± 0.3
	F_m/dex	-0.77 ± 0.04
	R_F/kpc	6.6 ± 0.1
	F_R/dex kpc^{-1}	-0.055 ± 0.004
	$k_{\rm halo}/4.27 \times 10^{-4}$	2.4 ± 0.3

We assume the recovered distributions are uncorrelated Gaussians

increasing R_d reduces the number of hot stars from inside. Finally, we see a weak anti-correlation between $\mathcal{F}$ and σ_z for the thick disc, and between $k_{\rm halo}$ and both σ_0 for the thick disc. Lowering the weight of the thick disc ($\mathcal{F}$) requires σ_z to increase to fill the wings of the v_z distribution, whilst lowering $k_{\rm halo}$ has the same effect.

References

Foreman-Mackey D, Hogg DW, Lang D, Goodman J (2013) emcee: The MCMC Hammer. PASP 125:306–312
Nelder JA, Mead R (1965) Comput J 7, 308
Schönrich R, Binney J (2009) MNRAS 396, 203

Curriculum Vitae

Jason Sanders
Institute of Astronomy
University of Cambridge, CB3 0HA, UK
jls@ast.cam.ac.uk

I am a Postdoctoral Research Associate in the Institute of Astronomy at the University of Cambridge. I recently completed my DPhil at the University of Oxford reading Theoretical Astrophysics under the supervision of Prof. James Binney. I am interested in the dynamics of the Milky Way. In particular, I research how tidal streams can be used to constrain the shape of the Galaxy and what the combination of chemistry and dynamics of stars reveals about the structure and history of the Galaxy.

Education & Employment

October 2014–present	**Postdoctoral Research Associate**, *Institute of Astronomy, University of Cambridge* funded by Science and Technology Facilities Council (STFC) rolling grant for 3 years. Member of the Local Group research group
October 2011–September 2014	**DPhil Astrophysics**, *Merton College, University of Oxford* supervised by Prof. James Binney, funded by STFC grant. Thesis title: *Dynamics of the Milky Way: Tidal Streams and Extended Distribution Functions for the Galactic Disc.*

- December 2014: Thesis chosen as **best thesis examined in Rudolf Peierls Centre for Theoretical Physics 2014** and to be **published by Springer**.
- October 2013: Awarded **Merton College Prize Scholarship** for academic distinction.
- April 2013: Awarded Merton College Graduate Research Grant to attend Gaia First Science Conference in Nice, France.

J. Sanders, *Dynamics of the Milky Way*, Springer Theses,
DOI 10.1007/978-3-319-18772-3

October 2007–June 2011 **B.A. (hons) MSci Experimental and Theoretical Physics**, *University of Cambridge.*

- Awarded **Dr. William Hodge Prize** for achievement in Part II Natural Sciences
- Awarded **Dr Stevens Prize** for Part IB Natural Sciences and a **Foundation Scholarship**.
- Awarded the **Foundress Prize** (for firsts in all four Part IA exams) and **Materials Prize** for a report on a manufactured article.

2005–2007 **A level: 5 A grades, 3 Advanced Extension Award Distinctions**, *Richard Huish College*, Somerset.
Awarded the **Rutt Memorial Prize** for highest A level mark in Chemistry in the college.

Publications

2014 **Extended Distribution Functions for the Galactic disc**, *Sanders & Binney, **submitted** to Monthly Notices of the Royal Astronomical Society (MNRAS)*, An initial presentation of the ideas behind using extended distribution functions for the Galaxy and comparison with data from the Geneva-Copenhagen survey and SEGUE.

2014 **A fast algorithm for estimating actions in triaxial potentials**, *Sanders & Binney, MNRAS, 2014*, Generalized the procedure of Binney (2012) to estimate angles and actions for a general triaxial potential using the Stäckel potential formulation.

2014 **Actions and angles from numerically integrated orbits**, *Sanders & Binney, MNRAS, 2014*, Devised a procedure to accurately find actions and angles in triaxial potentials by constructing the appropriate transformation from a toy system to the target system from an orbital time series.

2014 **Probabilistic model for constraining the Galactic potential using tidal streams**, *Sanders, MNRAS, 2014*, Constructed a fully probabilistic formulation for constraining the potential of the Galaxy using tidal stream observations. The formulation relied on insights from the expected stream structure in the angle-frequency space of the correct potential.

2013 **Stream-orbit misalignment II: a new algorithm to constrain the Galactic potential**, *Sanders & Binney, Monthly Notices of the Royal Astronomical Society (MNRAS), 2013*, Proposed a new algorithm to find the potential of the Milky Way using observations of tidal streams. This improved on previous approaches by correctly including the physics of the formation of streams.

2013 **Stream-orbit misalignment I: the dangers of orbit-fitting**, *Sanders & Binney, MNRAS, 2013*, One method for using tidal streams to find the potential of the Galaxy is to fit the stream with an orbit. In this paper we demonstrated that this is wrong, and can lead to large biases in estimating the potential.

2012 **Determining the velocity dispersion of the thick disc**, *Sanders, MNRAS, 2012*, A recent work had used a particular method to find the velocity dispersion of the Galactic thick disc, and from these results had derived a very low estimate of the local dark matter density. Here we demonstrated that the method used was incorrect, and how one could improve it.

2012 **Angle-action estimation in a general axisymmetric potential**, *Sanders, MNRAS, 2012*, Angle-action coordinates are useful quantities in galactic dynamics, but their calculation is tricky. I developed a method for estimating these coordinates which can be used in any axisymmetric potential.

Conference Proceedings

2013 **Stream-orbit misalignment and a new algorithm for constraining the Galactic potential with streams**, *Sanders & Binney*, International Astronomical Union Symposium 298 Conference Proceedings (IAUS 298 Conf. Proc.), 2013.

2013 **Dynamical models and Galaxy surveys**, *Binney & Sanders*, IAUS 298 Conf. Proc., 2013.

Research Talks

November 2014 **Extended distribution functions for the Galactic disc**, *Gaia-ESO Second Science*, Porto, Portugal.

November 2014 **Dynamics of the Milky Way**, *Lunchtime Newcomer talks* University of Cambridge, UK.

April 2014 **Extended Distribution Functions & Tidal Streams**, *2nd Oxford-Heidelberg meeting*, University of Oxford, UK.

February 2014 **Dynamics of the Milky Way: Tidal streams**, *Galaxy Evolution meeting*, University of Oxford, UK.

August 2013 **Streams in angle-action space**, *Gaia Challenge meeting*, University of Surrey, UK.

August 2013 **Tidal streams and the Galactic potential**, *Milky Way Workshop* MPIA, Heidelberg, Germany.

July 2013 **Modelling RAVE with Extended Distribution Functions**, *RAVE meeting 2013*, Oxford, UK.

May 2013 **Stream-orbit misalignment**, *International Astronomical Union Symposium 298*, Lijiang, China.

April 2013 **Chemodynamical distribution functions**, *Gaia-ESO First Science*, Nice, France.

Teaching

2014 **1st year Undergraduate Physics Supervisions**, *Pembroke College, University of Cambridge.*
Taught two small groups of students, marked weekly work.

2013–2014 **3rd year Undergraduate General Relativity Tutorials**, *Magdalen College, University of Oxford.*

2012–2013 **3rd year Undergraduate Special Relativity Tutorials**, *Magdalen College, University of Oxford.*
Taught two small groups of students, set and marked fortnightly work, took revision classes, and marked collection papers in preparation for the end-of-year exams.

2012–2013 **3rd year Undergraduate Astrophysics Lab Demonstrating** *University of Oxford.*
Supervised students whilst they performed practical experiments, answered any questions they had, and discussed the experiments with them.

2012 **2nd year Quantum Mechanics marking**, *University of Oxford.*
Marked the work of 12 students taking Quantum Mechanics.

Research Experience & Other Employment

Dec 2010–May 2011 **Masters Research Project**, *Open Fermionic Systems*, University of Cambridge, Supervised by Dr Emilio Artacho.

- Investigated how one could use a Hamiltonian of mean force to investigate the properties of open fermionic systems.
- Wrote a final report, examined by viva, and presented the results to a small group.

Jun–Sept 2010 **Summer Undergraduate Research Fellowship**, California Institute of Technology, USA, supervised by Prof. Marc Kamionkowski.

- Selected by Pembroke College to undertake a 10 weeks summer research project in the Theoretical Astrophysics including Relativity (TAPIR) group.
- Derived a theoretical formalism for the detection of cosmological birefringence from active galactic nucleus (AGN) maps. Implemented method computationally, tested using mock data, and applied to data from the AGN 3C47.
- Wrote a technical research proposal, progress reports and an extended final report. Concluded the project with a presentation suitable for an audience with a non-scientific background.

MIX
Papier aus verantwortungsvollen Quellen
Paper from responsible sources
FSC® C105338

If you have any concerns about our products,
you can contact us on
ProductSafety@springernature.com

In case Publisher is established outside the EU,
the EU authorized representative is:
Springer Nature Customer Service Center GmbH
Europaplatz 3, 69115 Heidelberg, Germany

Printed by Libri Plureos GmbH
in Hamburg, Germany